Stress and Human Performance

SERIES IN APPLIED PSYCHOLOGY

Edwin A. Fleishman, George Mason University
Series Editor

Teamwork and the Bottom Line: Groups Make a Difference
Ned Rosen

Patterns of Life History: The Ecology of Human Individuality
Michael D. Mumford, Garnett Stokes, and William A. Owens

Work Motivation
Uwe E. Kleinbeck, Hans-Henning Quast, Henk Thierry, and Hartmut Häcker

Psychology in Organizations: Integrating Science and Practice
Kevin R. Murphy and Frank E. Saal

Human Error: Cause, Prediction, and Reduction
John W. Senders and Neville P. Moray

Contemporary Career Development Issues
Robert F. Morrison and Jerome Adams

Justice in the Workplace: Approaching Fairness in Human Resource Management
Russell Cropanzano

Personnel Selection and Assessment: Individual and Organizational Perspectives
Heinz Schuler, James L. Farr, and Mike Smith

Organizational Behavior: The State of the Science
Jerald Greenberg

Police Psychology Into the 21st Century
Martin I. Kurke and Ellen M. Scrivner

Benchmark Tasks for Job Analysis: A Guide for Functional Job Analysis (FJA) Scales
Sidney A. Fine and Maury Getkate

Stress and Human Performance
James E. Driskell and Eduardo Salas

Stress and Human Performance

Edited by

James E. Driskell
Florida Maxima Corporation
Winter Park, Florida

and

Eduardo Salas
Naval Air Warfare Center Training Systems Division
Orlando, Florida

NEW YORK AND LONDON

First publishedby Lawrence Erlbaum Associates,Inc., Publishers
10 Industrial Avenue Mahwah, New Jersey 07430

This editions first Published in 2009 by Psychology press

Published 2016 by Routledge
711 Third Avenue, New York, NY 10017
2 Park Square, Milton Park, Abingdon, Oxfordshire OX14 4RN

First issued in paperback 2016

Routledge is an imprint of the Taylor and Francis Group, an informa business

cover design by Gail Silverman

Library of Congress Cataloging-in-Publication Data

Stress and human performance / edited by James E. Driskell and Eduardo Salas.
p. cm.
Includes bibliographical references and index.
ISBN 0-8058-1182-6 (alk. paper)
1. Job stress. I. Driskell, James E. II. Salas, Eduardo.
HF5548.85.S737 1996
158.7—dc20 96-3785
CIP

ISBN 13: 978-1-138-98304-5 (pbk)
ISBN 13: 978-0-8058-1182-7 (hbk)

Contents

Foreword

Edwin A. Fleishman

There is a compelling need for innovative approaches to the solution of many pressing problems involving human relationships in today's society. Such approaches are more likely to be successful when they are based on sound research and applications. This *Series in Applied Psychology* offers publications that emphasize state-of-the-art research and its application to important issues of human behavior in a variety of societal settings. The objective is to bridge both academic and applied interests.

This volume, entitled *Stress and Human Performance,* deals with an important area of concern to researchers and practitioners in many fields. The concept of stress has been difficult to pin down and has been loosely defined in a number of different contexts. In the present volume, James Driskell and Eduardo Salas have brought leading researchers together from a variety of different contexts; their common focus is the influence of stress on various aspects of human performance. Thus, the book provides a coherent and comprehensive treatment of the subject that will be of interest to everyone concerned with factors affecting human performance.

Specifically, *human performance* refers to people doing things—performing a task, carrying out a procedure, solving a problem, or doing some type of work or activity. Some tasks are relatively simple, such as typing on a computer keyboard, and some tasks are more complex, such as piloting an aircraft. It is difficult enough for people to learn to perform these types of tasks well in a normal setting, as the literature on skill acquisition and training attests. However, it is far more difficult to perform these tasks under real-world, high stress conditions that may include time pressure, noise, novel or threatening events, various people making demands or requests, and other distractions.

This book examines human performance under stress. It attempts to balance two important concerns. The first is theoretical—the desire to understand how human performance is impacted by stress. Accordingly, this volume highlights current research on topics such as stress and decision making, stress and team performance, and factors that moderate stress effects. The second concern is more applied—to examine how stress impacts performance in real-world settings, such as aviation or military enviroments, and how we may intervene through selection, training, and system design to avoid future stress-related errors and accidents.

Stress research is a wide-ranging, almost foreboding area that overlaps the fields of applied psychology, medicine, clinical psychology, physiology, and biology. The naive reader should be cautioned against wading into this literature without a high tolerance for ambiguity. However, the present volume is unique in that it presents a nonclinical focus on stress and performance. Thus, the focus of this book is on the effects of stress on performance in applied task settings. Moreover, this book brings together a set of authors who are not only prominent researchers within this field, but also are actively involved in the applicaiton of this research to real-world settings.

The content of this book reflects a recent resurgence of interest in stress and performance. This has been stimulated in part by the almost commonplace occurrence of accidents or near-accidents involving ships, trains, industrial plants, airplanes, and even spacecraft in which increased enviromental demand inevitably plays a role. Whereas these types of accidents are often broadcast on the evening news, on a smaller scale we all are, at one time or another, faced with having to perform under the pressure of deadlines while juggling multiple tasks in the face of various distractions. To the extent that we can better understand these incidents, both large and small, we may be able to avoid the often dire consequences that result.

Stress and Human Performance will prove to be a valuable resource for researchers in the areas of applied psychology, human factors, training, and industrial/organizational psychology. This book will also provide an important source of information for practitioners in industry, the military, aviation, medicine, law enforcement, and other areas in which effective performance under stress is required.

Preface

On July 3, 1988, the USS Vincennes detected an approaching aircraft in the Persian Gulf, attempted verification, and then fired on and shot down the aircraft, all within a critical 4-minute span. The aircraft turned out to be a civilian Iranian airliner with 290 passengers and crew members aboard. The investigative panel who evaluated this accident found that although major hardware systems functioned correctly, the humans who operated these systems were under severe time pressure and operational stress that contributed to the Vincennes tragedy.

On February 22, 1994, USAir Flight 565, a DC-9 carrying 62 passengers from Washington to Boston, was forced to divert and make an emergency landing at New York's LaGuardia Airport because the airplane was about to run out of gas. Behind schedule in Washington, workers had forgotten to refuel the plane, and the flight crew had not checked fuel readings before takeoff.

In Tampa Bay during the predawn hours of August 10, 1993, a phosphate freighter collided with two barges carrying millions of gallons of jet fuel and oil, touching off a fiery explosion. With the outboard freighter and the inbound barges negotiating the bay's 600-foot-wide shipping channel, one party turned directly into the path of the other (exactly who did what remains a bit murky) and collided.

There were a number of contributing factors that led to these accidents. However, they are all similar in that they represent demanding task environments in which individuals must often perform under high stress. The crew of the Vincennes was under intense time pressure to identify a potential threat before it could endanger the ship, while the ship was lurching, lights were flickering, and communications were coming in simultaneously over left and right channels of operators' headphones. Commercial aviation has a strong safety record, yet airways are becoming more crowded, airlines have increased the pressure for quick turnarounds in airport terminals, and coordination among the flight crew, cabin crew, maintenance, air traffic control, and others is difficult and demanding. Operating an oil tanker costs $5,000 to $20,000 a day, whether it is at sea or sitting in port, and thus there are tremendous pressures for captains to take risks, run as fast as possible, and run under almost any conditions.

This book is about stress and performance. It is probably quite apparent to the reader that the term *stress* can refer to a broad range of topics, and that this statement does little to narrow the scope of this book. It is perhaps more informative to describe what this book is *not* about. It is not about stress-related disorders, clinical interventions, or coping. It is not about psychopathology, illness, or biochemical responses. The primary characteristic that distinguishes this volume from other related texts is its specific focus: We attempt to present basic and applied research that addresses the effects of acute stress on performance.

There are several implications of this focus. First, we do not intend to depreciate the extensive work that has been conducted on mental health, stress, and coping. However, clinical perspectives on stress and coping are more than well-represented in the research literature. In fact, the dominant, if unstated, psychoanalytic paradigm underlying much stress research has led to an emphasis on coping and defense mechanisms and to a preoccupation with disordered behavior and illness. Almost any text with *stress* in the title will invariably devote a considerable amount of pages to these issues. The bulk of stress research has almost ignored effective task performers in normal populations. By contrast, we intend this book to focus more directly on stress and performance from an applied perspective.

Second, we focus primarily on acute stress: intense, novel stress of limited duration that requires a proximate response—what we typically think of as emergency conditions. Thus, we are less interested in general life stress, chronic anxiety, daily hassles, and the like. Furthermore, we are interested in the effects of acute stress on performance, somewhat broadly defined. Thus, the effects of stress on task performance, the effects of stress on decision making, and the effects of stress on team interaction are all relevant topics. In keeping with our applied focus, we are further interested in not only how stress impacts performance, but in interventions to overcome these effects.

Third, we believe that the emphasis on stress and performance embodied in this book is especially timely. The pace of life in our high technology world has quickened. With the advent of electronic mail, fax machines, and computer-mediated communication systems, information can be received from anyone in almost any location in an instant; however, those on the other end are likely to want a response just as quickly. Industries that do not become more efficient, often by requiring a faster production turnaround with less slack, are superseded. As technology expands, as airplanes become faster and airways more crowded, as Naval battleships become more lethal, as decisions to launch or not launch a space shuttle become more complex, the demands imposed by these systems increase, taxing the capabilities of those who must operate them. These workers face an environment in which they must perform under more time pressure and under greater task load, in which stress is more prevalent, and in which the consequences of poor performance are more critical than ever before.

We begin this volume with an introductory chapter that includes an overview and general introduction to the topic of stress and performance. The first part of the volume, "Stress Effects," provides a presentation of current concerns, perspectives, and research on stress and performance, including reviews of stress and decision making, stress and aircrew performance, stress and military performance, and factors that moderate stress effects. Part II, "Interventions," addresses selection, training, and system design approaches to overcoming the impact of stress on performance.

James E. Driskell
Eduardo Salas

List of Contributors

Patricia Backer. Division of Technology, San Jose State University, San Jose, California.

Clint A. Bowers. Department of Psychology, University of Central Florida, Orlando, Florida.

Janis A. Cannon-Bowers. Naval Air Warfare Center Training Systems Division, Orlando, Florida.

James E. Driskell. Florida Maxima Corporation, Winter Park, Florida.

Nehemia Friedland. Department of Psychology, Tel Aviv University, Tel Aviv, Israel.

Joyce C. Hogan. Department of Psychology, University of Tulsa, Oklahoma.

Sandra Hughes. Florida Maxima Corporation, Winter Park, Florida.

Joan Hall Johnston. Naval Air Warfare Center Training Systems Division, Orlando, Florida.

Barbara G. Kanki. Aerospace Human Factors Research Division, NASA-Ames Research Center, Moffett Field, California.

Giora Keinan. Department of Psychology, Tel Aviv University, Tel Aviv, Israel.

Gary Klein. Klein Associates Inc., Fairborn, Ohio.

Michael Lesser. Department of Psychology, University of Tulsa, Tulsa, Oklahoma.

Ben B. Morgan, Jr. Department of Psychology, University of Central Florida, Orlando, Florida.

Judith M. Orasanu. Aerospace Human Factors Research Division, NASA-Ames Research Center, Moffett Field, California.

Eduardo Salas. Naval Air Warfare Center Training Systems Division, Orlando, Florida.

Jeanne L. Weaver. Department of Psychology, University of Central Florida, Orlando, Florida.

Christopher D. Wickens. Aviation Research Laboratory, University of Illinois at Urbana-Champaign, Savoy, Illinois.

1 Introduction: The Study of Stress and Human Performance

Eduardo Salas
Naval Air Warfare Center Training Systems Division

James E. Driskell
Sandra Hughes
Florida Maxima Corporation

> *A wonderful concept is "stress"—*
> *What it means is anyone's guess.*
> *Though it's fun to be clinical*
> *and rude to be cynical,*
> *operationally it is a mess!*
> *—Parsons, cited in Weitz (1966)*

The topic of stress is approached by many with trepidation because it is a difficult and often confusing subject. Stress is a psychological concept and as such is not concrete—it cannot be touched or perceived directly. Therefore, many who deal with "real-world" matters day in and day out may prefer not to deal with the topic of stress at all. During a meeting on stress and military performance attended by one of the present authors, one officer stood up and said, "There is no stress in my organization—I will not allow it!" Another officer said, "I want my people to be stressed, it keeps them on their toes." After hearing several such statements, one of those present who was responsible for training remarked, "Let's just make sure our personnel know how to perform their jobs, and forget about stress."

Sitting in that meeting (having chosen, wisely, not to launch into a speech on the merits of scientific inquiry), several thoughts came to mind. First, it was obvious that stress means many things to many people. For every one person who says that stress helped him storm the beach at Iwo Jima, there is another person who says that stress almost caused her to miss a landing approach while piloting an aircraft. Second, some discussed stress in positive terms, some discussed stress in negative terms, and some obviously preferred not to discuss stress at all. Third, it seemed that the trainer's hesitance not to tackle such an amorphous subject, to focus on performance and just forget about stress, was understandable given the circumstances. Clearly, the trainer had decided that with so little consensus on what stress was, it was preferable to dispense with the topic altogether. It seemed then, and still does now, that the reasons behind this decision were warranted; we as scientists had provided the trainer with little practical or usable information on stress and performance. However, although the reasons behind the trainer's desire to avoid discussing stress were understandable, the decision itself would appear to be a grave error. An understanding of stress and stress effects is of direct relevance not only to trainers, but to researchers, applied scientists, and the general public for several reasons.

First, the impact of stress on performance is perhaps greater now than at any time in our history. We live in an increasingly complex, high-technology world in which the potential for catastrophic error has greatly increased. Ships have always collided or run aground with alarming frequency, but these were not the 1,000-foot-long supertankers that now carry millions of gallons of crude oil along our coasts. These carriers (when they reach this size, they are no longer called ships) are fitted with the latest in electronics and navigational wizardry, but this only seems to allow them to be pushed out of port faster and travel under more hazardous conditions. The worst oil spill in U.S. history, the Exxon Valdez accident that dumped 11 million gallons of crude oil into Alaska's Prince William Sound, does not even place among the world's top 25 tanker oil spills. Given the state of the marine industry (ship captains are under enormous production pressures, crews are often undertrained, and regulation on an international level is notoriously inconsistent), it is quite likely that this accident will be eclipsed in the future by an even greater and more devastating mishap. Perrow (1984) called the shipping industry an *error-inducing system*, one in which so many pressures are present that lend themselves to lapses, miscues, and errors that attempts to address any one area are almost self-defeating.

By contrast, the commercial aviation community has an enviable safety record, yet the airways are becoming increasingly crowded, pilots are under strong pressure to maintain schedules, and accidents are likely to involve large numbers of people in the air or on the ground. On July 19, 1989, United Airlines Flight 232 experienced the failure of an engine and complete loss of hydraulic pressure, leaving the airplane with no flight controls. The airplane crashed during an attempted landing at Sioux City, Iowa, and 111 of the 296 passengers and crew members were fatally injured. In this case, the fact that the flight crew was able to bring the airplane down under some measure of control was seen by most experts as just short of a miracle (National Transportation Safety Board, 1990).

The military provides numerous illustrations of extremely complex and demanding task environments. Military antiair warfare (AAW) systems aboard modern Navy ships allow military personnel to detect aircraft at great distances. On the other hand, the amount and complexity of information that must be processed in a short period of time once a target has been identified is enormous.

This informational complexity, task load, and time pressure increases the potential for error, such as the 1988 downing of an Iranian commercial aircraft by the USS Vincennes (see chapter 4).

Complex, high-technology systems pervade the aerospace, military, petrochemical, mining, maritime, transportation, and nuclear industries. The humans that operate these systems are increasingly taxed to make critical decisions under extreme pressures and demands. Emergency or crisis conditions occur suddenly and often unexpectedly, operators must make critical decisions under extreme stress, and the consequences of poor performance are immediate and often catastrophic. Furthermore, when accidents occur, investigators all too frequently cite "operator error" as a primary cause. Although the requirement for effective performance under stress has been present since our ancestors were first chased with a club, it is likely that modern high-technology systems have increased both the stress under which we must perform and the consequences of poor performance.

By noting this, our intention is not to join the Luddites in wishing for a return to simpler times. The rapid technological advances of the past several decades have yielded many benefits. The commercial aviation system provides rapid and almost unfettered travel, and advanced military systems aboard Naval ships allow operators to detect distant threats and protect those on board. The point we wish to make is that with the greater benefit comes a greater destructive potential. The more "efficient" an industry becomes (e.g., commercial airliners are pushing for a "20-minute turnaround" to spend less downtime in terminals and more time in the air), the less margin for error exists. Therefore, the impact of stress on performance has become a primary concern in industry (Spettell & Liebert, 1986), the military (Driskell & Salas, 1991a), aviation (Prince, Bowers, & Salas, 1994), and other applied settings in which effective performance under stress is required.

A second reason that we believe stress is a growing topic of interest is that stress impacts both unique high-demand performance environments and everyday settings. On one hand, there are a large number of applied settings that share the commonality of a potentially high-stress, high-demand performance environment. These include aviation (Prince et al., 1994), military operations (Driskell & Olmstead, 1989), emergency medicine (Mackenzie, Craig, Parr, & Horst, 1994), mining (Perrow, 1984), diving (Radloff & Helmreich, 1968), parachuting (Burke, 1980; Hammerton & Tickner, 1969), bomb disposal (Rachman, 1982), police work (Yuille, Davies, Gibling, Marxsen, & Porter, 1994), firefighting (Markowitz, Gutterman, Link, & Rivera, 1987), and so on. We think of these stereotypically high-stress environments as settings that impose a particularly high demand on those who work or play in them and in which there is a substantial potential for risk, harm, or error. It is in the best interests of all concerned that pilots, ship captains, police, and firefighters perform their jobs effectively under the high demands to which they are exposed. On the other hand, even in more everyday settings such as working in an office or driving home, we may be subjected to various stressors that have been shown to disrupt task performance. These include noise (Cohen & Weinstein, 1981), performance pressure (Baumeister, 1984), anticipatory threat (Paterson & Neufeld, 1987; Wachtel, 1968), time pressure (Wright, 1974), task load (McLeod, 1977), group pressure (Mullen, 1991), and other stressors.

Finally, the deleterious effects of stress on performance are profound and pervasive. Stress may result in *physiological changes* such as increased heartbeat, labored breathing, and trembling (Rachman, 1983); *emotional reactions* such as fear, anxiety, frustration (Driskell & Salas, 1991b), and motivational losses (Innes & Allnutt, 1967); *cognitive effects* such as narrowed attention (Combs & Taylor, 1952; Easterbrook, 1959), decreased search behavior (Streufert & Streufert, 1981), longer reaction time to peripheral cues and decreased vigilance (Wachtel, 1968), degraded problem solving (Yamamoto, 1984), and performance rigidity (Staw, Sandelands, & Dutton, 1981), changes in *social behavior* such as a loss of team perspective (Driskell, Salas, & Johnston, 1995) and decrease in prosocial behaviors such as helping (Mathews & Canon, 1975); and even lowered immunity to disease (Jemmott & Locke, 1984). Data show that performance stress alone may increase errors on operational procedures threefold (Villoldo & Tarno, 1984). Similarly, Idzikowski and Baddeley (1983) found that the time taken to complete manual tasks doubled under stress conditions.

In brief, stress is an increasingly salient factor in our technologically complex society, the potential for stress-induced error spans both unique high-demand settings such as aviation as well as everyday work environments, and effects of stress on physiological reactions, cognitions, emotions, and social behavior are manifold. On the other hand, stress is a diffuse concept, lampooned in limerick by some authors (as noted at the start of this chapter, the concept is indeed a "mess") and maligned by others. In the following, we examine why stress has gotten such a bad rap, and define more clearly our interests in stress and performance.

WHAT IS STRESS, AND WHY ARE PEOPLE SAYING SUCH BAD THINGS ABOUT IT?

Part of the difficulty surrounding the concept of *stress* has to do with the different meanings that have been attached to the term. It is perfectly legitimate to define a general term such as stress for various purposes. To a mechanical engineer, stress is the effect of some action on a body that produces strain, and is usually measured in pounds per square inch. To a linguist, stress is the relative force with which a sound or syllable is spoken. To a physician, stress may mean damage to the body. To someone interested in stress and performance, it may mean something entirely different. To assist the reader in determining what *we* mean when we speak of stress, we first present some of the more general definitions that have been offered, and then attempt to develop a working definition that is applicable to our present concern with stress and performance.

It is informative to consider the etymology of the term stress, which stems from the Latin *stringere,* to draw tight (other derivatives of *stringere* include the words *strain,* to exert or tax, and *strait,* affording little space or room). Thus, even from the early Latin, we have a strong hint of what stress means—it taxes, it strains, and it restricts.

The term was first used as a topic of scientific inquiry in the early part of this century in the work of Cannon (1935) and Selye (1936). Selye (1980) provided the early direction for stress research by defining stress as the "nonspecific result of any demand upon the body" (p. vii). Selye was a physician (an endocrinologist) and interested in what type of general syndrome could produce such diverse observed bodily changes as enlargement of the adrenal cortex, atrophy of the thymus-lymphatic system, and the appearance of peptic ulcers. Selye (1980) thus called this general syndrome of nonspecific bodily reactions *stress* and unabashedly claimed that this formulation brought the subject "out of the stage of vague cocktail party chitchat into the domain of science" (p. viii). Whether or not this was the case, Selye did provide a general conceptualization of stress as a reaction to noxious events, which he termed *stressors.* For our purposes, however, his aforementioned definition of stress is a statement so broad as to be almost meaningless.

Stress gradually became less of an exclusively medical domain and became a topic of psychological inquiry, partly fueled by wartime concerns of the effects of combat on soldier performance. Prefacing their study of the stress of war on Air Force combat crews, Grinker and Spiegel (1945) noted that, "Never in the history of the study of human behavior has it been so important to understand the psychological mechanisms of 'normal' individuals in situations of stress" (p. vii). Janis (1951) examined emotional reactions to air-raid attacks, as well as reactions to peacetime disasters (Janis, 1954). Other research began to document effects of stress on human performance (Basowitz, Persky, Korchin, Grinker, 1955; Lazarus, Deese, & Osler, 1952; Postman & Bruner, 1948). Gradually stress research came out from under the umbrella of medicine and into the field of psychology. The more psychologically oriented definitions of stress that evolved include the following:

> A stressful event is any change in the environment that typically induces a high degree of unpleasant emotion (such as anxiety, guilt, or shame) and affects normal patterns of information processing. (Janis & Mann, 1977, p. 50)

> [Stress is] an adaptive response, mediated by individual characteristics and/or psychological processes, that is a consequence of any external action, situation, or event that places special physical and/or psychological demands upon a person. (Ivancevich & Matteson, 1980, p. 8)

> A potential for stress [exists] when an environmental situation is perceived as presenting a demand which threatens to exceed the person's capabilities and resources for meeting it, under conditions where he expects a substantial differential in the rewards and costs from meeting the demand versus not meeting it. (McGrath, 1976, p. 1352)

Note that these definitions vary in the extent to which they emphasize the *stimulus* environment, the *response* of the individual, or the *relationship* between the person and environment. A stimulus-based definition, such as that provided by Janis and Mann (1977), tends to be couched in terms of the environmental events that impact the individual. A response-based definition, as provided by Ivancevich & Matteson (1980), emphasizes the response of the individual to stress. Selye's definition of stress as the body's response to impinging stimuli is the embodiment of a response-based definition.

Dissatisfaction with primarily stimulus or response definitions of stress led researchers to develop a more relational perspective, as provided by McGrath (1976), emphasizing the relation between environmental demands and individual response. Lazarus and Folkman (1984) have provided one of the most generally accepted relational definitions of stress: "Psychological stress is a particular relationship between the person and the environment that is appraised by the person as taxing or exceeding his or her resources and endangering his or her well-being" (p. 19). In the following, we use the term stress in a similar manner. We define stress as a process by which certain environmental demands (i.e., performing in front of others, taking an examination, industrial noises) evoke an appraisal process in which perceived demand exceeds resources and results in undesirable physiological, psychological, behavioral, or social outcomes. There is nothing "magical" about this definition. Our working definition adapts the relational approach taken by Lazarus and Folkman (1984) and McGrath (1976) to our current interests in understanding stress and performance. However, we intend to further qualify our interest in stress by noting certain limiting characteristics.

First, we are primarily interested in *acute* stress—that which is sudden, novel, intense, and of relatively short duration, disrupts goal-oriented behavior, and requires a proximate response. Acute stress is illustrated by the prototypical "emergency" situation, in which the scenario unfolds rapidly, the task must be dealt with in a short time period, and the consequences of poor performance are immediate. The pilot's loss of a landing gear, the sudden appearance of an unidentified blip on a military radar operator's screen, and the plant operator's observance of an aberrant reading on a critical indicator are examples, but so is the more mundane example of a work team deliberating a task under tight deadline pressure.

This emphasis on relatively sudden demands that require a near-term response differentiates the present perspective from research on other types of stress, such as work dealing with chronic stress, stressful life events, and daily hassles. *Chronic stress* refers to stress factors that are in the background of our everyday activities, and includes job stress, family stress, and the stresses imposed by organizational requirements. Sometimes no one single source of chronic stress may seem that consequential, but the combined or cumulative effects of these stressors can lead to degraded performance over time. By placing the primary emphasis of the current analysis on acute stress, rather than on more persistent, chronic, stressors, we do not intend to minimize their relevance. In fact, Schaubroeck and Ganster (1993) noted that chronic stress may reduce the individual's capacity to respond effectively to acute stress demands. Certainly, personnel that have been on a rotating swing shift for several weeks may respond more poorly when an emergency situation suddenly arises. Although chronic stress is an important topic, attempts to account for the effects of chronic stress, as well as interventions to overcome negative effects, are likely to differ from that required to understand short-term, intense, acute-stress conditions.

Furthermore, we may distinguish the current emphasis on acute stress from research on *stressful life events*—major events in a person's life such as death of a loved one, marriage, divorce, illness, job loss, moving, and so on. These events are unpleasant and can clearly impact performance, and to the extent that the occur suddenly, disrupt goal oriented behavior, and require an immediate response, they may overlap somewhat our conceptualization of acute stressors. However, these events are relatively rare in a person's life and the impact is likely to be felt over weeks and months rather than require an immediate adaptive response. *Daily hassles* refers to more common events that occur day to day: traffic hassles, arguments, barking dogs, and other irritants. The point here is not to be too restrictive in our interests, but to clarify that there are a number of types of stressors that are discussed in the research literature, and to define the primary focus of our interest in acute stress.

Second, we are primarily interested in the *negative* consequences of stress. Certainly there are conditions under which stress can enhance performance. For example, too little stress, in the sense of operating in a stimulus-impoverished environment, can be understimulating, and a low-to-moderate level of stimulation is often required to keep the individual alert and on-task (this is a variant of the inverted U hypothesis, that there is an optimal level of stress required for effective functioning). However, the facilitative effects of low-level stress on enhancing performance is of less interest for our current purposes than the negative impact of acute stressors.

Furthermore, we are interested in *threat* situations in which the perception of danger exceeds one's resources to respond. Lazarus and Folkman (1984) noted that environmental events may be appraised as *threatening* or *challenging*. Individuals evaluate an event as threatening when the perceived danger exceeds the capacity to respond. Thus, threat appraisals focus on the potential harm of an event and are associated with primarily negative emotional reactions. Individuals evaluate an event as challenging when they perceive the possibility for gain to outweigh the potential for harm. Thus, challenge appraisals focus on the potential for gain or profit from an event and are associated with positive emotions such as eagerness. (Note that the term *challenge* is used in different ways in the research literature. Schaubroeck and Ganster (1993) defined a challenge situation as one in which sudden crisis events require a strong response to maintain successful task performance, a formulation more closely related to our concept of acute stress.)

Tomaka, Blascovich, Kelsey, and Leitten (1993) examined the distinction between threat and challenge appraisals by presenting research subjects with a mathematics test. Each individual's appraisal of the situation was assessed by asking them how threatening they expected the task to be (primary appraisal) and how able they were to cope with the task (secondary appraisal). Faced with this relatively innocuous task, the researchers expected some individuals to perceive the situation as a threat and some to perceive it as a challenge. Those who reported a high perceived threat relative to perceived coping ability were placed in the *threat* group. Those who reported a low perceived threat relative to perceived coping ability were placed in the *challenge* group. In general, results indicated that those who appraised the task as a challenge reported less subjective stress and performed better than those who appraised the task as a threat (subjects' physiological response was more complex: Tomaka et al. found that the challenged group evidenced greater cardiac activity [i.e., heart rate] and decreased vascular [i.e., blood pressure] resistance, whereas the threatened group showed relatively less cardiac reactivity and greater vascular resistance.)

It is important to note, as demonstrated in this study, that environmental events can be perceived as either a challenge or as a threat, and that people respond differently to challenge versus threat situations. For example, it is a common tendency for a researcher to attempt to study stress by examining a particular environment, such as students taking final examinations. It is quite likely that for a large proportion of these populations, for those who are prepared for final exams, these situations constitute a challenge rather than a threat. On the other hand, there are many situations that most people would agree involve very high demand. For example, when an air traffic controller suddenly finds several airplanes in the same air space, or when a fire breaks out aboard a ship, relatively few individuals would describe the task as a challenge or report it as a positive opportunity. Nevertheless, there are variations in how people respond to environmental events, and the extent to which an event is stressful is not inherent in the environment itself. In the present context, we are primarily interested in threat situations in which demands exceed resources; however, we should be cognizant of the central role that appraisal plays in the evaluation of environmental events.

In summary, what do we mean when we speak of stress and performance? Stress is a high-demand, high-threat situation that disrupts performance. It is time-limited; stress conditions occur suddenly and often unexpectedly; quick and effective task performance is critical; and consequences of poor performance are immediate and often catastrophic. This perspective is relevant to many applied settings that share the commonalities of high demand, high-risk performance conditions, such as aviation, military operations, and nuclear, chemical, and other industrial settings, as well as everyday situations in which individuals face stressors such as noise, time pressure, and task load that disrupt goal-oriented behavior.

Problems With the Concept of Stress

As we noted at the beginning of this chapter, stress means many things to many people. The term is not unique in this regard—many psychological constructs are somewhat ambiguous. However, Hogan and Hogan (1982) noted that the stress literature is "awash in a sea of terminology" (p. 153), and that there is so much conceptual confusion that researchers lack a common vocabulary. Therefore, every author must define his or her terms anew, and the reader must check each article on stress to make sure that he or she understands what subject the author is addressing. This lack of a consistent and uniform vocabulary serves to increase the ambiguity of the field.

Hogan and Hogan (1982) argued that one problem leading to the current fractionalization and confusion in the stress literature is the traditional emphasis on psychoanalysis as the dominant paradigm for studying stress, as evidenced in the work of Janis, Lazarus, and others. One consequence of this view is the prevalent emphasis on "stress and coping." An implication of the psychoanalytic perspective is that people are seen as constantly stressed (by chronic "background" stressors, daily hassles, etc.) and that those who are able to deal with this more effectively have better developed defenses or coping mechanisms to bring to bear. There are several problems with this view. First, it places an emphasis on that relatively small portion of the population that suffer under chronic stress, and the goal of research becomes discovering what coping mechanisms or defense mechanisms that these distressed persons *lack*. Almost ignored are normal people in normal circumstances for whom stress can be disruptive but is not a constant companion, and who are better characterized as actively pursuing a goal rather than actively building defenses.

A second consequence of the psychoanalytic view is that it leads to a preoccupation with illness. The zenith of the medical/biological view is represented by Selye (1980), who used the unfortunate term *eustress* to refer to curative stress and *distress* to refer to disease-producing stress. The study of illness is only marginally related to the understanding of stress and performance in a normal population.

Therefore, the predominant, if implicit, viewpoint of most stress research over the past several decades has been psychoanalytic, with its emphasis on coping, defense mechanisms, and the restoration of ego equilibrium. Accordingly, most treatises on stress invariably devote a considerable amount of pages to disordered behavior, coping, and treatment, and yet there are typically precious few pages devoted to performance, effectiveness, and productivity. The bulk of stress research has almost ignored effective task performers in real-world work environments.

Because of understandable concerns with the myriad of ways in which stress has been envisioned, many wonder whether the term carries so much excess baggage as to be better off dismissed with entirely. Accordingly, many have questioned whether the construct of stress is a useful one. First, a reasonable case can be made that the term stress is not a unidimensional construct at all. Stress is a term that describes a process by which certain environmental stimuli evoke an appraisal process that determines certain observable outcomes (i.e., physiological changes, emotional reactions, cognitive changes, etc.). Thus, the term describes a general process, from which researchers may study any number of issues. Accordingly, the term may be more appropriately used as a broad descriptor to refer to a loosely defined body of research. This research may include work that focuses on stimuli such as noise stress or crowding, responses such as physiological changes or performance outcomes, or the appraisal process itself and psychological moderators that determine appraisal such as personality or experience.

However, some have made a more specific argument, that there is little value in assuming a coherent body of stress literature, and that one would be better off by examining specific variables, such as the effect of noise on performance, without evoking the idea of stress at all. Willer and Webster (1970) have drawn a distinction between *empirical concepts,* which are observable and *theoretical constructs,* which are not. Concepts are drawn from everyday experience; they describe objects or events that are directly observable. For example, we may make statements about the effects of noise or the presence of others on performance, and these terms will be readily understood by most observers. In other words, we know what noise or group pressure are because they are observable in everyday situations. A great deal of valuable empirical research has been dedicated to examining the relationship of specific observable concepts, such as noise, to other concepts, such as distraction.

By contrast, a theoretical construct is abstract rather than concrete—it is not directly observable and in fact does not "exist" in a real sense. Constructs are abstract properties of objects or events that are defined for the purpose of explaining a broad class of phenomena. For example, we may define the construct "acute stressor," and argue that noise and group pressure are both examples of acute stressors. We use the construct of acute stressor for the purpose of developing theoretical statements, and noise and group pressure are then taken as indicators of this construct. Theoretical research in this instance takes a broader perspective; we are interested in relating certain constructs to other constructs (rather than relating certain observables to other observables). In this manner, when we observe the effects of noise in producing increased errors, and we observe the effects of group pressure in producing increased errors, we can offer a more general or abstract statement as an explanation for these findings. Therefore, our complaints that the term stress is too ambiguous or diffuse notwithstanding, constructs such as acute stress are intended to be abstract and open-ended. The primary function that we ask of such constructs is that they are convenient things to talk about and useful for explaining concrete phenomena.

A MODEL OF STRESS AND PERFORMANCE

In the following, we present a heuristic model of the stress process. This model provides a basic frame of reference for discussing the determinants and consequences of stress. Figure 1.1 illustrates a simple four-stage model of stress and performance: (a) An environmental stimulus becomes salient, (b) it acquires a positive or negative valence through the appraisal process, (c) this leads to the formation of performance expectations, (d) and these in turn determine a number of physiological, cognitive, emotional, and social consequences.

The process presented in Fig. 1.1 is activated by the introduction of specific environmental stimuli such as noise, threat, time pressure, task load, group pressure, or other potential stressors. The presence of a potential threat leads to the second stage of this model—the activation and operation of the appraisal process. *Appraisal* refers to the process by which the meaning of the environmental event is evaluated. Lazarus and Folkman (1984) distinguished between primary and secondary appraisal. *Primary appraisal* involves evaluation of the extent of threat that an event poses. *Secondary appraisal* is an evaluation of perceived capacity or resources to meet this threat. Traditional models of stress presume that appraisal is a function of the degree of discrepancy between demand and capacity.

Environmental stimuli become salient, and through the appraisal process become evaluated as positive (and seen as a challenge) or negative (and see as a

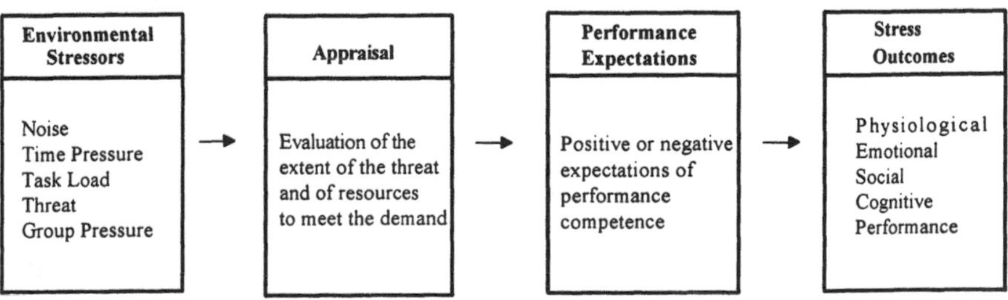

FIG. 1.1. A four-stage model of stress and performance.

threat). The appraisal process leads to the formation of performance expectations—feelings of self-efficacy or mastery. If demands exceed perceived resources, negative performance expectations are formed. If the perception of available resources exceeds the perceived threat, positive performance expectations result. Performance expectations are similar to the concepts of self-efficacy (Bandura, 1982), perceived mastery (Pearlin, Menaghan, Lieberman, & Mullan, 1981), and performance confidence (Rachman, 1982). The development of positive performance expectations is a crucial factor in preparing personnel to operate under high-demand conditions. An examination of specialized hazardous duty training given to British military bomb-disposal experts showed those who developed positive performance expectations reported relatively little fear during operations (Rachman, 1983). Other research has shown performance expectations to be a strong predictor of actual performance (Bandura, Reese, & Adams, 1982; Locke, Frederick, Lee, & Bobko, 1984).

Finally, stress results in a number of types of outcomes of interest, including physiological reactions, cognitive effects, emotional reactions, social behavior, and performance outcomes. *Physiological* reactions to stress have been assessed in various studies by a host of measures, including skin conductance, pulse rate, heart rate, heart rate variability, electrocardiograph (EKG) and impedance cardiographic (ZKG) signals, pulse transit time (PTT, the interval between the EKG R-wave and rise of the finger pulse wave), salivary immunoglobulin A (IgA), systolic and diastolic blood pressure, catecholamine (e.g., adrenaline and noradrenaline) output, glucocorticoid (e.g., cortisol) output, EMG level, blood glucose level, palmar sweating, and P300 evoked potential response, muscle tension, eye blink and eye blink duration, respiration rate, and a number of other measures. For those who are not psychophysiologists, this is a daunting list. There is no doubt that stress results in various types of physiological response. However, physiological measures of stress are often inconsistent and difficult to interpret. Furthermore, the relationship of physiological state to performance is unclear. Nevertheless, these reactions are, at the least, a source of distraction to the task performer. Individuals under stress may often overinterpret or assign heightened importance to novel bodily sensations (see Clark, 1988), and this attentional conflict may impair task performance. Worchel and Yohai (1979) found that individuals who were able to label or identify novel physiological reactions (i.e., individuals who were able to attribute their physiological reactions to some reasonable cause) were less distressed or aroused by those reactions.

Emotional reactions to stress may include subjective feelings of fear or anxiety, annoyance, tension, frustration, and increased concern for the well-being of self and others. Measures of self-reported or subjective stress typically assess state or transitory anxiety, by asking the extent to which the person in the current situation feels excited, tense, nervous, pressured, or anxious. Other measures of subjective stress include trait anxiety (although we would expect acute stressors to have little impact on trait measures), and specific measure of performance anxiety such as speech anxiety, test anxiety, or computer anxiety.

Cognitive effects of stress may include distraction, narrowing of attention, tunnel vision, decreased search activity, response rigidity, longer reaction time to peripheral stimuli, increased errors, and memory deficits. One of the more well-established findings in the stress literature is that as stress or arousal increases, the individual's breadth of attention narrows (Combs & Taylor, 1952; Easterbrook, 1959). Perhaps the earliest statement of this phenomena was James' (1890) belief that the individual's field of view varied, from a broader perspective under normal conditions to a more narrow, restricted focus under stress. For complex tasks in which the individual must attend to a relatively larger number of salient task cues, this narrowing of attention may result in the elimination of relevant task information and task performance will suffer. Thus, stress may result in degraded overall performance on complex tasks as attention is narrowed in response to overload.

Research has also documented other stress-induced cognitive changes. For example, individuals tend to scan solution alternatives less effectively when under stress. Wright (1974) found that fewer data dimensions were considered when decision makers were under time pressure. Keinan (1987) found that the scanning of solution alternatives was less organized and led to poorer decision outcomes when task performers were under stress. Cohen (1952) found that stressful conditions lead to greater problem-solving rigidity, a tendency to persist with a set method of problem solving when it ceased to provide a direct task solution. Dorner (1990) found that individuals under stress were prone to "ballistic" decision making; making decisions without checking the consequences of their decision. Dorner concluded that ballistic decision making tends to increase concreteness of behavior, because evaluating the consequences of one's actions is an essential means for adapting responses to changing environmental contingencies.

Social effects of stress may include a reduction in the tendency to assist others, increased interpersonal aggression, neglect of social or interpersonal cues, and less cooperative behavior among team members. Cohen (1978, 1980) noted that the narrowing of attention that occurs under stress may include a restriction of social cues as well, and that stress may lead to a neglect of social or interpersonal cues and decreased sensitivity to others. In fact, Driskell et al. (1995) found that team members were less likely to maintain a broad team perspective under stress and were more likely to shift to a more individualistic self-focus, resulting in poorer overall team performance. Other research has examined the effects of stressors on group decision making (Driskell & Salas, 1991b; Janis, 1982a; Mullen, Anthony, Salas, & Driskell, 1994), and the performance of groups in exotic environments such as space (Harrison & Conners, 1984).

Performance outcomes that are typically examined in the research literature include performance accuracy (usually assessed by the number of errors incurred on a task), performance speed (the time required to perform a task), and performance variability (variability in accuracy or speed).

In the following sections, we examine the effects of stress in greater depth by reviewing research on specific stressors, including noise stress, group pressure, task load, threat, and time pressure.

NOISE STRESS

Noise is defined as sound that is unwanted by the listener because it is unpleasant, bothersome, interferes with task activity, or perceived as being potentially harmful (see Cohen & Weinstein, 1981; Kryter, 1985). We are primarily interested in the *nonauditory,* or psychological, effects of noise. This excludes research that examines the effects of noise on sensory functions such as hearing loss, depth perception, or visual acuity as well as that which examines the role of noise in directly masking desired sound. This approach is similar to that taken by other researchers interested in the effects of noise on human performance (see Cohen & Weinstein, 1981; Grether, 1971; Kryter, 1985; Loeb, Jones, & Cohen, 1976).

Noise researchers commonly make a distinction between *sound* and *noise*. Sound is a physical phenomenon, referring to changes in air pressure detected by the ear. Noise, on the other hand, is a psychological concept, defined as "unwanted sound" (Cohen & Weinstein, 1981), or as an auditory stimulus that bears no task-related information (McCormick & Sanders, 1982). Noise cannot be defined strictly in physical terms because, for example, a particular engine sound may provide useful information to a technician, whereas it may simply represent an unwanted irritant to someone attempting to read a book.

There are several major theories offered to explain the effects of noise on performance. The classic arousal perspective argues that stress results in heightened arousal, and that arousal leads to a narrowing of attention (see Broadbent, 1971; Easterbrook, 1959). As attention narrows, peripheral (less relevant) task cues are first ignored, followed by further restriction of central or task relevant cues. To the extent that task relevant cues are neglected, performance suffers. Accordingly, complex tasks that demand attention to a wide range of cues are more susceptible to degradation under stress than simple tasks.

Poulton (1978) argued that the effects of noise on performance may be related to arousal, but that arousal subsides quickly after the onset of noise (he also argued that this initial arousing effect is often beneficial to performance). Poulton maintained that decrements in task performance, particularly under continuous noise conditions, are a function of the masking of acoustic cues, or even the masking of the "inner speech" of a task performer. In other words, people either can't hear subtle task relevant cues in the presence of noise, or they literally can't "hear themselves think." Poulton further argued that detrimental effects of intermittent noise on performance are often simply the result of distraction.

According to the distraction–arousal theory (Teichner, Arees, & Reilly, 1963), noise has two primary effects; it can distract the task performer or it can increase the level of arousal. For example, Teichner et al. found that research subjects performed more poorly when a baseline (81 dB) level of noise was either increased *or* decreased, suggesting that the change in the stimulus itself was distracting to the performers. Furthermore, Schoenberger and Harris (1965) demonstrated that the effects of change in stimulus level were lessened (i.e., performers were less distracted) when the task was well-practiced.

General Effects

From the earliest studies of the effects of noise on performance (e.g., Cassel & Dallenbach, 1918) to the more recent, the noise literature has been marked by an apparent inconsistency of results. Some research has found that noise degrades performance. For example, Finkelman, Zeitlin, Romoff, Friend, and Brown (1979) found that noise increased the incidence of errors on a short-term memory task. On the other hand, Kirk and Hecht (1963) discovered that variable noise facilitated the performance of a vigilance task. What Coates and Alluisi (1975) referred to as the schizophrenic nature of findings in this area is illustrated by the following statements:

> There is a growing body of knowledge which has demonstrated both negative and positive effects of noise. (Repko, Brown, & Loeb, 1974, p. 2)
>
> The findings have ranged from adverse effects through no effects to beneficial effects on performance. (Thackray & Touchstone, 1979, p. 1)
>
> No generalizations can be made; noise may or may not have a disruptive influence upon behavior. (Wilbanks, Webb, & Tolhurst, 1956, p. 1)
>
> Perhaps the only conclusion one can reach from reading reviews of the effects of noise on human performance is that there are effects. Whether these effects are detrimental or facilitative . . . remain largely undetermined. (Harris, 1968, p. 16)

These comments reveal that the state of knowledge regarding noise and performance is hardly settled. This may reflect the wide range and diversity of research being conducted in this area. However, for those interested in practical applications, and particularly in the effects of noise on performance, this ambiguity may be discomforting.

Performance Effects. A number of studies have examined the effects of noise on performance accuracy. In general, results indicate that noise tends to increase task errors. For example, Burger and Arkin (1980) found that uncontrollable noise blasts resulted in increased errors on a free-recall memory task. Hartley (1981) reported greater errors on a pursuit-tracking task when subjects worked under 95 dB noise. Finkelman et al. (1979) found a significant effect of noise on information processing performance: Subjects performing under noise exhibited more errors on a delayed digit-recall task. Eschenbrenner (1971) found that intermittent noise resulted in greater errors on a complex psychomotor task. In general, the available evidence suggests that noise tends to result in degraded accuracy on a variety of tasks.

Some studies have also assessed the effects of noise on speed of performance (i.e., speed of target detection, speed of response, etc.). For example, von Wright and Vauras (1980) found that noise resulted in increased errors on a recall task, but that response latencies (the time it took to respond) were somewhat shorter under noise stress. The existence of a speed/accuracy tradeoff, in which subjects tend to maintain speed of performance under stress at the expense of increased errors, has been noted by Broadbent (1957), Repko et al. (1974), and others.

Cohen and Weinstein (1981) noted that noise often produces highly variable performance. The typical pattern of effects is for noise (especially random or intermittent noise) to produce cycles of performance decrement and normal performance. This variability in performance can result in critical errors on tasks that require sustained high-level performance.

Cognitive Processes. A number of studies have documented a shift in attentional focus under noise stress. For example, Hockey and Hamilton (1970) found that on a memory-recall task, subjects under noise stress were able to recall words accurately, but were not able to accurately recall the location of the words in the word list. This suggests that under noise stress, attention is focused on the dominant part of a task to the detriment of more peripheral cues. Hockey (1970) found that, under noise stress, the detection of lights displayed on the periphery of a display was impaired, whereas the detection of centrally located lights was not affected. Studies of dual-task performance also indicate that individuals under noise stress tend to devote more effort to primary task performance at the expense of secondary tasks (Boggs & Simon, 1968; Finkelman & Glass, 1970).

Other studies suggest that individuals become more confident in their judgments about high-probability events when under noise stress. Broadbent (1979) examined individuals' confidence in their judgments in detecting signals when the probability of a signal was high and when the probability of a signal was low. He found that when under noise stress, and when the probability of a signal was high, subjects were much more likely to assert that they were confident that a signal was present or that a signal was not present.

Subjective Stress. Most studies show a relatively strong negative relationship between noise and perceived stress (Burger & Arkin, 1980; Lovallo &

Pishkin, 1980; Sherrod, Hage, Halpern, & Moore, 1977; Wohlwill, Nasar, De-Joy, & Foruzani, 1976). In a meta-analysis of the effects of noise on performance, Driskell, Mullen, Johnson, Hughes and Batchelor (1992) found a strong overall negative relationship between noise and self-reported stress (mean $r = -.56$). In fact, the relationship between noise and self-reported stress was much stronger than between noise and performance accuracy (mean $r = -.14$). This suggests that, in general, those experiencing noise stress are more likely to report being stressed than they are likely to exhibit performance decrements.

Social Behavior. A number of studies suggest that noise may have significant effects on social or group behavior. Mathews and Canon (1975) examined the effects of noise on the tendency to help or assist others. They found that under high noise conditions, individuals were less likely to come to the aid of others (in this case, to help another person pick up accidentally dropped materials). Sherrod and Downs (1974) found that following exposure to noise, subjects were less likely to help the experimenter by volunteering to perform a requested task. Cohen (1978) has attempted to explain these findings by noting that stress results in a narrowing of attentional cues, including social cues. Therefore, those under stress may be less likely to notice interpersonal cues such as another's request for assistance as well as less likely to render assistance to others. Others have also suggested that the decrease in cooperative behavior noted under stress may stem from the negative affective state induced by noise (Moore, Underwood, & Rosenhan, 1973).

Cohen, Conrad, O'Brien, and Pearson (1974) concluded that the effects of noise are difficult to predict, and are dependent on a number of factors, including intensity of the noise, temporal characteristics such as intermittency and duration, and the nature of the task. We briefly discuss these factors in the following sections.

Intensity. Most research indicates that the effects of noise on performance increase in severity as a function of loudness. Research further suggests that loud noise in which the higher frequencies predominate is generally more distracting than lower frequency noise (Broadbent, 1957). However, there is considerable ambiguity over the precise level of intensity at which noise produces a decrement in performance.

Variability. Variable noise is noise that changes in intensity within a specified range (i.e., from 80 dB to 60 dB). Research has shown that *changes* in noise level (both from loud to quiet and from quiet to loud) produce increases in noise stress (see Teichner et al., 1963). In general, noise stimuli that are novel or unusual (in terms of regularity of the presentation of the noise or variability of the noise) are distracting and tend to degrade task performance.

Duration. There are several temporal characteristics of noise of interest, including duration, intermittency, periodicity, and rise time. Duration is the time period over which the noise occurs. For example, a 100 dB noise may be presented for 100 seconds, or the noise may be presented for a duration of 10 seconds. In general, longer exposure to noise leads to increased decrements in performance. However, Driskell et al. (1992) also found that as exposure to continuous noise increased, individual's subjective feelings of stress decreased. Apparently, for continuous noise a longer noise duration may lead to decreasing stress effects, as the task performer becomes habituated to the noise. For random intermittent noise, it is likely that the longer the noise is presented, the greater the noise effects.

Intermittency. Intermittency is determined by the on–off time of the noise event. A noise event with an on ratio of 100% (with on time being expressed as a percentage of total time) is defined as *continuous,* whereas an on ratio of 0% represents no noise. Any point along this continuum represents *intermittent* noise: For example, a 50% on–off ratio may be represented by 1 second of noise followed by 1 second of silence. Because intermittent noise may be more distracting (Poulton, 1978) and impose more of an information load on the performer (Coates & Alluisi, 1975) than continuous noise, many researchers argue that intermittent noise has a more negative effect on performance (Eschenbrenner, 1971; Theologus, Wheaton, & Fleishman, 1974).

Periodicity. Periodicity is determined by the repetition rate, or the time from the beginning of one noise episode to the beginning of another. *Periodic* intermittent noise is noise that is patterned; for example, a noise of a 5-second duration that repeats every 5 seconds. *Aperiodic* or *random* intermittent noise is noise that is presented at varying repetition intervals; for example, a 5-second noise that is followed by a random duration of silence. Research suggests that random or aperiodic intermittent noise produces a greater decrement on task performance (Glass & Singer, 1972; Percival & Loeb, 1980).

Rise Time. A noise event that is transient and has a short rise time and rapid decay is termed an *impulse noise* or *burst.* Sound of high amplitude and sudden onset has been shown to produce a startle response. For example, May and Rice (1971) found that motor performance was impaired in the period immediately following a pistol shot. The performance impairment accompanying the startle response is quite transient, generally limited to a period of 1–30 seconds following the impulse stimulus.

Type of Task. Almost all reviews of the effects of noise on performance make claims that the effect of noise is dependent on the specific type of task. For example, McCormick and Sanders (1982) stated that cognitive tasks are most sensitive to disruption by noise. Theologus, Wheaton, Mirabella, Brahley, and Fleishman (1973) suggested that psychomotor tasks may be less sensitive to noise effects than cognitive tasks. Coates and Alluisi (1975) concluded that noise has no effect on vigilance performance. In a meta-analytic integration of noise stress research, Driskell et al. (1992) found empirical evidence to support these overall claims. They found that noise resulted in a significant decrease in performance accuracy for cognitive tasks ($r = -.224$) and psychomotor tasks ($r = -.160$), but had no significant impact on vigilance tasks ($r = -.076$).

GROUP PRESSURE

In one of the earliest observations of the effects of others on individual performance, Triplett (1898) observed that cyclists performed better (i.e., faster) when racing with other riders than when paced mechanically or when racing alone. Triplett followed this observation up with what is generally considered one of the first social psychological experiments, and again found that individuals who performed in the presence of others performed better than when alone. This research documented what has been termed *social facilitation:* the enhancement of performance when others are present. However, other early research produced contradictory findings. For example, the work of Moore (1917) and others found that in many instances, the presence of an audience degraded performance. In fact, nearly as many studies demonstrated *social impairment,* or the impairment of performance when others are present, as demonstrated social facilitation. Research on the effects of others on performance stalled until Zajonc published in 1965 an integrative rationale for these inconsistent findings. Research conducted since 1965 has been an extension of or reaction to this theory.

Drive Theory

The earliest comprehensive theoretical perspective on the effects of the presence of others on task performance was offered by Zajonc (1965). His "drive arousal" theory argues that the presence of others increases an individual's drive (a generalized state of activation or arousal), which increases the probability that the individual will emit a dominant response (a dominant response is a well-learned response, or one that is at the top of an individual's response hierarchy).

Zajonc further argued that the effects of this increased drive on performance would depend on the nature of the task. When the task was simple, the dominant response was likely to be the correct one; therefore, the increased drive generated by the presence of others should improve performance (social facilitation). When the task was complex, the dominant response was more likely to be incorrect; therefore, the increased drive generated by the presence of others should degrade performance (social impairment). In summary, Zajonc's approach provided a parsimonious explanation for those studies that showed performance impairment as well as facilitation.

Zajonc (1965) explained the effects of social facilitation/impairment on the basis of the "mere presence" of others. In other words, he argued that the mere presence of others was sufficient to increase a task performer's drive level. A number of other authors have adopted the drive perspective, but argue that factors other than mere presence moderate the effects of the presence of others on drive. Two such perspectives include the evaluation apprehension and the distraction/conflict theories.

Evaluation Apprehension

Cottrell (1972) argued that it was not the "mere presence" of others that caused increase drive, but rather the presence of others who have the potential to *evaluate* that person's performance that increases arousal. Cottrell proposed that audiences produce evaluation anxiety. The presence of an evaluative audience increased drive because of the individual's anticipation of positive or negative rewards. In other words, evaluation apprehensive stems from fear of failure before an evaluative audience. In summary, according to this perspective, it is evaluator apprehension, not the mere presence of others, that causes increased drive or arousal.

Distraction/Conflict

Baron (1986) presented another drive-related explanation of social facilitation/impairment by arguing that the presence of others distracts attention from the task. This creates attentional conflict, or conflict between the attentional demands of the task and the demands of being distracted by others. This conflict between competing response tendencies is a source of increased drive. Sanders (1981) also noted that this attentional conflict may have two distinct effects on performance: (a) It may serve to impair task performance in that it reduces attention to the task, and (b) it may serve to impair complex tasks and facilitate performance in simple tasks through the general effect of increasing drive. Therefore, the distracting effect of the presence of others should always impair complex task performance; however, it should facilitate the performance of simple tasks only to the extent that the beneficial effects of increased drive outweigh the degrading effects of reduced attention.

In summary, drive theories argue that the presence of others increases drive, with drive representing a nondirective increase in response. However, other theorists have proposed explanations which imply that the presence of others causes individuals to behave in a certain manner. Rather than others simply serving a motivational role (intensifying drive), in these theories, the presence of others causes the individual to attempt to behave in a specific way. In the following, we examine three such approaches: self-attention, self-presentation, and information processing.

Self-Attention

According to self-attention theory (Carver & Scheier, 1981), the presence of others causes increases in self-attention. More specifically, the presence of others causes the individual to become more self-aware, both of his or her own present behavior and of the salient standards of behavior. This in turn sets in motion a self-regulation process whereby the individual attempts to match their present behavior to whatever established standards of behavior are prevalent. The discrepancy between current behavior and expected behavior drives the individual to initiate action to match the standard. However, Carver and Scheier argued that this increase in self-directed attention causes performance to be facilitated in some circumstances and to be impaired in others. If the task is relatively easy, the discrepancy between current performance and salient standards is likely to be relatively small, the task performer should develop a positive outcome expectancy for resolving this discrepancy, and the result should be the facilitation of performance as the task performer attempts to improve performance to meet the standard. However, if the task is relatively difficult or complex, the discrepancy between current performance and salient standards is likely to be large, the task performer may develop a negative outcome expectancy for resolving the discrepancy, and he or she may simply withdraw from the task, resulting in performance impairment.

Self-Presentation

Self-presentation theorists have presented a nondrive explanation of social facilitation/impairment that retains the emphasis on evaluation, but holds that subject's *cognitions* related to performance mediate the effects of others on performance (Bond, 1982; Geen, 1979; Sanna & Shotland, 1990). This perspective argues that individuals who expect to do well on a task develop positive task expectations; for them, the presence of an evaluative audience results in improved performance. Individuals who expect to do poorly on a task develop negative task expectations; for them, the presence of an evaluative audience impairs performance. The implication of this approach is that for demanding tasks, individuals are more likely to develop negative performance expectations and perform more poorly in the presence of evaluative others, whereas for easy tasks, individuals are more likely to develop positive expectations of success at the task and perform more competently in the presence of others.

Information Processing

Information processing approaches assume that the presence of others leads to attentional overload which interferes with efficient task performance. Baron (1986) presented an information-processing explanation of the effects of others on performance that is an update of the distraction/conflict model. This approach holds that the presence of others is distracting and leads to attentional conflict as the task performer allocates attention between the task and the observers. Because individuals have a finite amount of attentional capacity, this increased demand can lead to a condition of cognitive overload; this in turn leads to a narrowing of attention to a smaller number of central task cues. (Note that this perspective draws from Easterbrook, 1959, and others who have postulated a narrowing in the range of cue utilization as a response to stress.)

The reduction in the utilization of task information affects task performance according to whether the task is simple or complex. An individual performing a simple task attends to a relatively small number of central task cues and a relatively large number of peripheral or distracting cues. As the informational overload stemming from the presence of others in the task environment leads to a narrowing of attention, it is likely that more of the peripheral cues will be eliminated first. Therefore, this may lead to an enhancement of performance as attention is focused on the remaining central task cues. For complex tasks, the individual must attend to a relatively larger number or wider range of central task cues. Accordingly, as attention is narrowed, it is more likely that central task relevant cues will be eliminated, leading to impairment of performance.

Geen (1989) noted that the effects of reduced attention on performance may depend on whether the task involves automated versus controlled processing of information. Automated processing of information occurs as tasks become well-rehearsed and performance becomes routinized. Automated processing requires less active attention to task cues. Controlled processing of information is required on tasks that are not well-learned, and places more of an active demand on attentional processes. Generally, tasks that are novel, complex, or variable require controlled processing. Because controlled processing places heavy demands on attentional capacity, the overall attentional overload imparted by the presence of others should be greater, leading to greater task impairment. Conversely, tasks that are routine, familiar, or easy require less active attention, and are less subject to disruption by increased attentional demands.

Performance Effects

Research generally supports the proposition that the presence of others facilitates the performance of a simple task and impairs the performance of a difficult task. The debate continues as to *why* this occurs: because of the mere presence of others, because of evaluation apprehension and fear of failure, because of distraction, or for other reasons. Some studies have shown support for the "mere presence" position (Schmitt, Gilovich, Goore, & Joseph, 1986; Worringham & Messick, 1983). For example, Schmitt et al. (1986) found that the presence of others, whether or not they actually observed the performers, affected performance; thus they concluded that the concern with evaluation need not be present for performance impairment. Other studies support the evaluation apprehension position, finding that the more evaluative the audience, the greater the performance impairment (Cottrell, Wack, Sekerak, & Rittle, 1968; Paulus & Murdoch, 1971). A key assumption of the distraction/conflict model is that the presence of others is distracting; a number of studies have reported greater distraction when in the presence of others (Baron, 1986; Sanders, 1981).

We can conclude that the basic effect of the presence of others on performance is robust and substantial—indeed, it is one of the most well-documented phenomena in social psychology. Sometimes this effect occurs in the mere presence of others. In a highly evaluative setting, evaluation apprehension and fear of failure may also become salient. Sometimes the presence of others may be distracting; in other cases, the presence of others may lead to reduced attention to the task. Which of these theoretical perspectives is more informative will depend on the nature of the task situation.

THREAT

We define threat in broad terms as the *anticipation or fear of physical or psychological harm*. Thus, a threat-provoking situation is one in which dangerous and novel environmental events pose the potential for pain or discomfort. Evidence from a broad range of studies indicates that the threat of dangerous or novel environments may result in increased subjective stress and impaired performance. For example, studies of parachuting report increases in subjective stress before and during early jumps (Basowitz et al., 1955; Powell & Verner, 1982). Both Weltman, Christianson, and Egstrom (1970) and Mears and Cleary (1980) found psychological stress to be a significant factor in degrading underwater (diving) performance.

There are three primary settings in which threat has been investigated. The first setting includes *real-world events* such as combat, aircraft emergencies, impending surgery, or natural disasters. Although these real-world events rarely provide the opportunity for controlled research experiments, a number of valuable studies have examined reactions to combat or to natural disasters. A second environment for studying threat includes *real-world simulations* such as realistic exercises or simulated emergencies. These simulations are manmade, yet designed to be as realistic as possible. For example, Keinan (1988) examined the performance of military personnel in a training setting using live artillery fire. A third environment for studying threat includes *laboratory studies*, in which threat is manipulated in an experimental laboratory setting. For example, Lee (1961) studied the effects of threat of shock on the performance of a verbal learning task. Representative research within each of these settings is presented in the next sections.

Real-World Events

A number of military studies have examined the effects of stress in combat. Many of these studies were conducted during World War II and documented the impact of battlefield threats on military performance. For example, Reid (1945) analyzed the calculation and plotting errors involved in measuring wind vectors by navigators on operational sorties during World War II. He found that compared to errors made over England, errors increased significantly once bombers crossed the enemy coast and increased even further as the bombers approached the target. When the bombers crossed the coast on the return journey, errors declined (suggesting that the initial dip in performance was probably not attributable to fatigue). Cox and Rachman (1983) examined the self-reported fear of British military bomb disposal operators during a 19-week tour of duty in Northern Ireland. The majority of personnel reported having experienced some degree of fear during their tour of duty, especially early in the tour. However, Cox and Rachman found few significant factors that predicted the extent of psychological reaction.

Other investigators have examined demanding environments such as parachuting. Not surprisingly, most investigators have found increases in subjective anxiety and impaired performance either preceding or during early jumps. Hammerton and Tickner (1967) found that novice military parachutists showed a decrement in a tracking task immediately before and after their descent from a balloon. MacDonald and Labuc (1982) observed military personnel at various stages in parachuting training. They measured the performance of novice and experienced parachutists on four tasks (logical reasoning, tracking, visual search, and decoding map references) during a baseline period in advance of parachute training, before jumping, and after jumping. Significant decrements in performance on the tests were found in nearly all phases of training. Burke (1980) examined Army jumpmaster training in order to search for variables predictive of ability to perform under threat. He found that the variable most highly correlated with jumping performance was perceived stress ($r = -.38$).

Deep diving is another area that has received considerable research attention. Research on divers has shown that manual dexterity is impaired underwater, and that the more dangerous the dive, the greater the impairment (Baddeley, 1966; Baddeley & Flemming, 1967; Baddeley, de Figueredo, Hawkswell-Curtis, & Williams, 1968). However, in a number of deep-diving studies, it is difficult to differentiate the effects of stress from the effects of nitrogen narcosis on performance.

Real-World Simulations

A defining characteristic of real-world simulations is that they attempt to provide a simulated and controllable, yet realistic, setting for examining the effects of threatening environments. Berkun, Bialek, Kern, and Yagi (1962) developed a realistic series of stress studies for the Army, all of which involved contrived emergencies and were designed to make subjects feel that they were actually in danger of losing their lives or had endangered others. In one scenario, research subjects were military basic trainees who were taken up in a DC-3 aircraft. In the experimental "threat" group, a severe inflight emergency was simulated: The plane lurched, one propeller was stopped by the pilot, an emergency was declared, and subjects were told to prepare for ditching. In addition, as the aircraft passed by the airfield, subjects could see fire trucks and ambulances on the airstrip in apparent expectation of a crash landing. After several minutes, the experimenter (posing as a steward) administered two questionnaires, both of which were designed to appear as plausible emergency information, but were actually cognitive tests. (One of the forms was given under the pretext that it would be used to furnish proof to insurance companies that emergency procedures had been properly followed.) A short time later, the aircraft made a safe landing, the subjects filled out more questionnaires, and were subsequently debriefed.

Berkun et al. found that the experimental (threat) group expressed greater subjective stress and performed more poorly on the performance measures than a control group who were not exposed to the threat. Berkun et al. implemented several other contrived emergency situations, including one in which research subjects were led to believe that they had detonated a charge of TNT and seriously injured another person. Over the next 45 minutes, as medical assistance was being summoned and the Military Police were on their way to question the subject (who thought he had apparently hurt someone), the subject completed several performance tasks. Performance was significantly poorer for the experimental (threat) subjects than for the controls. Not too surprisingly, these studies drew considerable professional censure for the unethical treatment of research participants.

More recent research has examined aircrew performance in realistic flight simulations. In one classic simulator study, B-747 flight crews were observed on a simulated trip from New York's Kennedy Airport to London (Ruffell Smith, 1979). During this tightly scripted scenario, an oil-pressure problem forced the crew to shut down an engine. The crew had to decide where to land the plane; however, this decision was further complicated by a hydraulic system failure, bad weather, poor air-traffic control, and a cabin crew member who demanded attention at the worst possible moments. Researchers recorded a total of 450 procedural errors in the 20 flight crews studied. Many of the errors were related to poor coordination among flight crew members. The demonstration of the adverse impact of high-stress conditions on interpersonal behavior, and the resulting breakdowns in crew coordination, provided the impetus for the development of Crew Resource Management (CRM) training within the commercial and military aviation community.

Laboratory Investigations

Ethical problems arise when subjects are put into situations designed to cause intense fear without their prior knowledge of the scope and purpose of the study. Most laboratory research, typically conducted in university laboratories with undergraduate research participants, incorporates more moderate levels of stress. (There are few institutional review boards that would allow a researcher to fire artillery over a student's head, even a freshman's.) However, whatever the researcher may lose in realism he or she gains in increased precision and greater control over experimental variables.

By far, the most common laboratory manipulation of threat is the threat of electrical shock. Most studies show that the threat of shock results in a significant increase in subjective stress. For example, Cox (1984) challenged a control group of research subjects to improve their performance on a pursuit rotor task, and challenged an experimental group by telling them they could avoid electrical shock by improving their performance. Those subjects threatened with shock reported significantly greater subjective stress. Most studies also indicate that the threat of shock leads to significant decrements in task performance. For example, Lee (1961) found that the threat of shock led to a decrease in the mean number of correct responses on a verbal learning task. Ryan (1962) found that the threat of shock impaired performance of a difficult motor task (balancing on a stabilometer, a horizontally pivoted platform).

Threat and Controllability. A number of researchers have argued that control can reduce the aversiveness of a threatening event (see Thompson, 1981). Control refers to the capability of the individual to alter or modify the environment. Two primary types of control that have been identified are *behavioral* and *cognitive control* (Averill, 1973; Thompson, 1981). Behavioral control implies that one has the behavioral capacity to terminate an event, shorten its duration, or otherwise avoid an aversive event. Cognitive control refers to the belief that one has a cognitive strategy available that can lessen the aversiveness of a threatening event.

Many laboratory studies of the threat of shock have found that subjects report less anticipatory stress in a situation in which they believe they have behavioral control (i.e., they could terminate or avoid the electrical shock) as well as when subjects believe they have some degree of cognitive control over the threatening event (Bowers, 1968; Champion, 1950; Corah & Boffa, 1970; Houston, 1972; Szpiler & Epstein, 1976). The moderating effect of controllability on stress reactions has been further demonstrated in studies of crowding, (Baum & Paulus,

1987; Epstein, 1982), noise (Cohen & Weinstein, 1981), air pollution (Evans & Jacobs, 1982), and heat (Bell & Greene, 1982).

Several concepts that are related to controllability include *temporal uncertainty, ambiguity,* and *unpredictability.* Temporal uncertainty refers to not knowing *when* an event is going to happen. An already threatening situation is more stressful if the individual knows that an unpleasant event is imminent, but does not know exactly when it will occur (Monat, Averill, & Lazarus, 1972). *Ambiguity* (a lack of situational clarity) is a type of uncertainty that may intensify threat by limiting the individual's sense of control and/or increasing a sense of helplessness (Lazarus & Folkman, 1984). Ambiguity is increased when task stimuli are difficult to distinguish, present events that disconfirm expectancies, or present conflicting information. *Predictability* is a concept closely related to controllability, although one may be able to predict the occurrence of an event and yet not control it. Predictability refers to the extent to which an event or task sequence can be predicted or anticipated. Evidence suggests that predictable stressors are less aversive than unpredictable stressors. For example, threatening stimuli that are predictable by way of a warning signal (a signal which occurs consistently prior to the stress event) are less aversive than unpredictable or unsignaled stressors (Badia & Culbertson, 1970; Weinberg & Levine, 1980).

TASK LOAD

Often, individuals are forced to perform more than one task at a time. Sometimes these are overt tasks (e.g., an air traffic controller talking to a pilot while scanning a visual display), whereas other times they may be covert (problem solving while driving a car). However, many studies show that performing multiple tasks carries a penalty, and individuals may perform more poorly due to increased task load. We define *task load* as performing two or more tasks concurrently. Task load is also referred to as divided attention or dual task performance. Broadly speaking, task load refers to the pressure or demand of performing multiple tasks.

Two major sets of theories (structural and capacity theories) have been offered to explain performance under high task load. Capacity theories assume that there is a limited pool of attentional resources, or capacity, that can be divided across tasks. Structural theories assume that the human information processing system is parallel, capable of processing separate channels, but at some point will narrow to a serial system that must handle only one channel at a time. An overview of these theories is provided in the following.

Structural Theories

Structural theorists seek to answer the question, "At what stage of processing does a parallel system, capable of processing separate channels concurrently 'narrow' to a serial system that must handle one channel at a time?" Broadbent (1958) and Treisman (1969) theorized that the bottleneck occurred at the stage of perception. This idea became known as *early selection theory.* In contrast, *late selection* theorists (e.g., Deutsch & Deutsch, 1963; Keele, 1973; Norman, 1968) postulated that the bottleneck occurred at the stage at which decisions were made to initiate a response (either an overt motor response or a covert response, such as storing material in long-term memory, or rehearsing it). *Late selection theory* assumes that a dedicated decision-making response-selection mechanism must be available in order for an individual to perform a task.

Kerr (1973) proposed a theory which accommodates both early and late selection theory, arguing that there is not a single stage of information processing that becomes the bottleneck. Instead, he postulated, there is a single limited capacity central processor (LCCP). The LCCP is said to act as a single server queue that must be engaged to complete certain mental operations, such as selecting a response, performing a mental transformation, or rehearsing. Thus, when the LCCP is in operation for one task, it is unavailable for a second concurrent task that might also require it. Therefore, performance on the second task will deteriorate. This view, which postulates that a number of mental operations require the LCCP in order to proceed, allows the possibility of more than a single bottleneck within the processing system.

Structural theorists acknowledge the role of task difficulty (a capacity concept) in generating interference by assuming that more difficult tasks occupy the bottleneck or LCCP for a relatively longer period of time. However, the emphasis of these theories is on structure, with either a stage or processing or a LCCP assumed to service only one process or task at a time.

Capacity Theories

Capacity theories typically examine how an individual's performance trades off between two tasks as task demands change. As a primary task (e.g., driving on a crowded interstate highway) demands more of a person's resources, fewer resources are available for a concurrent secondary task (e.g., speaking on a cellular phone), and performance on the latter task will be disrupted. In a similar manner, as the demands of the phone conversation increase, performance on the primary task, driving, may suffer. Thus, capacity theorists maintain that capacity can be allocated in graded quantity between separate activities. In 1967, Moray drew an analogy between human processing resources and the limited capacity of a general purpose computer, which can apply its limited capacity interchangeably to widely different classes of processing. Given such flexibility, Moray argued that it was not necessary to assume a given locus of task interference (or attentional bottleneck). The cause of interference would depend merely on the capacity demands at any particular stage of processing.

During the 1970s, the concept of capacity or resources as an intervening variable in dual task performance was developed from a loose concept to a quantitative theory with testable predictions by Kahneman (1973), Norman and Bobrow (1975), and Navon and Gopher (1979). Wickens (1984) has summarized and elaborated on these formulations, which he refers to as *resource theory.*

General Effects

Most studies of dual task performance indicate that the addition of a second, concurrent task tends to impair performance on the primary task (Kahneman, 1975; McLeod, 1977; Neisser & Becklan, 1975). Research further suggests that the negative effect of task load on performance is greater when the tasks performed are similar and when the tasks are novel, unfamiliar, or difficult.

Task Similarity. Some everyday activities can be performed simultaneously without difficulty, such as talking on the phone while drawing, or reading and listening to music. However, our experience also teaches us that some tasks, such as listening to a request from someone in the room while conversing on the phone, are more difficult to perform concurrently. There is a substantial body of research suggesting that the degree of similarity between two tasks is of great importance in determining the impact of task load on performance. Eysenck (1984) distinguished between three types of similarity: (a) similarity of stimuli involved in the task; (b) similarity of internal processing operations; and (c) similarity of responses.

Most of the research on *stimuli similarity* in dual task performance has focused on the sense modalities to which the task stimuli are presented. It seems to be easier to continuously divide attention between two inputs in different modalities than it is to divide attention between two inputs in the same modality. For example, it appears that is it difficult for an individual to handle two concurrent auditory inputs or two concurrent visual inputs. Allport, Antonis, & Reynolds (1972) found that research subjects who attempted to verbally shadow (repeat back) prose passages while learning auditorily presented words performed more poorly than when they read back passages while learning words presented visually.

Treisman and Davies (1973), Eijkman and Vendrik (1965), Moore and Massaro (1973), and Tulving and Lindsay (1967) found little or no decrement in individuals' accuracy in simultaneously detecting tones (auditory stimuli) and lights (visual stimuli), compared to their accuracy in detecting only a single target. In fact, Treisman and Davies discovered that two monitoring tasks interfered with each other to a greater extent when the stimuli for both tasks were presented in the same sense modality, whether it was visual or auditory.

Response similarity has also been shown to determine the impact of task load on performance. In a study by McLeod (1977), research subjects performed a continuous tracking task requiring manual response. Subjects also performed a concurrent tone identification task; half of the subjects were required to respond verbally, whereas the other half responded manually with the hand not involved in the tracking task. Performance on the tracking task was worse under conditions of high response similarity (manual responses on both tasks) than under low response similarity (manual response on one task and verbal response on the second).

In examining *processing similarity,* Wickens (1984), Navon & Gopher (1979) and Wickens and Flach (1988) described how tasks compete for specific processing resources within the brain. There are three dichotomous resource dimensions: (a) processing modalities (auditory vs. visual), (b) processing codes (verbal vs. spatial), and (c) processing stages (working memory vs. response). Wickens and Flach asserted that to the extent that two tasks share common levels on any of the three dimensions, timesharing will be less efficient, and to the extent that an increase in resource demand occurs at the level of the dimension shared by another task, there will be increasing interference between the two. Note that the first dichotomous resource dimension is analogous to Eysenck's (1984) concept of stimuli similarity discussed earlier. The second two resource components, processing codes and processing stages, are similar to Eysenck's concept of internal processing operations.

The dimension of processing codes (verbal vs. spatial) distinguishes information that is mostly spatial and analog in nature from that which is verbal and linguistic. This dichotomy suggests that a mixture of graphics and digital or verbal displays provides a more optimal format for presenting multiple task information than a homogeneous display.

The dimension of processing stages distinguishes between working memory and response. Wickens (1980) reviewed data which suggested that two tasks both demanding either response processes or perceptual or cognitive processes (e.g., decision making, working memory, information integration) will interfere with each other to a greater extent than will two tasks, one of which requires perceptual or cognitive processes and the other requiring response processes.

Driskell et al. (1992) conducted a meta-analysis of the research literature on dual task performance, specifically examining studies that assessed how the addition of a secondary task affected performance accuracy on a primary task. Results of this analysis indicated that the addition of a secondary task led to a significant and moderately large decrease in the accuracy of performance on the primary task. Moreover, the more similar the two tasks in requiring similar processing modalities, the more performance on the primary task suffered when the secondary task was added.

Practice. A number of studies have shown that highly practiced tasks can be performed jointly with little interference. For example, Spelke, Hirst, and Neisser (1976) and Hirst, Spelke, Reaves, Caharack and Neisser (1980) found that when subjects were asked to read aloud prose while taking dictation, performance dropped dramatically. However, with substantial dual task practice—over 50 hours—subjects could more readily read while taking dictation, achieving reading and comprehension rates similar to that of the single-task control subjects.

Fisk and Schneider (1983) found that task consistency plays an important role in determining how practice affects dual task performance. Task consistency has been manipulated in these and other studies by comparing consistent mapping (CM) versus varied mapping (VM) tasks. Mapping tasks consist of presenting subjects with a memory set of letters or numbers and then presenting a set of numbers or letters in a series of successive trials, with the subjects' task being to determine whether or not those items are in the memory set. With consistent mapping, the memory items never appear in the visual display except as targets. With variable mapping, memory set items presented in one trial may serve as distractors in further trials. In the Fisk and Schneider study, subjects performed a digit-span task, coupled with either a consistent visual category search task, or a variable visual category search task. When the digit-span task and consistent task were first performed together, detection accuracy dropped by about 10% from the single task detection baseline. After 90 additional trials of dual-task practice, however, category detection accuracy reached a level comparable to single-task accuracy. In contrast, when the digit-span task and the variable task were first performed together, detection accuracy dropped by about 25% from single-task levels. More importantly, additional dual-task practice of the digit-span and variable-search tasks did not improve detection accuracy.

Schneider and Detweiler (1988) proposed that automatic/controlled processing theory accounts for the observed changes with consistent practice. Automatic processing occurs in well-practiced, consistent tasks and is characterized as fast, parallel, relatively effortless, and not under subjective control. Automatic processing allows a person to carry out a task in an essentially resource-free fashion, avoiding interference with concurrent tasks. Conversely, controlled processing usually occurs in novel and inconsistent tasks and is characterized as slow, generally serial, effortful, capacity-limited, and largely under subjective control. Controlled processing is likely when the individual must vary his or her response from trial to trial. Most complex tasks require a combination of automatic and controlled processing.

Task Difficulty. It seems intuitively obvious that performing two simple tasks together should be easier than performing two difficult tasks together. Unfortunately, the concept of task difficulty is somewhat ambiguous, and there are a number of ways in which task difficulty has been operationalized. Hitch and Baddeley (1976) had research subjects perform a verbal-reasoning task in conjunction with rapid overt rehearsal of six digits; the digits were either ordered "one, two, three, four, five, six" or in a random sequence. When the more difficult random-digit sequence was rehearsed, subjects took considerably longer to perform the verbal reasoning task. Degradation of dual task performance due to task difficulty was also found by Sullivan (1976). Subjects performed two auditory tasks: shadowing (repeating) a spoken message represented to one ear and detecting target words in a second spoken message presented to the other ear. When the shadowing task was made more difficult by using a less redundant message, fewer targets were detected on the nonshadowed message.

TIME PRESSURE

Much of the world's work (and sometimes it seems like all of one's own work) consists of tasks that have some degree of temporal urgency. Many complex real-world tasks must be performed under extreme time pressure. For example, the crew of the USS Vincennes, cruising in the Persian Gulf, had approximately 3 minutes following initial radar contact of an unidentified aircraft to evaluate the threat posed by the approaching aircraft to the ship and take action to protect the ship (the Vincennes incident is described in detail in chapter 4). Just 4 minutes into the 1978 accident at the Three Mile Island nuclear power plant, operators were faced with a loss of reactor coolant (the most feared consequence in a nuclear plant in that it means the core can melt); reactor coolant pumps cavitating, thumping, and shaking; three audible alarms sounding; and many of the 1,600 annunciator lights on or blinking (Perrow, 1984). The members of the NASA and Morton Thiokol management team who made the ill-fated decision to launch the Challenger shuttle on January 28, 1986, had deliberated under the time pressure of three prior postponements of the scheduled launch date, a scrubbed launch attempt on the previous morning, and a narrowing window of opportunity for the Challenger to deliver a payload to observe Halley's Comet.

We may define *time pressure* as the restriction in time required to perform a task. Research suggests that time pressure may degrade performance because of the cognitive demands, or information overload, imposed by the requirement to process a given amount of information in a limited amount of time (Wright, 1974). Under the demand of time pressure, individuals may resort to several types of strategies to reduce information overload. The first strategy—*acceleration*—refers to the process whereby the individual increases information integration in order to "match" the speed with which information is being presented. This strategy assumes that the task performer is processing information at some suboptimal rate initially and is thus able to increase processing activity in order to reduce the impact of time stress. It is unlikely that individuals performing complex tasks will be operating at this suboptimal level to begin with, so this strategy may have limited application for many real-world tasks.

A second type of strategy involves a change or adjustment in information processing strategy. Under time pressure, individuals may alter the means by which they process information. For example, if an individual is evaluating each available alternative sequentially before attempting a task solution, he or she may adopt a strategy in which all alternatives are considered according to a specific criteria prior to decision making. These adjustments in decision-making strategy may provide an adaptive response to the constraints imposed by time pressure, but in some cases may lead to poorer performance if an inappropriate strategy is adopted.

A third strategy—*filtration*—refers to the general tendency for individuals to restrict information processing when under stress. By filtering or reducing the amount of information to be processed, the individual may reduce the overload imposed by time pressure. However, this restriction of environmental information may have detrimental effects on task performance if attention to task-relevant information is also restricted.

Research has documented the effects of time pressure on performance accuracy and speed. The typical finding is that under time pressure, subjects tend to work more quickly, but performance accuracy declines. In a study by Lulofs, Wennekens, and Van Houtem (1981), time pressure was manipulated by varying the interstimulus interval in a two-choice reaction time task, performed while cycling on a bicycle ergometer. Increasing time pressure induced subjects to respond more rapidly, however, at the expense of decreased accuracy. This speed–accuracy tradeoff has been noted in other similar studies (Link, 1971; Pachella, Fisher & Karsh, 1968).

Danev, de Winter, and Wartna (1972) also reported detrimental effects of time pressure on performance accuracy. This study required subjects to add 10 pairs of three-digit numbers per trial at their own pace or with a time limit which was slightly shorter than that required in the self-paced condition. Under time constraint, more errors, omissions, and "mental blocks" (unusually long addition times) were noted. McBride (1988) examined group performance under time pressure. This investigation involved groups of three members performing a dynamic group choice task. Time stress was induced by increasing the rate at which information was presented. Certain teams were taught a set of decision rules (heuristics), including what task events to ignore and what events to note. These groups reported lower levels of stress and performed better under moderate time pressure than under high time pressure. McBride concluded that with moderate time pressure, subjects were able to use these heuristics to enhance performance; however, high time pressure precluded subjects' ability to incorporate these strategies.

Decision Making. A number of studies have demonstrated that time pressure may impact the way people make decisions. Janis and Mann (1977) presented a model of decision making in which they distinguished between *vigilant* and *hypervigilant* decision-making patterns. Vigilant decision making was described as a thorough, systematic, rational pattern of decision making. According to Janis (1982b), the vigilant decision maker "searches painstakingly for relevant information, assimilates information in an unbiased manner, and appraises alternatives carefully before making a choice" (p. 73). This pattern of thorough consideration of all available alternatives and systematic, organized information search, Janis concluded, generally results in high-quality decision making.

Hypervigilant decision making is described as an impulsive, disorganized pattern of decision making. According to Janis and Mann (1977), the hypervigilant decision maker tends to "search frantically for a solution, [consider] a limited number of alternatives, and then latch onto a hastily contrived solution" (p. 51). In contrast to vigilant decision making, a hypervigilant pattern of decision making is characterized by consideration of limited alternatives, nonsystematic information search, rapid evaluation of data, and selection of a solution without extensive review or reappraisal.

Janis and Mann (1977) argued that stress may lead to the use of hypervigilant decision strategies and that this strategy often led to maladaptive actions and poor judgment. On the other hand, they noted that, under certain conditions, a hypervigilant decision making strategy may in fact be *adaptive*. Janis (1982b) noted that the combination of "sudden, unexpected threat, with extreme time pressure to avert the danger" (p. 81) leads to hypervigilance. Furthermore, under these conditions (that typify many real-world situations of interest), decision makers do not have the luxury to painstakingly search for information, weigh all available alternatives, and systematically eliminate each to arrive at a problem solution. In fact, under time pressure, Janis and Mann (1977) noted that vigilant decision making may be maladaptive: "Decision makers are often under severe pressure of time, which precludes careful search and appraisal" (p. 22).

Driskell, Salas, and Hall (1994) conducted a study to examine the use of vigilant and hypervigilant decision strategies. Results indicated that, on a naturalistic task simulation, subjects employing a hypervigilant decision strategy performed better than those employing a vigilant strategy. Furthermore, subjects employing a vigilant strategy performed more poorly under time pressure than under no-stress conditions, whereas the effectiveness of a hypervigilant strategy did not degrade under stress. These results suggest that the view of hypervigilant decision making as indicative of a performer "breaking down" under stress is not accurate. The results of this study provide empirical evidence to support Janis' (1982b) claim that under time pressure, a hypervigilant decision style can be an adaptive and effective strategy.

A number of other changes in decision making under time pressure have been documented. For example, Wright (1974) demonstrated that under time pressure individuals tend to accentuate negative evidence and use less information in judgment decisions. Rothstein and Markowitz (1982) gave research subjects 7 seconds to decide which of two sets of numbers had the larger mean. Subjects under time pressure simplified the task by using the larger sum as the indicator of the correct alternative, even when the set with the larger sum had the smaller mean. Ben Zur and Breznitz (1981) demonstrated that decision makers under time pressure tended to make choices with lower risk (where low risk is defined as choices of gambles with lower variance) and spend more time viewing negative dimensions (amount of loss and probability of loss). Rothstein (1986) studied the effects of time pressure on judgment, and found that although subjects were able to implement sound decision policies under time pressure, their behavior was much more erratic than under unstressed conditions.

Some investigators have studied how people make decisions under time pressure in negotiations. As Carnevale and Lawler (1986) noted, time pressure is a common feature of nearly all negotiations: Typically, the negotiators desire to end the negotiation quickly, the process of negotiation takes time away from other pursuits, the goods in dispute may have a limited shelf life (e.g., fruit beginning to spoil), and a specific deadline may be set for the negotiators to reach an agreement. Most research suggests that time pressure increases the level of cooperation of parties involved in negotiation because it facilitates concession making (see Pruitt, 1981; Rubin & Brown, 1975). However, Walton and McKersie (1965) argued that time pressure is likely to inhibit joint problem solving and, specifically, the search for integrative agreements. Pruitt and Carnevale (1982) distinguished between integrative agreements and compromises. Integrative agreements reconcile the two parties' divergent interests and are highly beneficial to both parties (Carnevale & Lawler, 1986). Compromises involve concessions to a middle point on some dimension of value: A compromise only partly satisfies each party. It seems likely that integrative agreements require more time to develop than nonintegrative or compromise agreements, and that high time pressure may interfere with their development. For example, Yukl, Malone, Hayslip, and Pamin (1976) manipulated time pressure in a two-issue negotiation task. Research subjects vied for points, which they could attain by reaching an agreement within 30 minutes. Under high time pressure conditions, the negotiators were given 30 extra points and told that they would lose 1 point for each minute spent in bargaining. Under low time pressure conditions, no extra points were given and there was no cost for the time spent in negotiation. Yukl et al. reported that under high time pressure, subjects reached agreements sooner, but they made fewer offers and reached poorer joint outcomes.

SUMMARY

At the beginning of this chapter, we argued that stress plays a defining role in our complex, high-technology society than is unprecedented in history. This type of claim is always somewhat suspect. Certainly, each period of history has presented unique demands on its inhabitants, from the time in which one's daily work was likely to be interrupted by a marauding tiger to the relatively lawless times of the Wild West. However, it is likely that those who populate the latter half of the 20th century and who will command the 21st century, from pilot to fireman to office worker, are likely to face an environment in which they must perform under more time pressure and under greater task load, in which stress is more prevalent, and in which the consequences of poor performance are more critical than ever before.

There are several reasons why we believe that a concern with the effects of stress on performance will continue to play an increasingly salient role in both professional and practical debate. First, advances in equipment and hardware are fast exceeding the capability of those who must operate them. For example, the demands placed on a military jet pilot by modern high technology aircraft are enormous. Furthermore, the push to design more sophisticated, more complex, and more demanding technological systems shows no signs of abating. Engineering "fixes," rather than reducing task complexity, often contribute to it. For example, any gains achieved through automation, ostensibly a means to lessen task load by offloading some tasks from the operator to the machine, often simply give the engineer the leeway to add more capabilities to the system. The result is little net reduction and often an increase in overall task load. Engineers have also designed a head-up display (HUD) that projects aircraft information onto the cockpit windshield. This allows the pilot to maintain a direct line of vision outside the cockpit with less scanning of the instrument panel. The function of a head-up display is not to make the pilot's job any *easier,* but to allow the pilot to maintain control of aircraft that are becoming faster and more complex, and to fly them under more demanding conditions.

A second reason we believe the topic of stress and performance will continue to gain in interest is that the scale and destructive potential of technology is ever increasing. Our airways are becoming more crowded, tankers are becoming larger (and older—industry statistics reported by Enders, 1993, indicate that 65% of the world's tankers are 15 years old, considered a typical lifespan), Navy battleships are more lethal, and high-speed rail systems will travel faster. Perrow (1984) presented the thesis that advanced technologies involved in military systems, petrochemicals, recombinant DNA technology, and other "high-tech" areas are becoming so complex that accidents must be considered a normal or inevitable feature of their operation. In brief, because of the increasing scale and destructive potential of these systems, failures or accidents are more likely to occur, to be substantial, and involve greater damage to life and property.

Finally, we as a consuming public seem unwilling to accept errors; that is, we want fast and inexpensive air travel but with fewer accidents, we want cheap oil but pristine waters, we want the most extensive and exotic emergency medical treatment available but are intolerant of any mistakes. Therefore, we want the advantages provided by advanced technology, but we do not accept that the increased demands imposed by these systems on those who operate them lead to increased potential for error. We seem to have developed a "zero risk" mentality—we are tolerant of very little risk in our daily lives.

Compounding this general societal aversion to risk is the fact that we live in a worldwide community in which news is nearly instantaneous. We hear of the crash of a commuter aircraft in central Europe almost as quickly as we hear of an automobile accident on a local highway. The extensive media coverage of worldwide incidents such as airplane accidents, industrial accidents, and railway mishaps makes these dangers *seem* much larger. Therefore, we have become highly sensitized to the apparent dangers of air travel, marine accidents, rail travel, industrial mishaps, medical mistakes, and military accidents that are portrayed on the daily news. The concern with how individuals perform in these demanding, stressful environments is likely to continue to be of growing theoretical and practical interest.

REFERENCES

Allport, D. A., Antonis, B., & Reynolds, P. (1972). On the division of attention: A disproof of the single-channel hypothesis. *Quarterly Journal of Experimental Psychology, 24,* 225–235.

Averill, J. R. (1973). Personal control over aversive stimuli and its relationship to stress. *Psychological Bulletin, 8,* 286–303.

Baddeley, A. D. (1966). The influence of depth on the manual dexterity of free divers: A comparison of open sea and pressure chamber testing. *Journal of Applied Psychology, 50,* 81–85.

Baddeley, A. D., de Figueredo, D., Hawkswell-Curtis, H., & Williams, A. N. (1968). Nitrogen narcosis and preformance underwater. *Ergonomics, 11,* 157–164.

Baddeley, A. D., & Flemming, N. C. (1967). The efficiency of divers breathing oxy-helium. *Ergonomics, 10,* 311–319.

Badia, P., & Culbertson, S. (1970). Behavioral effects of signalled vs. unsignalled shock during escape training in the rat. *Journal of Comparative and Physiological Psychology, 72,* 216.

Bandura, A. (1982). Self-efficacy mechanism in human agency. *American Psychologist, 37,* 122–147.

Bandura, A., Reese, L., & Adams, N. E. (1982). Microanalysis of action and fear arousal as a function of differential levels of perceived self-efficacy. *Journal of Personality and Social Psychology, 43,* 5–21.

Baron, R. S. (1986). Distraction-conflict theory: Progress and problems. In L. Berkowitz (Ed.), *Advances in experimental social psychology* (Vol. 19, pp. 1–40). New York: Academic Press.

Basowitz, H., Persky, H., Korchin, S., & Grinker, R. (1955). *Anxiety and stress: An interdisciplinary study of a life situation.* New York: McGraw-Hill.

Baum, A., & Paulus, P. (1987). Crowding. In D. Stokols & I. Altman (Eds.), *Handbook of environmental psychology* (pp. 533–570). New York: Wiley.

Baumeister, R. F. (1984). Choking under pressure: Self-consciousness and paradoxical effects of incentives on skillful performance. *Journal of Personality and Social Psychology, 46,* 610–620.

Bell, P., & Greene, T. (1982). Thermal stress: Physiological comfort, performance, and social effects of hot and cold environments. In G. W. Evans (Ed.), *Environmental Stress.* New York: Cambridge University Press.

Ben Zur, H., & Breznitz, S. J. (1981). The effects of time pressure on risky choice behavior. *Acta Psychologica, 47,* 89–104.

Berkun, M., Bialek, H., Kern, R., & Yagi, K. (1962). Experimental studies of psychological stress in man. *Psychological Monographs, 76*(15, Whole No. 534).

Boggs, D. H., & Simon, R. S. (1968). Differential effect of noise on tasks of varying complexity. *Journal of Applied Psychology, 52,* 148–153.

Bond, C. F. (1982). Social facilitation: A self-presentational view. *Journal of Personality and Social Psychology, 42,* 1042–1050.

Bowers, K. (1968). Pain, anxiety, and perceived control. *Journal of Clinical and Consulting Psychology, 32,* 596–602.

Broadbent, D. E. (1957). Effects of noises of high and low frequency on behavior. *Ergonomics, 1,* 21–29.

Broadbent, D. E. (1958). *Perception and communication.* London: Pergamon.

Broadbent, D. E. (1971). *Decision and stress.* New York: Academic Press.

Broadbent, D. E. (1979). Human performance and noise. In C. M. Harris (Ed.), *Handbook of noise control* (pp. 17.1–17.20). New York: McGraw-Hill.

Burger, J. M., & Arkin, R. (1980). Prediction, control, and learned helplessness. *Journal of Personality and Social Psychology, 38,* 482–491.

Burke, W. P. (1980). *Development of predictors of performance under stress in Jumpmaster training* (Research Report No. 1352). Ft. Benning, GA: U.S. Army Research Institute.

Cannon, W. B. (1935). Stresses and strains of homeostasis. *American Journal of Medical Science, 189,* 1–14.

Carnevale, P. J. D., & Lawler, E. J. (1986). Time pressure and the development of integrative agreements in bilateral negotiations. *Journal of Conflict Resolution, 30,* 636–659.

Carver, C. S., & Scheier, M. F. (1981). Self-attention induced feedback loop and social facilitation. *Journal of Experimental Social Psychology, 17,* 545–568.

Cassell, E. E., & Dallenbach, K. M. (1918). The effect of auditory distraction upon the sensory reaction. *American Journal of Psychology, 29,* 129–143.

Champion, R. A. (1950). Studies of experimentally induced disturbance. *Australian Journal of Psychology, 2,* 90–99.

Clark, D. M. (1988). A cognitive model of panic attacks. In S. Rachman & J. D. Maser (Eds.), *Panic: Psychological perspectives* (pp. 71–89). Hillsdale, NJ: Lawrence Erlbaum Associates.

Coates, G. D., & Alluisi, E. A. (1975). *A review and preliminary evaluation of methodological factors in performance assessments of time-varying aircraft noise effects.* Hampton, VA: National Aeronautics and Space Administration, Langley Research Center.

Cohen, E. L. (1952). The influence of varying degrees of psychological stress on problem-solving rigidity. *Journal of Abnormal and Social Psychology, 47,* 512–519.

Cohen, H. H., Conrad, D. W., O'Brien, J. F., & Pearson, R. G. (1974). *Effects of noise upon human information processing* (Research Report No. NASA CR–132469). Washington, DC: National Aeronautics and Space Administration.

Cohen, S. (1978). Environmental load and the allocation of attention. In A. Baum, J. E. Singer, & S. Valins (Eds.), *Advances in environmental psychology* (Vol. 1, pp. 1–29). Hillsdale, NJ: Lawrence Erlbaum Associates.

Cohen, S. (1980). Aftereffects of stress on human performance and social behavior: A review of research and theory. *Psychological Bulletin, 88,* 82–108.

Cohen, S., & Weinstein, N. (1981). Nonauditory effects of noise on behavior and health. *Journal of Social Issues, 37,* 36–70.

Combs, A. W., & Taylor, C. (1952). The effect of the perception of mild degrees of threat on performance. *Journal of Abnormal and Social Psychology, 47,* 420–424.

Corah, N. L., & Boffa, J. (1970). Perceived control, self-observation, and response to aversive stimulation. *Journal of Personality and Social Psychology, 16,* 1–4.

Cottrell, N. B., Wack, D. L., Sekerak, G. J., & Rittle, R. H. (1968). Social facilitation of dominant responses by the presence of an audience and the mere presence of others. *Journal of Personality and Social Psychology, 9,* 245–250.

Cottrell, N. B. (1972). Social facilitation. In C. G. McClintock (Ed.), *Experimental social psychology* (pp. 185–236). New York: Holt.

Cox, D., & Rachman, S. (1983). Self-reported fear during a nineteen-week tour of duty. *Advances in Behavior Research and Therapy: An International Review Journal, 4*(3), 141–152.

Cox, R. H. (1984). Consolidation of pursuit rotor learning under conditions of threat of electrical shock. *International Journal of Sports Psychology, 15,* 1–10.

Danev, S. G., de Winter, C. R., & Wartna, G. F. (1972). Information processing and psychophysiological functions in a task with and without time stress. *Activitas Nervosa Superior, 14,* 8–12.

Deutsch, J. A., & Deutsch, D. (1963). Attention: Some theoretical considerations. *Psychological Review, 70,* 80–90.

Dorner, D. (1990). The logic of failure. In D. E. Broadbent, J. Reason, & A. Baddeley (Eds.), *Human factors in hazardous situations* (pp. 463–473). Oxford, England: Clarendon Press.

Driskell, J. E., Mullen, B., Johnson, C., Hughes, S., & Batchelor, C. (1992). *Development of quantitative specifications for simulating the stress environment* (Report No. AL–TR–1991–0109). Wright–Patterson AFB, OH: Armstrong Laboratory.

Driskell, J. E., & Olmstead, B. (1989). Psychology and the military: Research applications and trends. *American Psychologist, 44,* 43–54.

Driskell, J. E., & Salas, E. (1991a). Overcoming the effects of stress on military performance: Human factors, training, and selection strategies. In R. Gal & A. Mangelsdorff (Eds.), *Handbook of military psychology* (pp. 183–193). London: Wiley.

Driskell, J. E., & Salas, E. (1991b). Group decision making under stress. *Journal of Applied Psychology, 76,* 473–478.

Driskell, J. E., Salas, E., & Hall, J. K. (1994, April). *The effect of vigilant and hypervigilant decision training on performance.* Paper presented at the annual meeting of the Society for Industrial and Organizational Psychology, Nashville, TN.

Driskell, J. E., Salas, E., & Johnston, J. (1995). *Does stress lead to a loss of team perspective?* Manuscript under review.

Easterbrook, J. A. (1959). The effect of emotion on cue utilization and the organization of behavior. *Psychological Review, 66,* 183–201.

Eijkman, E., & Vendrik, A. J. H. (1965). Can a sensory system be specified by its internal noise? *Journal of the Acoustical Society of America, 37,* 1102–1109.

Enders, J. (1993, March 21). Aging tankers draw companies concern. *The Orlando Sentinel,* p. F-6.

Epstein, Y. (1982). Crowding stress and human behavior. In G. W. Evans (Ed.), *Environmental stress* (pp. 133–148). New York: Cambridge University Press.

Eschenbrenner, A. J. (1971). Effects of intermittent noise on the performance of a complex psychomotor task. *Human Factors, 13,* 59–63.

Evans, G. W., & Jacobs, S. V. (1982). Air pollution and human behavior. In G. W. Evans (Ed.), *Environmental stress* (pp. 105–132). New York: Cambridge University Press.

Eysenck, M. W. (1984). *A handbook of cognitive psychology.* Hillsdale, NJ: Lawrence Erlbaum Associates.

Finkelman, J. M., & Glass, D. C. (1970). Reappraisal of the relationship between noise and human performance by means of a subsidiary task measure. *Journal of Applied Psychology, 54,* 211–213.

Finkelman, J. M., Zeitlin, L. R., Romoff, R. A., Friend, M. A., & Brown, L. S. (1979). Conjoint effect of physical stress and noise stress on information processing performance and cardiac response. *Human Factors, 21,* 1–6.

Fisk, A. D., & Schneider, W. (1983). Category and word search: Generalizing search principles to complex processing. *Journal of Experimental Psychology: Learning, Memory, and Cognition, 9,* 177–195.

Geen, R. G. (1979). Effects of being observed on learning following success and failure experiences. *Motivation and Emotion, 4,* 355–371.

Geen, R. G. (1989). Alternative conceptions of social facilitation. In P. B. Paulus (Ed.), *Psychology of group influence* (pp. 15–51). Hilldale, NJ: Lawrence Erlbaum Associates.

Glass, D. C., & Singer, J. E. (1972). *Urban stress: Experiments on noise and social stressors.* New York: Academic Press.

Grether, W. F. (1971). *Noise and human performance* (Report No. AMRL–TR–70–29). Wright–Patterson Air Force Base, OH: Aerospace Medical Research Laboratory.

Grinker, R. R., & Spiegel, J. P. (1945). *Men under stress.* Philadelphia: Blakiston.

Hammerton, M., & Tickner, A. H. (1967). *Tracking under stress* (APRE Technical Report No. 67/CS 10a). Farnborough, England: Army Personnel Research Establishment.

Hammerton, M., & Tickner, A. H. (1969). An investigation into the effects of stress upon skilled performance. *Ergonomics, 12,* 851–855.

Harris, C. S. (1968). *The effects of high intensity noise on human performance* (Report No. AMRL–TR–67–119). Wright–Patterson Air Force Base, OH: Aerospace Medical Research Laboratory.

Harrison, A. A., & Conners, M. M. (1984). Groups in exotic environments. *Advances in Experimental Social Psychology, 17,* 49–87.

Hartley, L. R. (1981). Noise, attentional selectivity, serial reactions and the need for experimental power. *British Journal of Psychology, 72,* 101–107.

Hirst, W., Spelke, E. S., Reaves, C. C., Caharack, G., & Neisser, U. (1980). Dividing attention without alternation or automaticity. *Journal of Experimental Psychology: General, 109,* 98–117.

Hitch, G. J., & Baddeley, A. D. (1976). Verbal reasoning and working memory. *Quarterly Journal of Experimental Psychology, 28,* 603–621.

Hockey, G. R. J. (1970). Effect of loud noise on attentional selectivity. *Quarterly Journal of Experimental Psychology, 22,* 26–28.

Hockey, G. R. J., & Hamilton, P. (1970). Arousal and information selection in short-term memory. *Nature, 226,* 866–867.

Hogan, R., & Hogan, J. C. (1982). Subjective correlates of stress and human performance. In E. A. Alluisi & E. A. Fleishman (Eds.), *Human performance and productivity: Stress and performance effectiveness* (pp. 141–163). Hillsdale, NJ: Lawrence Erlbaum Associates.

Houston, B. K. (1972). Control over stress, locus of control, and response to stress. *Journal of Personality and Social Psychology, 21,* 249–255.

Idzikowski, C., & Baddeley, A. D. (1983). Fear and dangerous environments. In R. Hockey (Ed.), *Stress and fatigue in human performance* (pp. 123–144). Chichester: Wiley.

Innes, L. G., & Allnutt, M. F. (1967). *Performance measurement in unusual environments* (IAM Technical Memorandum No. 298). Farnborough, England: RAF Institute of Aviation Medicine.

Ivancevich, J. M., & Matteson, M. T. (1980). *Stress and work: A managerial perspective.* Glenview, IL: Scott, Foresman.

James, W. (1890). *The principles of psychology* (Vol. 1). New York: Holt.

Janis, I. L. (1951). *Air war and emotional stress.* New York: McGraw-Hill.

Janis, I. L. (1954). Problems of theory in the analysis of stress behavior. *Journal of Social Issues, 10,* 12–25.

Janis, I. L. (1982a). *Groupthink* (2nd ed.). Boston: Houghton Mifflin.

Janis, I. L. (1982b). Decision-making under stress. In L. Goldberger & S. Breznitz (Eds.), *Handbook of stress: Theoretical and clinical aspects* (pp. 69–80). New York: The Free Press.

Janis, I. L., & Mann, L. (1977). *Decision Making: A psychological analysis of conflict, choice, and commitment.* New York: The Free Press.

Jemmott, J. B., & Locke, S. E. (1984). Psychosocial factors, immunologic mediation, and human susceptibility to infectious diseases: How much do we know? *Psychological Bulletin, 95,* 78–108.

Kahneman, D. (1973). *Attention and effort.* Englewood Cliffs, NJ: Prentice-Hall.

Kahneman, D. (1975). Effort, recognition and recall in auditory attention. *Attention and Performance, 6,* 65–80.

Keele, S. W. (1973). *Attention and human performance.* Pacific Palisades, CA: Goodyear.

Keinan, G. (1987). Decision making under stress: Scanning of alternatives under controllable and uncontrollable threats. *Journal of Personality and Social Psychology, 52,* 639–644.

Keinan, G. (1988). The effects of expectations and feedback. *Journal of Applied Social Psychology, 18,* 355–373.

Kerr, B. (1973). Processing demands during mental operations. *Memory and Cognition, 1,* 401–412.

Kirk, R. E., & Hecht, E. (1963). Maintenance of vigilance by programmed noise. *Perceptual and Motor Skills, 16,* 553–560.

Kryter, K. D. (1985). *The effects of noise on man.* (2nd ed.) New York: Academic Press.

Lazarus, R. S., Deese, J., & Osler, S. F. (1952). The effects of psychological stress upon performance. *Journal of Experimental Psychology, 43,* 100–105.

Lazarus, R. S., & Folkman, S. (1984). *Stress, appraisal, and coping.* New York: Springer.

Lee, L. C. (1961). The effects of anxiety level and shock on a paired-associate verbal task. *Journal of Experimental Psychology, 61,* 213–217.

Link, S. W. (1971). Applying RT deadlines to discrimination reaction time. *Psychonomic Science, 25,* 355–358.

Locke, E. A., Frederick, E., Lee, C., & Bobko, P. (1984). Effect of self-efficacy, goals, and task strategies on task performance. *Journal of Applied Psychology, 69,* 241–251.

Loeb, M., Jones, P. D., & Cohen, A. (1976). *Effects of noise on non-auditory sensory functions and performance* (NIOSH–76–176). Cincinnati, OH: National Institute for Occupational Safety and Health.

Lovallo, W. R., & Pishkin, V. (1980). Performance of Type A (coronary-prone) men during and after exposure to uncontrollable noise and task failure. *Journal of Personality and Social Psychology, 38,* 963–971.

Lulofs, R., Wennekens, R., & Van Houtem, J. V. (1981). Effect of physical stress and time pressure on performance. *Perceptual and Motor Skills, 52,* 787–793.

MacDonald, R. R., & Labuc, S. (1982). *Parachuting stress and performance* (Memorandum 82m511). Farnsborough, England: Army Personnel Research Establishment.

Mackenzie, C. F., Craig, G. R., Parr, M. J., & Horst, R. (1994). Video analysis of two emergency tracheal intubations identifies flawed decision making. *Anesthesiology, 81,* 4–12.

Markowitz, J. S., Gutterman, E. M., Link, B., & Rivera, M. (1987). Psychological responses of firefighters to a chemical fire. *Journal of Human Stress,* 84–93.

Mathews, K. E., & Canon, L. K. (1975). Environmental noise level as a determinant of helping behavior. *Journal of Personality and Social Psychology, 32,* 571–577.

May, D. N., & Rice, C. G. (1971). Effects of startle due to pistol shot on control precision performance. *Journal of Sound and Vibration, 15,* 197–202.

McBride, D. J. (1988). *An exploration of team information processing in a dynamic group choice task involving uncertainty.* Unpublished Doctoral Dissertation, University of Minnesota.

McCormick, E. J., & Sanders, M. S. (1982). *Human factors in engineering and design.* New York: McGraw-Hill.

McGrath, J. E. (1976). Stress and behavior in organizations. In M. D. Dunnette (Ed.), *Handbook of industrial and organizational psychology* (pp. 1351–1395). Chicago: Rand McNally.

McLeod, P. (1977). A dual task response modality effect: Support for multiprocessor models of attention. *Quarterly Journal of Experimental Psychology, 29,* 651–667.

Mears, J. D., & Cleary, P. J. (1980). Anxiety as a factor in underwater performance. *Ergonomics, 23,* 549–557.

Monat, A., Averill, J. R., & Lazarus, R. S. (1972). Anticipatory stress and coping reactions under various conditions of uncertainty. *Journal of Personality and Social Psychology, 24,* 237–253.

Moore, B., Underwood, B., & Rosenhan, D. L. (1973). Affect and altruism. *Developmental Psychology, 8,* 99–104.

Moore, H. T. (1917). Laboratory tests of anger, fear, and sex interests. *American Journal of Psychology, 28,* 390–395.

Moore, J. J., & Massaro, D. W. (1973). Attention and processing capacity in auditory recognition. *Journal of Experimental Psychology, 99,* 49–54.

Moray, N. (1967). Where is attention limited. A survey and a model. *Acta Psychologica, 27,* 84–92.

Mullen, B. (1991). Group composition, salience, and cognitive representations: The phenomenology of being in a group. *Journal of Experimental Social Psychology, 27*, 297–323.

Mullen, B., Anthony, T., Salas, E., & Driskell, J. E. (1994). Group cohesiveness and quality of decision making: An integration of tests of the groupthink hypothesis. *Small Group Research, 25*, 189–204.

National Transportation Safety Board. (1990). *Aircraft accident report: United Airlines DC-10-10 engine explosion and landing at Sioux City Iowa* (NTSB/AAR–90/06). Washington, DC: Author.

Navon, D., & Gopher, D. (1979). On the economy of the human processing system. *Psychological Review, 86*, 254–255.

Neisser, U., & Becklan, R. (1975). Selective looking: Attending to visually significant events. *Cognitive Psychology, 7*, 480–494.

Norman, D. (1968). Toward a theory of memory and attention. *Psychological Review, 75*, 522–536.

Norman, D. A., & Bobrow, D. J. (1975). On data-limited and resource-limited processes. *Cognitive Psychology, 7*, 44–64.

Pachella, R. J., Fisher, D. F., & Karsh, R. (1968). Absolute judgments in speeded tasks: Quantification of the tradeoff between speed and accuracy. *Psychonomic Science, 12*, 225–226.

Paterson, R. J., & Neufeld, R. (1987). Clear danger: Situational determinants of the appraisal of threat. *Psychological Bulletin, 101*, 404–416.

Paulus, P. B., & Murdoch, P. (1971). Anticipated evaluation and audience presence in the enhancement of dominant responses. *Journal of Experimental Social Psychology, 7*, 280–291.

Pearlin, L. I., Menaghan, E. G., Lieberman, M. A., & Mullan, J. T. (1981). The stress process. *Journal of Health and Social Behavior, 22*, 337–356.

Percival, L., & Loeb, M. (1980). Influence of noise characteristics on behavioral aftereffects. *Human Factors, 22*, 341–352.

Perrow, C. (1984). *Normal accidents: Living with high-risk technologies*. New York: Basic Books.

Postman, L., & Bruner, J. S. (1948). Perception under stress. *Psychological Review, 55*, 314–323.

Poulton, E. C. (1978). A new look at the effects of noise: A rejoiner. *Psychological Bulletin, 85*, 1068–1079.

Powell, F. M., & Verner, J. P. (1982). Anxiety and performance relationships in first time parachutists. *Journal of Sport Psychology, 4*, 184–188.

Prince, C., Bowers, C. A., & Salas, E. (1994). Stress and crew performance: Challenges for aeronautical decision making training. In N. Johnston, N. McDonald, & R. Fuller (Eds.), *Aviation psychology in practice* (pp. 286–305). Hants, England: Avebury.

Pruitt, D. G. (1981). *Negotiation behavior*. New York: Academic Press.

Pruitt, D. G., & Carnevale, P. J. D. (1982). The development of integrative agreements. In V. Derlega and J. Grzelak (Eds.), *Cooperation and helping behavior: Theories and research*. New York: Academic Press.

Rachman, S. J. (Ed.) (1983). Fear and courage among military bomb-disposal operators [Special issue]. *Advances in Behaviour Research and Therapy, 4*(3).

Rachman, S. J. (1982). *Development of courage in military personnel in training and performance in combat situations* (Research Report No. 1338). Alexandria, VA: U.S. Army Research Institute for the Behavioral and Social Sciences.

Radloff, R. & Helmreich, R. (1968). *Groups under stress: Psychological research in Sealab II*. New York: Appleton-Century-Crofts.

Reid, C. (1945). Fluctuations in navigator performance during operational sorties. In E. J. Dearnaley & P. V. Warr (Eds.), *Aircrew stress in wartime operations* (pp. 63–73). London: Academic Press.

Repko, J. D., Brown, B. R., & Loeb, M. (1974). *Effects of continuous and intermittent noise on sustained performance* (ITR-74-29). Arlington, VA: U.S. Army Organizations and Systems Research Laboratory. (AD-785740)

Rothstein, H. G. (1986). The effects of time pressure on judgment in multiple cue probability learning. *Organizational Behavior and Human Decision Processes, 37,* 83–92.

Rothstein, H. G., & Markowitz, L. M. (1982, May). *The effect of time on a decision strategy.* Paper presented to the meeting of the Midwestern Psychological Association, Minneapolis, MN.

Rubin, J. Z., & Brown, B. R. (1975). *The social psychology of bargaining and negotiation.* New York: Academic Press.

Ruffell Smith, H. P. (1979). *A simulator study of the interaction of pilot workload with errors, vigilance, and decisions* (NASA Technical Memorandum 78482). Moffett Field, CA: NASA-Ames Research Center.

Ryan, D. E. (1962). Effects of stress on motor performance and learning. *Research Quarterly, 33,* 111–119.

Sanders, G. S. (1981). Driven by distraction: An integrative review of social facilitation theory and research. *Journal of Experimental Social Psychology, 17,* 227–251.

Sanna, L. J., & Shotland, R. L. (1990). Valence of anticipated evaluation and social facilitation. *Journal of Experimental Social Psychology, 26,* 82–92.

Schaubroeck, J., & Ganster, D. C. (1993). Chronic demands and responsivity to challenge. *Journal of Applied Psychology, 78,* 73–85.

Schmitt, B. H., Gilovich, T., Goore, N., & Joseph, L. (1986). Mere presence and social facilitation: One more time. *Journal of Experimental Social Psychology, 22,* 242–248.

Schneider, W., & Detweiler, M. (1988). *The role of practice in dual task performance: Toward workload modeling in a connectionist/control architecture.* Unpublished manuscript.

Schoenberger, R. W., & Harris, C. S. (1965). Human performance as a function of changes in acoustic noise levels. *Journal of Engineering Psychology, 4,* 108–119.

Selye, H. (1936). A syndrome produced by diverse noxious agents. *Nature, 138,* 32.

Selye, H. (1980). The stress concept today. In I. L. Kutash & L. B. Schlesinger (Eds.), *Handbook on stress and anxiety: Contemporary knowledge, theory, and treatment* (pp. 127–143). San Francisco: Jossey-Bass.

Sherrod, D. R., & Downs, R. (1974). Environmental determinants of altruism: The effects of stimulus overload and perceived control on helping. *Journal of Experimental Social Psychology, 10,* 468–479.

Sherrod, D. R., Hage, J. N., Halpern, P. L., & Moore, B. S. (1977). Effects of personal causation and perceived control on responses to an aversive environment: The more control the better. *Journal of Experimental Social Psychology, 13,* 14–27.

Spelke, E., Hirst, W., & Neisser, U. (1976). Skills of divided attention. *Cognition, 4,* 215–230.

Spettell, C. M., & Liebert, R. M. (1986). Training for safety in automated person–machine systems. *American Psychologist, 41,* 545–550.

Staw, R. M., Sandelands, L. E., & Dutton, J. E. (1981). Threat-rigidity effects in organizational behavior: A multi-level analysis. *Administrative Science Quarterly, 26,* 501–524.

Streufert, S., & Streufert, S. C. (1981). *Stress and information search in complex decision making: Effects of load and time urgency* (Technical Rep. No. 4). Arlington, VA: Office of Naval Research.

Sullivan, L. (1976). Selective attention and secondary message analysis. A reconsideration of Broadbent's filter model of selective attention. *Quarterly Journal of Experimental Psychology, 28,* 167–178.

Szpiler, J. A., & Epstein, S. (1976). Availability of an avoidance response as related to autonomic arousal. *Journal of Abnormal Psychology, 85,* 73–82.

Teichner, W. H., Arees, E., & Reilly, R. (1963). Noise and human performance: A psychological approach. *Ergonomics, 6,* 83–97.

Thackray, R. I., & Touchstone, R. M. (1979). *Effects of noise exposure on performance of a simulated radar task* (FAA–AM–79–24). Washington, DC: Federal Aviation Administration, Office of Aviation Medicine.

Theologus, G. C., Wheaton, G. R., & Fleishman, E. A. (1974). Effects of intermittent, moderate intensity noise stress on human performance. *Journal of Applied Psychology, 59,* 539–547.

Theologus, G. C., Wheaton, G. R., Mirabella, A., Brahley, R. E., & Fleishman, E. A. (1973). *Development of a standardized battery of performance tests for the assessment of noise stress effects* (Rep. No. NASA CR–2149). Washington, DC: National Aeronautics and Space Administration.

Thompson, S. C. (1981). Will it hurt less if I can control it? A complex answer to a simple question. *Psychological Bulletin, 90,* 89–101.

Tomaka, J., Blascovich, J., Kelsey, R. M., & Leitten, C. L. (1993). Subjective, physiological, and behavioral effects of threat and challenge appraisal. *Journal of Personality and Social Psychology, 65,* 248–260.

Treisman, A. M. (1969). Strategies and models of selective attention. *Psychological Review, 76,* 282–299.

Treisman, A. M., & Davies, A. (1973). Divided attention to ear and eye. In S. Kornblum (Ed.), *Attention and Performance: Volume IV* (pp. 101–117). London: Academic Press.

Triplett, N. (1898). The dynamogenic factors in pacemaking and competition. *American Journal of Psychology, 9,* 507–533.

Tulving, E., & Lindsay, P. H. (1967). Identification of simultaneously presented simple visual and auditory stimuli. In A. F. Sanders (Ed.), *Attention and Performance: Vol. I.* Amsterdam: North-Holland.

Villoldo, A., & Tarno, R. L. (1984). *Measuring the performance of EOD equipment and operators under stress* (Technical Rep. No. 270). Indian Head, MD: Naval Explosive Ordnance Disposal Technical Center.

von Wright, J., & Vaurus, M. (1980). Interactive effects of noise and neuroticism in recall from semantic memory. *Scandinavian Journal of Psychology, 21,* 97–101.

Wachtel, P. L. (1968). Anxiety, attention, and coping with threat. *Journal of Abnormal Psychology, 73,* 137–143.

Walton, R. E., & McKersie, R. B. (1965). *A behavioral theory of labor negotiation: An analysis of a social interaction system.* New York: McGraw-Hill.

Weinberg, J., & Levine, S. (1980). Psychobiology of coping in animals: The effects of predictability. In S. Levine & H. Ursin (Eds.), *Coping and health* (NATO Conference Series III: Human factors). New York: Plenum.

Weitz, J. (1966). *Stress* (Research Paper P–251). Washington, DC: Institute for Defense Analyses. (AD–633–566)

Weltman, G., Christianson, R. A., & Egstrom, G. H. (1970). Effects of environment and experience on underwater work performance. *Human Factors, 12,* 587–598.

Wickens, C. D. (1980). The structure of attentional resources. In R. Nickerson (Ed.), *Attention and performance: Vol. VIII* (pp. 239–257). Hillsdale, NJ: Lawrence Erlbaum Associates.

Wickens, C. D. (1984). Processing resources in attention. In R. Parasuraman & D. R. Davies (Eds.), *Attention and performance: Volume IV* (pp. 63–102). London: Academic Press.

Wickens, C. D., & Flach, J. M. (1988). Information processing. In E. L. Wiener & D. C. Nagel (Eds.), *Human factors in aviation* (pp. 111–149). San Diego: Academic Press.

Wilbanks, W. A., Webb, W. B., & Tolhurst, G. C. (1956). *A study of intellectual activity in a noisy environment.* Pensacola, FL: U.S. Naval School of Aviation Medicine.

Willer, D., & Webster, M. (1970). Theoretical concepts and observables. *American Sociological Review, 35,* 748–757.

Wohlwill, J. F., Nasar, J. L., DeJoy, D. M., & Foruzani, H. H. (1976). Behavioral effects of a noisy environment: Task involvement versus passive exposure. *Journal of Applied Psychology, 61,* 67–74.

Worchel, S., & Yohai, S. M. L. (1979). The role of attribution in the experience of crowding. *Journal of Experimental Social Psychology, 15,* 91–104.

Worringham, C. J., & Messick, D. M. (1983). Social facilitation of running: An unobtrusive study. *Journal of Social Psychology, 121,* 23–29.

Wright, P. (1974). The harassed decision maker: Time pressures, distractions, and the use of evidence. *Journal of Marketing Research, 44,* 429–443.

Yamamoto, T. (1984). Human problem solving in a maze using computer graphics under an imaginary condition of "fire." *Japanese Journal of Psychology, 55,* 43–47.

Yuille, J. C., Davies, G., Gibling, F., Marxsen, D., & Porter, S. (1994). Eyewitness memory of police trainees for realistic role plays. *Journal of Applied Psychology, 79,* 931–936.

Yukl, G. A., Malone, M. P., Hayslip, B., & Pamin, T. A. (1976). The effects of time pressure and issue settlement order on integrative bargaining. *Sociometry, 39,* 277–281.

Zajonc, R. B. (1965). Social facilitation. *Science, 149,* 269–274.

I Stress Effects

2 The Effect of Acute Stressors on Decision Making

Gary Klein
Klein Associates Inc., Fairborn, Ohio

It is difficult enough to make decisions in operational settings where the stakes are high and data are ambiguous, but another problem can enter in—stress. Whether the stressors are time constraints, noise, workload, or threat, they can play havoc with the clear thinking needed in these settings. They can degrade the quality of judgments, prevent the use of rational decision strategies, and severely compromise performance; at least, that is a popular appraisal of stressors. The thesis of this chapter is that each of these assertions is either incorrect or misleading. Decision makers are adaptive in their reactions to stressors. The decision strategies used in the presence of stressors may be simpler, but they are rational and make powerful use of experience. Moreover, stressors do not necessarily degrade decision quality, and in some cases can improve it.

Because context helps to clarify issues, the issues in this chapter are framed around the context of a Naval destroyer or cruiser engaged in an antiair warfare mission, emphasizing the responsibilities of the Commanding Officer of the ship. The selection of this perspective is not accidental. On July 3rd, 1988, the USS Vincennes shot down an Iranian airliner, and that incident, perhaps more than any other, has highlighted the question of how stressors affect decision making. One of the many activities initiated by this tragic episode was a large-scale behavioral research project, entitled Tactical Decision Making Under Stress (TADMUS). My experience working on TADMUS, and learning from the many other investigators engaged in the project, helped me prepare this chapter.

This chapter focuses on what Driskell and Salas (1991) termed *acute* stressors—those that are sudden, novel, unexpected, and of short duration. These would include personal threat, time constraint, noise, task overload, and so on. This chapter does not address topics such as life stress, fatigue, prolonged exposure to heat or cold, boredom, or sleep loss.

The organization of the chapter is as follows. The first section discusses aspects of decision making, in order to provide a baseline for understanding which types of decision strategies are most severely affected by stressors. The second section presents a series of assertions about the ways that stressors influence decision making. The third puts these assertions into context by reviewing the Vincennes incident itself, to better understand the role played by stress.

JUDGMENT AND DECISION STRATEGIES

Decision strategies can rely on analytical comparisons between different courses of action, to select the best one, or on more intuitive strategies that may not rely on comparisons at all. Hammond, Hamm, Grassia, and Pearson (1987) have shown that analytical strategies are more effective for some sorts of decision tasks, whereas intuitive strategies are more useful for others. Payne (1976) and Beach and Mitchell (1978) also discussed the difference between analytical and nonanalytical decision strategies.

This distinction between analytical and nonanalytical strategies is important. For example, time pressure and noise may disrupt the analytical strategies, but may have little if any impact on nonanalytical strategies. The following two subsections describe the use of analytical strategies to select the best course of action from a set of alternatives and how people can adopt courses of action without making any comparisons.

Analysis of the Best Course of Action

Sometimes decision makers need to choose between several options or courses of action. This is the moment of choice that has been so carefully studied by decision researchers (e.g., Baron, 1988; von Winterfeldt & Edwards, 1986). Strategies such as decision analysis using subjective expected utility judgments and multiattribute utility analyses are best suited for cases in which there is less time pressure, more carefully collected data, multiple stakeholders, or generally lower levels of experience. The strategies are termed *compensatory* strategies, because they are designed to compensate for a small weakness on one or two attributes or evaluation dimensions if an option shows major strengths on other evaluation dimensions. Thus, in performing a multiattribute utility analysis (Janis & Mann, 1977) a decision maker would list all the options on one axis, all the evaluation dimensions on another, and would assign weights to the evaluation dimensions. After rating each option on each evaluation dimension, the decision maker would perform the necessary calculations to find the option with the highest score.

Compensatory decision strategies take time and require effort, and therefore are not well suited for stressful conditions. There are a set of *noncompensatory* decision strategies for selecting the best option from a set of alternatives that could be used in operational settings. Svenson (1979) has cataloged various possible strategies for making such a choice, and Zsambok, Beach, and Klein (1992) have expanded this catalog. For noncompensatory strategies, the strengths on one evaluation dimension do not balance against weaknesses on another. For example, the *lexicographic strategy* (Fishburn, 1974) is simply to select the option with the highest ratings for the most important dimension. In the *elimination-by-aspects* strategy (Tversky, 1972), the various options are evaluated using the most important dimension, and all that fail to meet a given criterion are rejected. The remaining options are evaluated using the next most important dimension, and so forth. All of these noncompensatory strategies presuppose a careful screening (Beach, 1990), and all of them share an additional characteristic—they are intended to select among options that are fairly well balanced so that it is not obvious which is best.

Because the options are so well balanced, it may not make a great deal of difference which is chosen—the ambiguity in the data may greatly overshadow the advantages and disadvantages of the options. Minsky (1986) has noted the paradox that the closer together the options, the harder the decision but the less importance it has.

Naturalistic Decision Making

Compensatory and even noncompensatory strategies require time and concentration. If decision makers are not relying on these types of strategies under conditions of stress, what other types of strategies are available? This question has opened up a field of inquiry known as *Naturalistic Decision Making* research. Klein, Orasanu, Calderwood, and Zsambok (1993) have presented a set of judgment and decision strategies that appear to be well suited to naturalistic settings. These differ from the classical strategies for making optimal decisions. The interest of Naturalistic Decision Making researchers is to describe what decision makers actually do, especially during nonroutine and critical challenges.

Orasanu and Connolly (1993) have described some critical features of naturalistic settings (see Table 2.1): The decision task itself is often unclear (ill-structured, with shifting and ill-defined goals); the available information may lead to uncertainty, due to missing, unreliable, or inaccurate data or conditions that keep changing; and stressors such as time pressure and high stakes are often present. The judgment and decision strategies that people use in such settings must conform to these situational features. For example, it makes little sense to adopt a decision strategy that requires comparable data across options if missing data make it impossible to compare options. And it makes little sense to choose a time-consuming, analytical strategy when faced with severe time pressure.

TABLE 2.1
Characteristics of Naturalistic Decision Settings

1. Ill-structured problems
2. Uncertain and dynamic environments
3. Shifting, ill-defined, or competing goals
4. Action/feedback loops
5. Time stress
6. High stakes
7. Multiple players
8. Organizational goals and norms

Decision making does not have to be restricted to finding the best course of action from a limited set of alternatives. Most of us make decisions throughout the day without having to compare the strengths and weaknesses of competing options. This is because we can draw on our experience to identify reasonable options as the first ones we consider. In operational settings, as opposed to laboratory settings, people encounter tasks with which they have some familiarity, so they can rely on experience for guidance. If a decision maker can use experience to generate adequate courses of action in this way, there is little marginal utility in generating large sets of options that must be evaluated and contrasted.

The Recognition-Primed Decision (RPD) model (Klein, 1989, in press) explained how people could make rapid but effective decisions by using experience to size up the situation and to generate and evaluate courses of action one at a time (as opposed to comparatively) (see Fig. 2.1). In the simplest case, Level 1 of Fig. 2.1, the decision maker sizes up a situation and reacts with the typical course of action. More complex situations require diagnosis, as illustrated by Level 2. Finally, sometimes the decision maker will need to make a more deliberate assessment of a course of action, as shown in Level 3. There appear to be different strategies used for decisions about diagnosing a situations versus selecting a course of action.

Diagnostic Strategies (Level 2). These strategies are intended to account for the way people size up a situation, including attempts to understand anomalies in a situation. When a person has initially identified an event or problem as "that kind of situation," the situation assessment usually clarifies which goals are possible to attain, what cues are important to monitor as the situation unfolds, what to expect next, and what courses of action are typical. In most cases, situation assessment requires little thought. When the situation assessment is instantaneous (e.g., Hintzman, 1986; Kahneman & Miller, 1986; Noble, 1993), there is rarely a need for conscious deliberation about what is happening. During

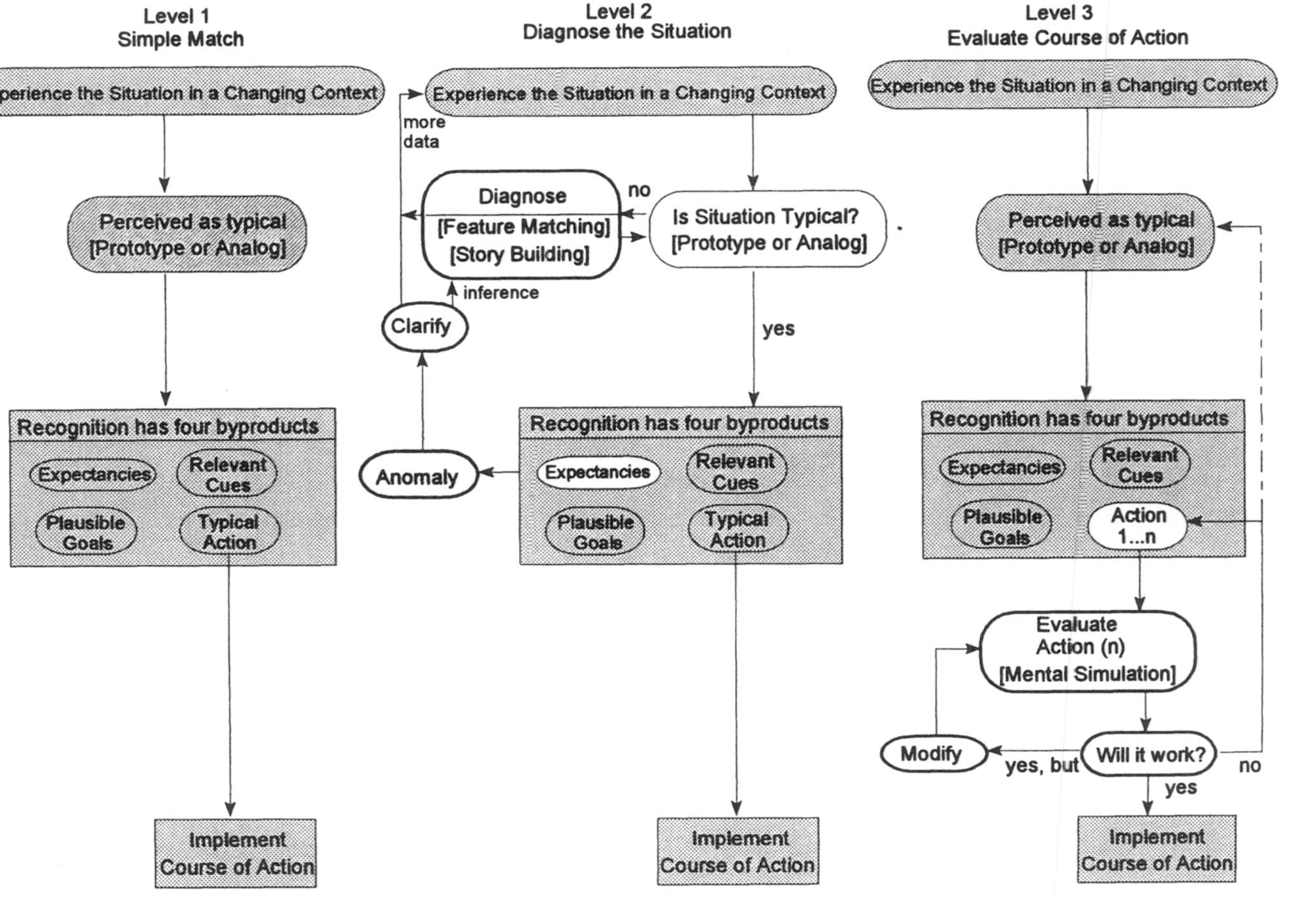

FIG. 2.1. Recognition-primed decision model.

situation assessment, the decision maker usually senses in a straightforward way what is going on.

Diagnostic deliberation is required when things are not straightforward, when a decision maker is uncertain how to classify a situation, or when one must choose from a set of possible situation interpretations. Zsambok et al. (1992) have identified several strategies that relate to situation diagnosis and fit the constraints of naturalistic decision making. These would all provide the diagnosis called for on Level 2 of the RPD model in Fig. 2.1.

The first of these strategies is feature matching (e.g., Noble, 1993). This applies to cases where these are competing hypotheses and a person consciously uses criterion features to analyze which of the competing hypotheses forms the better match to the observed characteristics of the situation. Feature matching corresponds to Rasmussen's (1985) Rule-Based Behavior, and also to Reason's (1987) GEMS model. Feature matching probably accounts for most of the diagnostic decisions encountered.

Kaempf, Klein, Thordsen, and Wolf (in press) collected a number of Navy incidents requiring complex decision making in AEGIS Combat Information Centers. Feature matching accounted for 87% of the diagnostic decisions about the nature of the situation. In one incident an unidentified surface track was spotted as coming from the direction of an unfriendly country, was not sending any identifying signal, was ignoring radio messages, and was sending military radar emissions. These features matched the profile of a hostile ship much more closely than they matched the profile of a neutral or friendly ship. This incident shows how a simple feature match was sufficient to diagnose the situation.

A more deliberative diagnostic strategy is to construct a story to try to account for the observations (Klein & Crandall, 1995; Pennington & Hastie, 1993). Here, the features are important but are not sufficient. The decision maker tries to construct a causal account that takes advantage of experience about what people might do in different situations, with different goals. In an incident similar to the one described in the last paragraph, an unidentified aircraft was flying very slowly, at approximately 80–100 knots, which ruled out a hostile fighter, but permitted the hypothesis of a small plane outfitted with weapons. The decision maker tried to imagine a sequence of events in which the unfriendly country actually would outfit such an airplane and send it off, without any support, without any fire control radar or other requirements for attack. The decision maker was simply unable to construct a plausible story, and rejected the hypothesis that the target was a threat. In contrast, there had been times when *friendly* helicopters had gotten lost, or had radio malfunctions, or had flown beyond the horizon and had broken off radar contact, so it was possible to construct a plausible story of how this particular track could be a friendly helicopter. Therefore, the decision maker decided that the aircraft should not be attacked, even though it fit many of the features of a hostile aircraft. In fact, the track turned out to be a friendly helicopter that had gotten lost and was zig-zagging between ships to find the one to which it was supposed to return.

Kaempf et al. (in press) reported that 12% of the judgments of situation awareness used a story-building strategy.

Strategies for Selecting a Course of Action. Once a person has reached a situation awareness, it is usually obvious how to respond to that situation. This is the thesis of the *Recognition-Primed Decision* (RPD) model. The simplest case, where a person recognizes a situation, recognizes the typical way of responding, and carries out that action, is Level 1 of the RPD model. In more complex cases, Level 3 of the RPD model, the decision maker needs to evaluate an option when no others have been generated for contrast. For these cases the RPD model posits the use of a deliberative mental simulation (Klein & Crandall, 1995) to envision how the course of action will be carried out, and to find potential difficulties that need to be avoided. Rarely is there a need to contrast options in naturalistic settings, even if the incidents are complex and demanding.

To summarize, this section has contrasted analytical and nonanalytical decision strategies, because these may be affected differentially by stressors. Analytical strategies include both compensatory and noncompensatory methods for weighing the strengths and weaknesses of different courses of action. Nonanalytical strategies refer to the way people use experience in naturalistic settings to avoid having to generate alternative courses of action. They do this by diagnosing the type of situation they are encountering and then recognizing a reasonable course of action.

This section has presented a brief orientation to the varieties of decision strategies. The next section examines the nature of stressors and the ways they can affect the types of decision strategies used.

EFFECTS OF STRESSORS ON DECISION STRATEGIES

At the beginning of this chapter we claimed that stressors do not necessarily degrade decision making. Our claim rests on a set of assertions:

- Stress is an intervening variable rather than a hypothetical construct.
- We can distinguish between the mediating and the cognitive reactions to stressors.
- Stressors may prevent the use of analytical decision strategies, but these strategies are rarely used in naturalistic settings.
- Naturalistic decision making is not very susceptible to stressors.
- Reactions to stressors are adaptive rather than dysfunctional.
- Stressors do not necessarily increase the likelihood of decision biases.
- Stressors do not necessarily degrade decision performance.

In this section we provide a rationale for each of these statements. Our discussion of the first two assertions about the nature of stress relies heavily on the reviews of stress research conducted by Edland (1989), Driskell et al. (1991), and Mross and Hammond (1990).

Proposition 1: Stress is an Intervening Variable Rather Than a Hypothetical Construct

An intervening variable is a convenient fiction that helps organize ideas and observations, but has no reality. For example, the concept of "force" in physics is an intervening variable. A hypothetical construct is one that actually exists as an unseen but real phenomenon that affects performance. In physics, subatomic particles may be postulated to explain certain events, and then researchers actually try to find evidence for the existence of these particles, which are hypothetical constructs.

In this chapter, stress is considered to be an intervening variable, not a hypothetical construct, as I have not seen compelling arguments for the existence of a general phenomenon of stress. One overriding reason for our reluctance to regard stress as a hypothetical construct is the lack of convergence on physiological concommitants of stress. Driskell and Salas (1991) described some of the physiological effects of stress as increased heartbeat, labored breathing, and trembling. If the same pattern of physiological effects was found for all manipulations considered to be stressors, there would be a justification for combining them into a common category. To date, there has been little success in demonstrating common physiological patterns.

Whenever possible, we try to specify which stressor we are discussing, rather than referring to a generic phenomenon. The variables studied as stressors, such as time constraint and noise, do affect decision making and these effects can be studied in themselves, without having to invoke an additional state.

We also do not want to become excessively specific "by listing endless putative environmental 'stressors' and cobbling up a list of matching responses." (Stokes, personal communication, June 12, 1992) A middle ground may be to identify common cognitive responses for different environmental stressors. Faced with stressors, people may adopt simpler decision strategies. The idea of common cognitive stress reactions treats stress as an intervening variable.

Proposition 2: We Can Distinguish Between the Mediating and the Cognitive Reactions to Stressors

To sort out the different effects of stressors, we must first clarify which stressors we are considering. Driskell and Salas (1991) reviewed the literature to identify a number of factors that can be considered to be acute stressors: crowding, noise, performance pressure, workload, anticipated threat of pain, anticipation of dangerous conditions, and emergency conditions.

Folkman (1984) defined some common features of stressors—they tax or exceed a person's resources, they create overload, and/or they threaten the person's well-being. Driskell and Salas (1991) added to this list the idea that stressors include overload conditions where demand is greater than resources. These categories correspond to the three major categories presented by Edland (1989): those linked to impending failure at a task, those linked to task overload, and those linked to various types of threats. Each of these three conditions is sufficient to be considered a stressor. Events that make demands greater than capacity (e.g., McGrath, 1970) need not be accompanied by threat in order to be considered stressors. Such events can include increased workload or distractors such as noise.

One must be careful not to confound manipulations with their effects. Both Driskell and Salas (1991) and Edland (1989) included "overload" as a stressor. If we create overload for a person performing a task, it is likely that performance will suffer; however, we cannot conclude that stress was the culprit. Overload, by definition, makes a task more difficult. Making a task more difficult, by definition, reduces performance. We have a tautology, not a finding. The same holds true for time pressure. As Edland (1989) and Keinan (1987) pointed out, decreasing the available time to a level below some minimum requirement will necessarily make a task more difficult, and will prevent a person from scanning all elements of the task. Therefore, finding that time pressure narrows attention (Edland, 1989; Wright, 1974) should not be surprising because reduced time will leave less opportunity to acquire new information.

To help sort these influences out, Fig. 2.2 presents a distinction between mediating and cognitive reactions to stress. Three mediating reactions are posited:

- Restricted ability to gather information, which commonly occurs when time pressure prevents the person from scanning all the cues.
- Interference with inner speech which commonly occurs when noise prevents echoic rehearsal of cues. Poulton (1976) has argued that noise can interfere with inner speech, and therefore can disrupt performance even on tasks that do not involve auditory stimuli.
- Imposition of a secondary task, as when the need to monitor symptoms such as rapid breathing and increased sweating and trembling, and to take these symptoms into account during task performance, may compete for attentional resources. Driskell and Salas (1991) noted that stressors can create a need for increased self-monitoring to detect and manage levels of exhaustion, increased rates of breathing, trembling, and so forth. For example, Zakay (1993) suggested

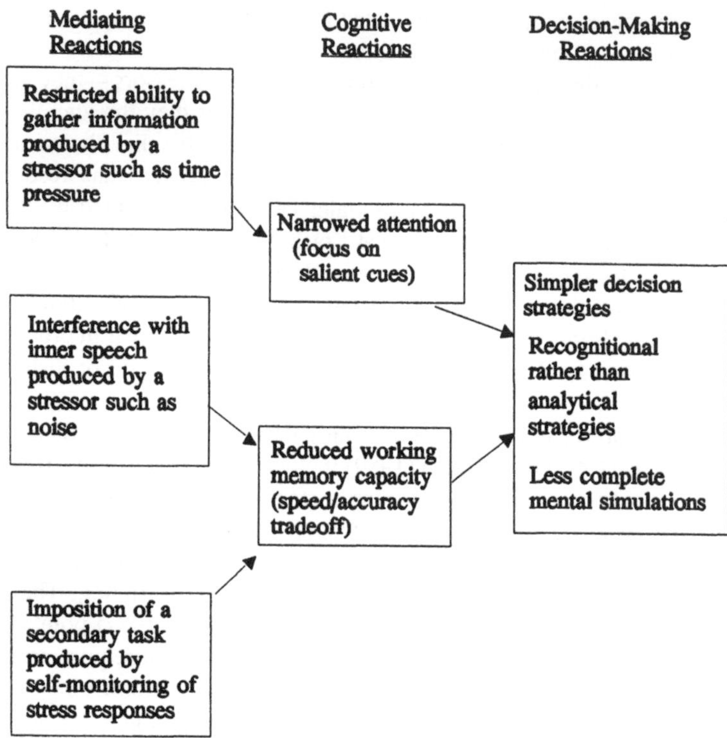

FIG. 2.2. Mediating, cognitive, and decision-making reactions to stress.

that time pressure creates the need to manage the task more carefully, forcing the decision maker to attend to time perception and reducing attentional resources for other task features. Similarly, overload may require more careful task management, thereby adding a secondary task.

Driskell and Salas (1991) also listed a fourth mediating reaction—motivational loss that can affect performance. However, it would seem that there are many situations in which the stressor increases motivation to perform well, such as when the consequences of poor performance increase the threat, or discomfort. At other times, the goals of managing or avoiding the stressors may serve to compete with task completion goals, and thereby reduce the motivation to successfully accomplish the task. We have not listed this fourth mediating reaction in Fig. 2.2 because it will depend on the context of the task and it is not directly tied to decision-making reactions.

One of the key implications of Fig. 2.2 is that different stressors can affect cognitive performance in the same ways. That is why we can treat stress as an intervening variable, because qualitatively different stressors can create the same types of mediating responses such as distracting secondary tasks.

For example, threat can be a stressor. The threat can involve impending shock, pain, or other forms of physical discomfort, or it can involve psychological consequences, such as making a poor result public. These types of manipulations serve as distracting *secondary* tasks. The salience of the threats makes it difficult to ignore them, and the severity of the threats can require mental stimulation to prepare for them. As with any secondary task, we would expect disruption of the performance of the primary task.

Now we can look at the second column in Fig. 2.2—the cognitive effects of the mediating reactions to stressors. To date, researchers have identified two primary cognitive reactions, narrowed attention and reduced capacity of working memory.

The most general class of cognitive reaction is on perception. Easterbrook (1959) and Combs and Taylor (1952) found that there was a narrowing of the perceptive field along with cue restriction. Edland (1989) cited evidence for a reduction in responsiveness to peripheral visual cues during emotional arousal. Wright (1974) found that time pressure led subjects to use fewer cues, Baddeley (1972) found that impending danger narrowed the range of attention, Streufert and Streufert (1981) reported decreased search behavior, and Wachtel (1968) found decreased vigilance. Figure 2.2 shows the narrowed attention to result from the physical restrictions often found in stressful tasks such as time pressure and overload.

Edland and Svenson (1993) reviewed the literature on time pressure and concluded that it results in a number of cognitive effects, including the following: greater selectivity of inputs and increase in salience of important cues. Therefore, the narrowed attention is not necessarily dysfunctional if experienced decision makers can select the most meaningful cues and spend less attention on irrelevant cues.

The second cognitive reaction listed in Fig. 2.2 is reduced working memory. Both Bacon (1974) and Hamilton (1982) also reported findings that showed that stressors reduced the capacity of working memory. Figure 2.2 postulates two ways that this can occur, through the interference with inner speech that would itself disrupt the active rehearsal needed for working memory (see also Hockey, 1984), and through the secondary task of managing the stress reactions. Reduced working memory can result in shifts in speed–accuracy tradeoffs, as the limited capacity requires the decision maker to either respond with less information and assessment or to wait longer in order to reach a given level of situational understanding.

In considering cognitive reactions, we must also take note of individual differences. For example, Ferrari (1994) has performed research suggesting individual differences in decisional procrastination, such as the purposive delay in making decisions despite unpleasant consequences. This characteristic seems related more to decision avoidance than to a need to gather increased information to reduce ambiguity. There may also be differences between people on a reflective–impulsive dimension (Ferrari, 1994; Ferrari & Emmons, 1994). For complex tasks where accuracy is important, a reflective style may be valuable, whereas dynamic tasks favoring speed over accuracy might call for an impulsive style. One hypothesis is that individuals who differ on this dimension might show different reactions to the need to narrow attention—those with a reflective style might be more uncomfortable with restricting attention than those with an impulsive style. Along the same lines, Stiensmeier-Pelster and Schurmann (1993) have built on Kuhl and Beckman's (1985) theory of action control, and have presented evidence that state-oriented individuals typically respond to time pressure with acceleration (doing everything more quickly), where action-oriented individuals respond with filtration (doing fewer things).

The third column of Fig. 2.2 shows the effects of the mediating and cognitive reactions on decision strategies: to select simpler and less analytical decision strategies, and to interfere with the mental simulation component of recognitional decision strategies. Consider the case where a person believes that capacity will be inferior to demand. This is one of the stressors identified by Driskell and Salas (1991). If it can be shown that this belief affects performance, without doing anything to actually reduce capacity or increase demand, then this would be an instance in which the stressor alone affected decision making. Such a result could be accounted for by arguing that the belief that resources will be inadequate for a task may itself promote a different decision strategy. Although the task has not changed, the person has changed the strategy used.

The next two propositions address the claim that stressors prevent the use of analytical decision strategies.

Proposition 3: Stressors May Prevent the Use of Analytical Decision Strategies, but These Strategies are Rarely Used in Naturalistic Settings

Zakay and Wooler (1984) found that under conditions of time pressure, subjects who were trained to use multiattribute utility analysis did not apply this strategy to a decision task. This finding confirms our belief that analytical strategies require sufficient time, data, quality, and attention to be carried out successfully. Keinan (1987) also found that the stressor of electric shock led subjects to select options without scanning all the alternatives.

Furthermore, Payne, Bettman, and Johnson (1988) found that time pressure led to the use of simpler analytical decision strategies. Payne et al. developed a computer simulation of different decision strategies and found that as time pressure increased, there was a shift in preferred strategy where preference was based on the expected mental effort required by the strategy and the expected power of the strategy for making a good decision. Payne et al. found a shift from compensatory to noncompensatory strategies, which paralleled the shift observed in actual subjects. In the computer simulation, the strategies that worked best under time pressure were noncompensatory strategies such as elimination by aspects and lexicographic strategies.

Therefore, the decision strategies that have been proposed for choosing between options can be inappropriate for naturalistic settings where there is time pressure or incomplete information. This blocking of analytical decision strategies is not a cause for concern because people rarely use analytical strategies in naturalistic settings even in the absence of stressors (e.g., Isenberg, 1984; Mintzberg, 1975).

Proposition 4: Naturalistic Forms of Decision Making Are Not Very Susceptible to Stressors

Naturalistic strategies appear to be less vulnerable to stress than the prescriptive decision strategies. For instance, recognitional strategies reduce vulnerability to time pressure. If you are a Commanding Officer faced with a difficult situation, and you rapidly size up the situation and see what to do, the time pressure may not be a factor. On the other hand, if you try to go through a process of identifying all the possible courses of action, and the most important dimensions for evaluating these, and the appropriate weights, and if you then rate each option on each dimension, but are interrupted in the middle of all this by the need to decide *NOW!*, before you have finished your deliberations, you will be unable to respond. The prescriptive decision strategies are of limited relevance in most operational settings. Prescriptive strategies such as Multiattribute Utility Analysis, Decision Analysis, and Bayesian Analysis are not designed for conditions of high time pressure, frequent interruptions, missing and ambiguous data, ill-defined goals, and frequent shifts in goals requiring fresh starts. This is not a criticism of these strategies, but it is a statement about situations in which they are applicable. There are many naturalistic settings, such as high-level policy planning, where these strategies are appropriate and helpful.

Because decision makers in operational settings such as the combat information center of an AEGIS cruiser are not trying to generate and evaluate multiple options, we need not worry that stressors may interfere with the analytical strategies. Kaempf et al. (in press) found minimal use of analytical decision strategies or of comparisons between courses of action in a study of actual antiair war incidents. This does not mean that people in these settings are making degraded decisions. The ill-defined qualities of these operational settings preclude analytical strategies. In addition, the more "intuitive" strategies enable decision makers to take advantage of experience in a way that is not possible using analytical techniques.

I use the RPD model as an example of an intuitive strategy, because it was derived from observations and interviews aimed at understanding how people make reasonably good decisions under time pressure. The RPD model asserts that experienced decision makers can recognize typical themes, and can rely on familiarity as the basis for quickly formulating a situation assessment. This type of recognitional process does not require much cognitive effort or time, and therefore would seem to be fairly robust to most stressors.

However, if the person cannot diagnose the situation easily and must devote effort to its assessment, then there can be a vulnerability to time pressure. One strategy we have seen is to fashion a story, using mental simulation to envision what events might have preceded the current state of affairs. This strategy is Level 2 of the RPD model, in Fig. 2.1. This process does appear to require cognitive resources, and does take time; therefore, stressors that compete for cognitive resources, or restrict the time available to use these resources, can degrade the quality of the situation assessment and can lead to more errors in sizing up the situation, although the degree of degradation may not be very high.

Similarly, for adopting a course of action, in most cases it will be obvious how to proceed as long as the situation is familiar, so stress should have little impact on decision making. However, for a complex course of action that includes novel elements, Level 3 of the RPD model asserts that the person may try to assess the consequences, looking for pitfalls. This type of mental simulation would also be vulnerable to stressors. The result would be a greater chance for blunders than if the person performed a comprehensive mental simulation.

Stressors would not prevent the use of recognitional decision strategies, but stress could interfere with the full use of mental simulation processes that require cognitive effort. Mental simulation is sometimes needed for situation assessment as well as for evaluating a course of action, and it is a cognitively demanding process that may well be affected by distraction and time pressure.

Proposition 5: Reactions to Stressors Are Adaptive Rather Than Dysfunctional

Wickens, Stokes, Barnett, and Hyman (1988) performed an elegant series of studies on stress and decision making. They have looked at the decision making of skilled and unskilled pilots, and have varied different types of stressors; that is, noise, pessimistic instructions that made failure seem likely, dual-task loading, time pressure, and financial risk. Stokes, Belger, and Zhang (1990) used a test battery of tasks sensitive to working memory and showed that noise and concurrent workload in combination as stressors had a strong effect on working memory. The subjects showed little decrement in a task measuring long-term memory, and in a dynamic flight task the set of stressors, applied concurrently, reduced the decision performance of novice pilots but slightly improved the performance of the experts. Presumably, the experts' reactions to stress were adaptive to the situation.

Research suggests that decision makers' reactions to stressors can be adaptive, in a number of ways. The adaptive reactions include use of simpler decision strategies, perceptual narrowing, and filtration. The results reported by Payne et al. (1988) show that effects such as perceptual narrowing and filtration of information, suggested by researchers such as Yates (1990) to be the most typical responses to stress, are in fact quite adaptive. Persisting in the use of the normative model, such as processing all of the attributes for all of the choices, would be suboptimal if time was short and information was missing or uncertain.

Perceptual narrowing may improve performance as it supports a more selective use of cues, especially when there is insufficient time to examine all of the information in the task. Wickens (1987) noted that the narrowing of attention in response to a stressor can be seen as adaptive. The narrowing of attention may represent a useful reaction to task conditions and to having to work with more limited resources, and in some situations may serve as an effective strategy for focusing attention. Particularly for experienced decision makers with the skill to prioritize information sources, narrowing of attention and filtration would seem to make sense.

Experienced decision makers performing difficult tasks in operational settings, when faced with various stressors, can use heuristics such as availability and representativeness. These heuristics describe means of accessing an experience base. Without these heuristics, decision makers might have to rely on inefficient prescriptive strategies. The use of heuristics is discussed in more detail later.

It is also claimed that people become more conservative under stress. This claim, as put forth by Edland (1989), is that time pressure leads to a greater avoidance of negative consequences. It has been disputed by Lieblich (1968), who found more risk taking under time pressure. Even if Edland's findings are replicated, this conservatism is not necessarily a disadvantage. If time pressure reduces a person's ability to perform mental simulations to anticipate pitfalls, then it might be a very reasonable strategy to avoid complex plans, because there may not be opportunities to carefully investigate potential pitfalls.

Stressors increase the likelihood that decision makers choose the first option they consider. Stokes, Kemper, and Marsh (1992), using a desktop microcomputer-based flight-decision simulator, found that more-skilled pilots showed better decision making under time pressure than less-skilled pilots and chose to carry out the first alternative they considered more often than less-experienced pilots. When multiple alternatives were considered, the pilots with 1000 or more flying hours chose their first response 71% of the time, whereas pilots with 400 flying hours or less chose their first response only 53% of the time. Actually, these data underestimate the effect because the more experienced pilots were generating more options. When option rate is taken into account, the rate of selection of first responses by more experienced pilots is almost double that of less-experienced pilots, and is highly significant ($p < .003$).

If decision makers under stress choose their first option, how good can this option be? Klein, Wolf, Militello, and Zsambok (1995) studied skilled chess players and found that they were able to use their experience to generate high-quality moves as the first ones they considered. Taken together, these findings show that experienced decision makers are able to generate reasonable options as the first ones they consider, and select these first options to carry out when performing a stressful task such as flying a complex mission in a simulator.

In research with fireground commanders, Klein, Calderwood, and Clinton-Cirocco (1986) heard again and again that a mediocre choice, initiated in time, was better than an optimal choice that came too late. The biggest sin was failure to act. In operational settings—ranging from police reaction to emergencies, to fireground command, to air-to-air combat, to command and control—there seems to be an individual difference in the readiness to act given incomplete data. Just as impulsiveness carries clear penalties, so does the reflective style that keeps asking for more information as the window of opportunity closes. This reflective style comes in for more concern and criticism than the impulsive readiness to take charge of a situation. From this context, rapid fixation on a course of action and "premature closure" do not seem so dysfunctional at all—delayed closure is perhaps a graver problem.

I am *not* arguing that decision quality is improved under stressors such as time pressure, noise, threat, and overload. Under certain conditions, this may be true (for example, threat often increases motivation to perform well), but generally decision quality should decrease. My point is that, given the inherent difficulties of performing a task with reduced time and attention and working memory, decision quality is usually satisfactory, although not necessarily optimal. Schneider and Detweiler (1988) have identified a set of compensatory strategies that enable people to adapt to overload and time-pressured conditions. These include the following: shedding, delaying, and preloading tasks; abandoning high-workload strategies; increasing efficiency in allocating attention; reducing information search and visual scan behaviors; and filtering out less-relevant items of information. These adaptive strategies, described as signs of effective coping, correspond nicely to the behaviors I have described earlier.

Proposition 6: Stressors Do Not Increase the Likelihood of Decision Biases

I assert that there is no compelling evidence to believe that stressors increase decision biases. Furthermore, there have been serious difficulties in applying the concept of decision biases to naturalistic settings; little is gained by linking stress to biases.

Janis and Mann (1977, p. 11) popularized the idea that decision making falls apart under stress. They presented an ideal, systematic strategy with a number of essential steps, as described in Table 2.2. Under stress, people were likely to fall

TABLE 2.2
Steps for Vigilant Decision Making

The decision maker, to the best of his ability and within his information processing capabilities:

1. Thoroughly canvasses a wide range of alternative courses of action;
2. Surveys the full range of objectives to be fulfilled and the values implicated by the choice;
3. Carefully weighs whatever he knows about the costs and risks of negative consequences, as well as the positive consequences, that could flow from each alternative;
4. Intensively searches for new information relevant to further evaluation of the alternatives;
5. Correctly assimilates and takes account of any new information or expert judgment to which he is exposed, even when the information or judgment does not support the course of action he initially prefers;
6. Reexamines the positive and negative consequences of all known alternatives, including those originally regarded as unacceptable, before making a final choice;
7. Makes detailed provisions for implementing or executing the chosen course of action, with special attention to contingency plans that might be required if various known risks were to materialize.

Note. Adapted from Janis and Mann (1977).

into a condition that Janis and Mann labeled *hypervigilance,* which included impulsive and disorganized behaviors resulting in the abbreviation or omission of one or more of the steps listed in Table 2.2.

This account runs into difficulty if the steps in Table 2.2 do not constitute the ideal. The steps in Table 2.2 are appropriate for well structured tasks, but do not apply to ill-structured tasks performed in operational settings.

Driskell, Salas, and Hall (1994) recently performed a study using experienced Navy technical school personnel in a simulation involving air threats, surface threats, and subsurface threats. The stressors were time pressure, auditory distraction, and task load, presented together to create conditions of high or low overall stress. In one condition, subjects were instructed to carefully scan all information items, in a sequential manner, devoting equal time to each item, and reviewing the scanned information before making a decision (vigilance condition). In another condition, subjects were asked to perform the task by scanning only the items needed to make a decision, in any sequence, devoting as much or as little time to each item as required, and performing reviews only where necessary (hypervigilance condition). Performance was measured by the number of information fields accurately identified or labeled. The subjects using the hypervigilant decision strategy performed better than those using the vigilant strategy. Moreover, the vigilant strategy led to poorer performance under stress, compared to no-stress conditions. In contrast, for subjects using the hypervigilant strategy, performance was not reduced by stress. Therefore, we may not have to worry that stress will prevent people from using the vigilant strategy shown in Table 2.2, because this strategy may not be very useful in many naturalistic conditions.

Now let us look more specifically at decision biases. The heuristics and biases paradigm was introduced by Tversky and Kahneman (1974) who performed a number of studies showing that under carefully controlled conditions, subjects committed a variety of errors in reasoning owning to their use of heuristics. More than two dozen heuristics/biases have been cataloged by Sage (1981), including availability, representativeness, anchoring and adjustment, and seeking confirmation. One hypothesis is that stressors potentiate these biases, resulting in poorer decision making. For example, Sheridan (1981) speculated that time pressure could lead subjects to treat all data sources as equally reliable, which would exaggerate representativeness biases. Sheridan also reported that under time pressure subjects tended to overlook data sources. Barnett (1993) interpreted these findings to suggest that a stressor such as increased task load would increase a number of biases (e.g., representativeness, availability, anchoring, contrast, and confirmation biases) because subjects would sample fewer cues.

The suggestion that stress increases decision biases runs into two difficulties: The *concept* of decision biases has come under criticism, and the *evidence* for degraded decision performance under stress has also been challenged. The concept of decision biases appears to rest on a misinterpretation of the findings reported by Tversky and Kahneman (1974). Lopes (1981) pointed out that Tversky and Kahneman set out to show that some heuristics were so strong, subjects would rely on them even under conditions where performance became degraded. This paradigm was necessary in order to demonstrate the strength of the heuristics, but it has led to the erroneous conclusion that outside of these carefully controlled conditions, people are inherently biased. In fact, their data do not lead to such a conclusion.

Other data suggest just the reverse, that the informal reasoning on which we all rely is not inherently biased, and does not result in degraded decision making. Gigerenzer (1987) found that even under laboratory conditions, when additional contextual information is presented the biases sharply decrease. Christensen-Szalanski (1986) and Fraser, Smith, and Smith (1992) demonstrated that the leading "biases" have little impact on the quality of decisions in naturalistic settings. Shanteau (1989) found little evidence for biases in a study of accountants. Smith and Kida (1991) found a minimal impact of biases in experienced auditors. Cohen (1993) argued that the concept of bias carries little meaning outside the controlled laboratory conditions needed to demonstrate the use of heuristics.

It therefore is difficult to maintain a hypothesis that stress increases the impact of biases, because the concept of biases is so questionable, the impact of biases outside the laboratory seems negligible, and there is no strong evidence that stressors systematically degrade decision quality.

Proposition 7: Stressors Do Not Necessarily Degrade Decision Performance

Stressors should strongly reduce subjects' abilities to carry out tasks, and indeed the majority of studies that have blasted subjects with noise, with shock, or with threatened shock have managed to disrupt performance. Likewise, Edland and Svenson (1993) reported that the research literature shows that time pressure decreases the accuracy of judgment.

Importantly, however, there are some studies showing that stress can *improve* performance. Edland (1989) also cited a range of studies showing improved performance with stressors such as threat and time pressure. Harvey, Hammond, Lusk, and Mross (1992) and Lusk (1993) found that increased stress heightened weather forecasters' predictive accuracy. Hockey (1970) found that noise as a stressor did reduce the range of cues attended to, but in doing so focused attention on the dominant part of the task. Hockey, Dornic, and Hamilton (1975) also found that noise led to higher recognition scores and faster reading of messages.

One of the most surprising findings has been that time stress can improve the performance of decision-making teams. Serfaty, Entin, and Volpe (1993) studied well-trained Navy Combat Information Center four-person teams using a task that simulates antiair warfare operations. When time pressure increased from low to moderate, error rates went up. However, when time pressure was increased further, error rates went down even below the levels for low time pressure. Serfaty et al. discovered that the teams were changing strategies. Ordinarily, the subordinates were responding to information requests from the commander. With moderate time pressure the teams accelerated this pattern, with less success. With high time pressure the subordinates stopped waiting for information requests and instead provided the commander with information they anticipated would be helpful. These data show how strategy shifts can lead to performance that is superior under stress than under nonstressed conditions.

Calderwood, Klein, and Crandall (1988) studied the effect of time pressure on chess performance. Two six-person groups of chess players, Masters and Class B players, were studied. The members of each group played a series of actual games against the other members, and the quality of the moves was evaluated by a Grandmaster who was not informed about the player's strength or about the time condition. The data analysis considered only the complex conditions where several possible choices were present, and excluded the simple cases where the choice of moves was obvious. Two conditions were used—regulation time, with approximately 2.6 minutes per move, and blitz games, with approximately 6 seconds per move. For the Masters, the proportion of blunders showed no effect of time pressure, 8% under regulation time and 7% under blitz conditions. For the class B players, the proportion of blunders did increase with time pressure, from 11% to 25%. Therefore, extreme time pressure seemed to have no impact on the chess Masters, although it did sharply reduce the performance of the class B players (see Table 2.3). Shanteau (1988) also suggested that experts were much less affected by time pressure than were nonexperts. Other researchers have shown that stress does not necessarily result in worse performance. Poulton (1976) cited a number of experiments showing that stressors such as heat, noise, and vibration can reliably improve performance, particularly in tasks that require speed or vigilance. Poulton argued that it could be a mistake to eliminate such stressors from an environment.

Time pressure and workload have been studied in regard to aviation weather forecasters' performance. Experienced forecasters made a series of predictions of thunderstorm severity at Stapleton International Airport. Time pressure was inferred from task overload, and the activity level (signifying degree of workload) was identified for each forecasting trial. Harvey et al. (1992) found that forecasters working under high time pressure achieved the same probability level using less evidence than forecasters working under low time pressure. Lusk (1993) looked at the same data set and found that the forecasters did *not* show greater sensitivity under low time pressure, but the reverse—sensitivity was higher under high time pressure, .057 versus .016 (although this trend was not significant). To summarize, stressors such as noise, threat, heat, vibration, and time pressure do not always reduce performance levels, and the narrowing of attention that is sometimes found to accompany stressors may appear to be adaptive.

Implications

In concluding this section, I summarize the major points thus far. Stressors may improve performance as well as disrupt it. The changes induced by stressors appear fairly adaptive—to trade off accuracy for speed, as when faced with a threat, and to narrow the focus of attention when faced with capacity limitations and attentional disruptions. Fear and anxiety may provoke intense physiological reactions that lead to self-monitoring, which itself serves as a distracting, sec-

TABLE 2.3
Percentage of Poor Moves in Complex Situations for Master and Class B Games Under Blitz and Regulation Conditions

	Game type	
	Blitz	*Regulation*
Skill	*M*	*M*
Master	.07	.08
Class B	.25	.11

ondary task of managing the anxiety reaction. Decision makers may show a range of adaptive reactions when faced with stressors: simpler decision strategies, perceptual narrowing, and the use of heuristics that take advantage of experience.

THE VINCENNES INCIDENT

In 1988, the USS Vincennes shot down an Iranian airliner, killing all those aboard. This event has been the subject of a great deal of investigation and analysis, and stands as a highly publicized example of faulty decision making under stress. There have been other famous accidents. Three Mile Island was the key incident for illustrating the importance of human factors for system design. Faulty crew coordination was exemplified by two landmark accidents. One was the crash of an L-1011 in the Florida Everglades in 1972 while its crew was preoccupied with a malfunctioning light bulb. The second accident was the 1982 crash of a B-737 in Washington, DC during a snowstorm, with the voice recorder subsequently revealing a first officer ineffectively trying to warn the captain that something was not right (in fact, ice had blocked a sensor, leading to false high-thrust readings, so the engines were set at less-than-normal takeoff thrust). The Vincennes shoot-down has taken its place with these accidents and has turned a neglected topic into a question of high interest for the research community; namely, how stressors can degrade decision making.

What, however, can we learn from the Vincennes incident? In some ways, it is a Rorschach test, a projective technique for applied researchers. To those of us interested in display design, it is an example of a poorly designed interface. To those studying biases, it is an example of decision biases in action. For those studying team performance, it is an example of what can go wrong during team decision making. For stress researchers, it is a case study of how stress clouds judgment. One cannot ignore the incident, because it is the defining exemplar of decision making under stress. By trying to clearly understand what happened and why it happened, one will be in a position to use the incident to evaluate the theses presented in the last few sections.

Overview[1]

On July 3rd, 1988, a U.S. Navy AEGIS cruiser shot down an Airbus 300 Iranian commercial airliner, Iran Air Flight 655, which had taken off from the Bandar

[1]The account of the Vincennes incident presented in this chapter is based on the U.S. Navy's official investigation as described in the Fogarty Report, as well as additional articles and reports, especially Roberts and Dotterway (1995), comments from knowledgeable sources such as Ted Kramer and Pete Edgar, and interviews with Capt. Will Rogers III, the Commanding Officer of the USS Vincennes at the time of the incident.

Abbas airport. The entire flight lasted 7 minutes and 8 seconds, from 0947 to 0954 (local times). Active monitoring began at 09:47:49, when the track was identified as "Assumed Enemy." The final segment, which lasted over 3 minutes, involved deliberation about the flight as a specific threat to the Vincennes. According to the Commanding Officer (CO) of the Vincennes there were approximately 189 seconds from the time the airliner became tactically significant until missile launch.

There were five primary states of situation awareness during the incident: (1) the initial battle between the Vincennes and Iranian surface ships, (2) detection that an unknown aircraft took off from Bandar Abbas airport, (3) identification of the aircraft as Iranian military, (4) reports that the aircraft was descending, and (5) final assessment that the aircraft was probably hostile and was a threat to the Vincennes.

Background

There were a number of previous events that shaped the mindset of the Vincennes' crew at the time of the incident:

- On April 18th, 1988, the U.S. military had conducted an operation "Praying Mantis" to attack Iranian forces disrupting Persian Gulf shipping; retaliation for this operation was expected. On April 18th, there was a battle in the same area in the Persian Gulf, in which missiles were fired by an Iranian F-4.
- Iranian F-14s had been shifted down to the Bandar Abbas airport two weeks before the incident.
- A recent incident had occurred in which an Iranian F-14 had flown towards a U.S. cruiser, had been warned, and had broken off after the cruiser directed fire control radar at it.
- In another recent incident, a commercial airliner had taken off from Bandar Abbas with an Iranian F-4 flying just below, tucked underneath it. This may have happened more than once.
- Iranian military aircraft were also observed squawking Mode III IFF (Identify Friend or Foe). The IFF system was designed to distinguish military from commercial, friendly from other. Mode III is supposed to be reserved for use by commercial aircraft.
- There was intelligence information that the Iranians intended to take some provocative action during the 4th of July weekend.
- On July 2, 1988, Iranian gunboats suddenly became very active.

Part of the background is the set of base rates involving radio challenges to potentially threatening aircraft in the Persian Gulf. From June 2nd, 1988 to July

2nd, 1988, 150 challenges were issued by the U.S. Navy to aircraft. Of these, two (1.3%) were to commercial flights COMAIR and 125 (83%) were to Iranian military aircraft. Also, showing the increasing tensions, no Iranian F-14s were challenged from June 2nd to June 17th, 1988, but seven Iranian F-14s were challenged from June 17th to July 2nd, 1988.

Events of July 3rd, 1988

On the morning of July 3rd, the USS Elmer Montgomery was surrounded by 13 Iranian gunboats. The Vincennes was called in to help out, and sent a helicopter to establish a communications link back to the Vincennes from the Montgomery location for further transmission. At 0930, the Vincennes was 35–40 miles away from the Montgomery.

The Vincennes' helicopter transited to the Montgomery, and was fired on. At this time, the Vincennes was 10 miles away. When the Vincennes headed in the direction of its helicopter, the Iranian gunboats appeared to close on the Vincennes on erratic courses. (They may also have just been milling around inside their own territorial waters.) This action was evaluated as demonstrating hostile intent. After receiving permission from higher authority, the Vincennes opened fire with deck guns. Almost simultaneously the gunboats commenced firing at Vincennes.

The surface battle began. Small craft are very maneuverable, fast, and difficult to hit; therefore, the Vincennes went to maximum rate of gunfire. These events were in progress at the time the Airbus began its trip.

When the Airbus took off, the Vincennes was 1 mile west of the center line of civilian airway Amber 59, the air land assigned to the Airbus. It should be noted that in view of traffic density, commercial flights in the Gulf rarely deviated from the airway center line. The air lane was 20 miles wide, 10 miles on each side of the center line. The AEGIS displays only show the center line and not the entire air lane.

The Airbus was attempting to reach an assigned altitude of 14,000 feet. It was still climbing through 12,000 feet when the Vincennes' crew decided to engage it. It was shot down at approximately 13,500 feet, at a speed of approximately 380 knots.

The Shootdown Incident

The following section lists events during the Vincennes incident. These crew members were involved:

CO: Commanding Officer. The officer in charge of the Vincennes.

AAWC: Antiair Warfare Coordinator. The crew member in charge of monitoring air threats and assets pertaining to the Vincennes itself.

FAAWC: Force Antiair Warfare Coordinator. The officer subordinate to the CO who was in charge of air defense for the Vincennes and also for the entire battle group.

IDS: Identification Supervisor, in charge of identifying air, surface, and subsurface tracks.

TIC: Tactical Information Coordinator, in charge of relaying critical data.

Identify Friend or Foe (IFF) refers to a system for electronically querying an unknown airplane triggering the transponder on that aircraft to respond information that would support identification. Mode II is for military aircraft and Mode III IFF is for use by commercial aircraft.

0630–0947 Situation Awareness #1: Fight With Gunboats. A mission to support the USS Elmer Montgomery. In the course of supporting the Montgomery, the Vincennes became involved in a fight with Iranian gunboats.

0947 Situation Awareness #2: Takeoff. Commercial airliner took off from Bandar Abbas airport, on a course directly over the Vincennes. Airliner was identified as "Unknown—Assumed Enemy" the designation for flights originating from potentially hostile joint use airfields. A crew member identified it as a single track, and gave it Track Number TN 4474. Seven different crew members all had TN 4474 in close control. The distance of the airliner from Vincennes at 0947 was 47 nautical miles. TN 4474 was given the new TN 4131. The reason for the switch in track numbers was that the USS Sides had identified the same air track and had given it the designation TN 4131. The AEGIS system determined that TN 4474 and TN 4131 were the same aircraft, so it used a single number for that aircraft.[2] By chance, the system selected the track number chosen by the USS Sides.

IDS checked the commercial air (COMAIR) schedule. It should be noted that the Airbus flight was 27 minutes late. Also, there are several different time zones in the Persian Gulf. Further, every time the Vincennes fired its five guns at the gunboats, the lights in the combat information center dimmed; IDS did not find the Iran Air 655 flight.

An Iranian P-3 in the area was challenged by the Vincennes over Military Air Distress (MAD) and International Air Distress (IAD), and acknowledged the request to stay clear. However, the P-3 was on an inbound course to Bandar Abbas and the Vincennes' crew worried that the P-3 could have been sending targeting information to the unknown aircraft, enabling it to fly cold-nose (without its radars on).

[2]The Naval Tactical Data System passed this type of data between ships and aircraft in a battle group, allowing correlation between tracks.

0948 The distance of the airliner from Vincennes was 44 NM.

0949 An attempt was made to contact the airliner via radio. The FAAWC then ordered TN 4131 to be challenged over Military (MAD) and commercial international (IAD) Air Distress radio frequencies.

0950 The FAAWC ordered the radio warnings to be repeated.

The CO of the Vincennes informed his superior, the commander of the Joint Task Force (who was on another ship), that if aircraft did not respond to warnings and closed within 20 miles Vincennes intended to engage (i.e., fire a missile at it). This was Standard Operating Procedure. The commander of the Joint Task Force agreed, and added that Vincennes should warn the aircraft again. This was done.

The first IAD warning, at 0950, was "Unknown aircraft on course 206, speed 316 position 2702N/05616E you are approaching U.S. Naval warship request you remain clear." The second IAD warning, at 0951, was "Unidentified aircraft on course 207, speed 360, altitude 7000. You are approaching U.S. Naval warship bearing 205, 30 miles from you. Your identity is not known, your intentions are not clear. You are standing into danger and may be subject to USN defensive measures. Request you alter course immediately to 270." A third IAD warning was issued at 0952 "Unknown aircraft on Course-210, Speed-360, Altitude 10,000. You are approaching USN warship Bearing 201, 20 miles from you. You are standing into danger and may be subject to USN defensive measures."[3] Approximately 12 warnings were transmitted, including MAD and IAD.

The aircraft was approximately 28–34 miles away.

0950 Situation Awareness #3: Aircraft Identified as Iranian Military AC. At 0950 the target was identified by many crew members as Iranian Military hostile. The IDS saw Mode II-1100, indicating an F-14, and reported so over the Vincennes' internal communications net. Other crew members also saw Mode II-1100. A crew member then tagged TN 4131 as an F-14 on the large screen display, for all to see; at the same time, one crew member told the CO that TN 4131 was possibly COMAIR. The CO acknowledged this information.

FAAWC requested permission to engage when TN 4131 got to 20 NM. The CO rejected this request, even though he had set it up as a trigger point, because the aircraft had not turned its radar on. The CO of the Vincennes did not feel he had enough information. He could not conceive that the pilot was going to continue to bore right at him, but could not figure out what the pilot was doing. He told the FAAWC to generate the radio warnings continuously. As described

[3]Subsequent to the shoot-down, this protocol was altered to include more careful geographical data so that commercial airline pilots could more easily determine who was the target of such a warning.

earlier, the warnings had gone from "Please state your intentions. . . ." to "We are prepared to take immediate defensive action if you do not respond."

Another factor was the CO's confidence in the AEGIS system. He felt they could down the aircraft at any time, without difficulty. The CO still had time left, although some crew members were getting anxious to engage.

The FAAWC requested and received permission from the CO to illuminate the aircraft (i.e., use fire control radar) at 20 NM. The purpose of illumination was to warn the aircraft to divert its course. By using fire-control radar, the Vincennes was locking on to the track, preparatory to firing a radar-guided missile.

The Airbus departed from the optimal course down the center line of the air lane. For the Persian Gulf, this was unusual; the departure was directly in line with the Vincennes. By the time of the shoot-down, the Airbus was 3.35 NM west of the centerline.

The CO again requested permission from his supervisors to engage the target. This was a standard procedure, giving the CO permission to engage at will, if necessary. The Vincennes received permission to engage in self-defense. At this time, the distance from Airbus to Vincennes was 28 NM.

0952 Situation Awareness #4: Aircraft is Descending. A crew member reported "descending" on one of the internal communications nets. Several other crew members also saw the descent. The CO did not verify this for himself—he was too busy with the surface battle. There was no indication of radar or other activity from TN 4131. The aircraft was within 22 nautical miles of the Vincennes at 0952—this decreased to 18 miles, then 17 miles, then 16 miles.

0953 The FAAWC kept asking if the CO wanted to engage. The CO kept refusing. At this point, the approaching aircraft had become the prime interest of the CO. The distance was 15–16 NM. When the aircraft reached 15NM, the TIC began announcing the decreasing range every open spot on the internal communications net. In fact, the TIC was dominating the net during this episode.

0954 The distance from the Vincennes was 12 NM.

Situation Awareness #5: The Aircraft was Beginning its Attack on the Vincennes. None of the data available to the Vincennes were inconsistent with the view that the aircraft had hostile intent, and some data were inconsistent with the hypothesis that the aircraft was COMAIR. Would an Iranian fighter pilot take off during a battle with surface ships, squawk Mode II IFF, change the transponder to squawk Mode III, fly directly at the Vincennes, ignore radio warnings to divert course, and descend towards the Vincennes? This was conceivable. Would a commercial airline pilot take off in the middle of a battle or be sent off by the control tower at Bandar Abbas, head directly into the middle of the battle, depart from the center line, ignore IAD notifications, squawk Mode II IFF, and descend towards the Vincennes? The answer is clearly "no." It is very difficult to make up a story to explain such behavior, so the CO rejected this hypothesis. In retrospect, we know that the last two information items were incorrect.

The aircraft reached ten miles. The CO's opportunity for self-defense was rapidly diminishing. Under 10 miles, the probability of Vincennes being able to make a successful missile intercept would be problematic. He had to shoot then, or not at all. The Vincennes had 400 people on board and was worth $400 million dollars. The decision to fire was made at about 9–10 NM, just about the minimal range for the missiles to be effective. The Airbus was hit at 8 NM.

Situation Awareness #6: Iranian F-4s. A section of Iranian F-4s came up and turned on their Fire Control radars. The Iranian F-4s harassed the Vincennes but did not attack it.

Situation Awareness #7: Iranian Hovercraft. A high speed surface contact approached at 55 knots. It was an Iranian hovercraft. The Vincennes fired a warning shot, and the hovercraft broke off.

Difficulties

Reason (1990) and Woods, Johannesen, Cook, and Sarter (1993) argued that we need to rethink examples of decision errors. When an accident such as the Vincennes shootdown occurs, we must resist the temptation to blame it on human error, and instead we should use this assessment as the starting point of the investigation, rather than the explanation. Figure 2.3 shows how the operator who appears directly responsible for the accident may be considered to be the sharp end of a range of causal factors, many of which existed in the system for long periods of time. Reason has termed these systemic flaws "latent pathogens," because they exist without effect until the wrong combination of events leads the human decision maker astray.

In this subsection, I review a range of difficulties that existed prior to the Commanding Officer's decision to engage Track Number 4131. I have identified 15 problems (see Table 2.4). Those problems attributable to training are marked as [T], those attributable to system design as [SD], and those attributable to organizational policies as [O].

Difficulty #1. The individual sitting at the AAWC console, although familiar with his duties, was not formally qualified [T, O. Impact: Low]. The person acting as AAWC had not yet completed his qualifications test and had difficulty in using the Fire Control System (also see #7 later). The AAWC did not commit

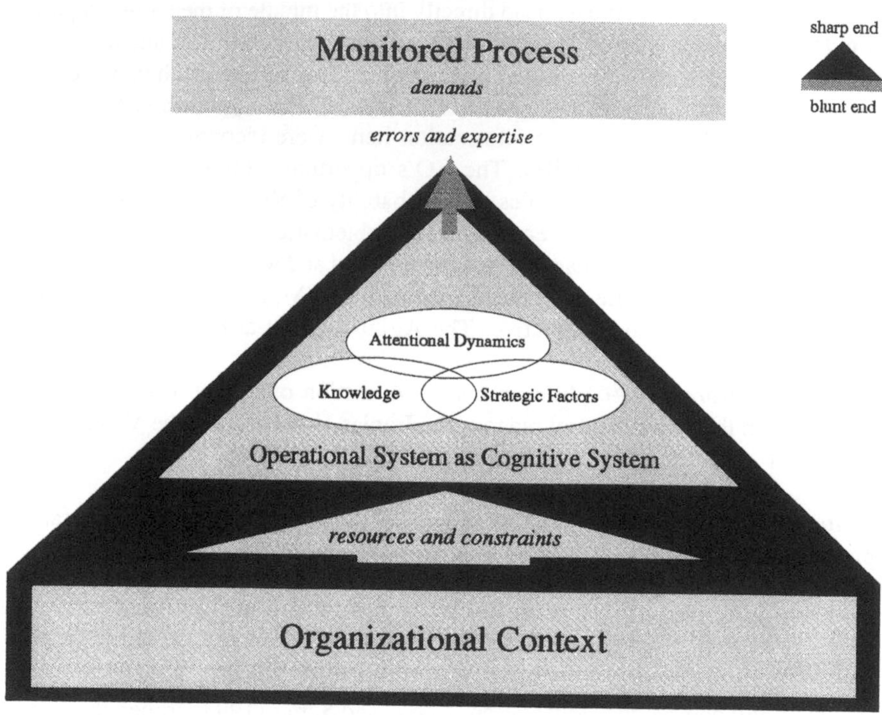

FIG. 2.3. The sharp and blunt ends of a large complex system. The interplay of problem demands and the resources of practitioners at the sharp end govern the expression of expertise and error. The resources available to meet problem demands are shaped and constrained in large part by the organizational context at the blunt end of the system. From Woods, Johannesen, Cook, and Sarter (1993). State-of-the-Art Report prepared for CSERIAC. Reprinted by permission.

any important errors, but did not catch any errors either. That is, he did not provide redundancy for independent verification of judgments such as altitude trends.

Difficulty #2. The Airbus was not located on the commercial flight schedule [T, SD. Impact: Medium]. The crew member responsible for identifying new tracks (IDS) attempted to locate the airbus, Iran Air 655, on the printed flight schedule for Bandar Abbas but did not. This was not a design flaw, but this type of problem would be easily remedied using a computer database.

Difficulty #3. The Airbus was identified as military [T, SD. Impact: High]. A crew member reported receiving Mode II IFF signal from the airplane, indicat-

TABLE 2.4
The Vincennes Shootdown: Errors and Remedies

		Impact	Remedy
1.	Unqualified AAWC	Low	Training
2.	Missed Airbus on schedule	Medium	Training
3.	Airbus observed to squawk Mode II IFF	High	Training & design
4.	CAP not used	High	Training & organization
5.	Vincennes failed to contact Airbus via radio	High	Training, organization, & design
6.	AAWC failed to illuminate Airbus, or report failure	Low	Training & HCI
7.	USS SIDES failed to inform Vincennes of COMAIR ID	Medium	Training
8.	Communications overload	Medium	Design
9.	USS Sides used wrong TN	Medium	Training & design
10.	Track Number switched	High	Design
11.	CO calls out wrong TN	High	Training & design
12.	Erroneous TN not corrected	High	Training
13.	Altitude trends hard to read	Medium	Design
14.	Vincennes Misread Airbus altitude	High	Training & design
15.	FAAWC overloaded	Medium	Organization

ing a military rather than a commercial aircraft. This report spread through the Vincennes' Combat Information Center and was the basis for tagging the aircraft as an F-14 on the Large Screen Displays. The most plausible speculation (which has never been officially confirmed) is that the manual control for tracking aircraft was left over the edge of a runway at Bandar Abbas, and in this position picked up the Mode II IFF signals from a military aircraft on the ground at Bandar Abbas airport. Because the Airbus was the only airborne track in the area, the crew member correlated these data and asserted that the Mode II IFF came from the track we now know was the Airbus.

Difficulty #4. Combat Air Patrol was not used [T, O. Impact: High]. American F-14s were in the air, not too far away, and could have been called in for a visual identification of the unknown aircraft. The Vincennes requested CAP from the USS Forrestal as soon as the helicopter was fired on. There were Combat Air Patrol on station approximately 80 miles away from the Vincennes at the time of the incident, a distance that could have been covered in about 7 minutes; some reports put CAP at 5 minutes away at the critical time. Therefore, CAP may have reached the area in time to identify the Airbus if the Vincennes had brought them over in time. Various reasons have been given for the failure to vector the F-14s to the site of the incident; for example, organizational factors such as not wanting to exacerbate the situation by bringing in

F-14s, and possible other reasons such as not realizing the need for the F-14s until it was too late. Here, a more experienced AAWC might have been more proactive about asking for CAP.

Difficulty #5. The Vincennes crew was unable to contact the crew of the Airbus via radio [O. Impact: High]. There were several difficulties here. First, the Vincennes did not have radio capabilities for simultaneously querying the aircraft and contacting the Bandar Abbas Air Traffic Controller, so it did not try to reach the Air Traffic Controller to sort out the problem. Also, the cryptic information the Vincennes sent out on the International Air Distress net might not have permitted the Airbus pilot to readily figure out that they were talking to him. The radio messages would have been difficult to interpret by a commercial pilot. The communications protocol was subsequently changed. The crew was following the standard procedures, and didn't realize the need to create a new protocol on the spot.

Difficulty #6. AAWC initially failed to complete the console sequence required to illuminate the aircraft with the fire control system. It is unclear if the AAWC ever successfully completed this action [T. Impact: Low]. This failure was unimportant because the Airbus was commercial and did not receive the radar signal from the Fire Control System. However, the AAWC may not have reported his failure back to the CO. The CO of Vincennes could not remember whether the illumination was successful, or, if unsuccessful, whether he had been so informed. To make matters worse, the Fogarty Report suggests that on the Vincennes' radar screen the aircraft seemed to "flinch" around the time it would have been illuminated. This could have been interpreted as evidence that the pilot was military, received the lockon signal, and reacted. If the Commanding Officer of the Vincennes had known that lockon never occurred, he would not have reached this conclusion. However, in an interview Capt. Rogers, the Commanding Officer, stated that he did not recall seeing the aircraft flinch.

Difficulty #7. The crew of the USS Sides failed to inform the Vincennes that the target appeared to be commercial air [T. Impact: Medium]. This omission was particularly serious because the Sides knew that the Vincennes had identified the airliner as an F-14. Note that the Sides identified the aircraft as COMAIR because it was not misled by two erroneous pieces of information—the Mode II IFF signal, and the report of descending altitude (see the following).

Difficulty #8. Communications limits [SD. Impact: Medium]. There was some degradation of the internal communication net, possibly due to overuse. The internal circuits were prone to overload, and there was a need for the Tactical Action Office to shift between internal communications nets. The Tactical Action Officer kept radio discipline very well, and switched circuits whenever he felt the net degrade, but this process added to the confusion level.

Difficulty #9. The USS Sides used the wrong Track Number [T. Impact: Medium]. When the Airbus first took off, the Vincennes spotted it and gave it TN 4474, while the USS Sides simultaneously tagged it as TN 4131.[4] In doing so, the Sides used a number from the block that had been reserved for the Vincennes. When the AEGIS system correlated TN 4474 and TN 4131 as the same aircraft, it had to select one of the two track numbers. It selected the one given by Sides—4131—which might have added to the confusion for the Vincennes' crew.

Difficulty #10. The AEGIS system reassigned TN 4474 to another aircraft in the area [SD. Impact: High]. Roberts and Dotterway (1995) determined that once the AEGIS system correlated TN 4474 and TN 4131 as the same aircraft, and settled on TN 4131 as the designator, it assigned TN 4474 to another aircraft in the Persian Gulf. It happened that this aircraft, the new TN 4474, was an American A-6, approximately 100 miles from the Vincennes. The problem was not in the design of the AEGIS system, but in the larger system and procedure used to provide intercommunications. Modifications have been made, and the system has subsequently been modified to prevent this type of reassignment from occurring.

Difficulty #11. The Vincennes CO called out the wrong track number [T. Impact: High]. In the middle of the incident, the Commanding Officer asked his crew "What is 4474 doing?" He mistakenly used the original track number, rather than the new one. Hutchins (in press) has found a number of communications errors such as this in a study of a simulated AEGIS cruiser combat information center. Note that all numbers in a block tend to start with the same numeral, for ease of database management.

Difficulty #12. No one corrected the wrong track number [T. Impact: High]. When the CO asked what TN 4474 was doing, none of the crew members informed him that he was using the wrong track number and should have been asking about TN 4131. Accordingly, some crew members may have punched in TN 4474 into their consoles, whereas others appear to have entered the correct TN 4131.

Difficulty #13. The altitude trends were hard to read [SD. Impact: Medium]. The display was a Character Read Out (CRO), an alphanumeric report of the current data including altitude for an aircraft of interest. Altitude is presented on the left-hand side of the CRO, which is small and hard to read. It is not an analog or graphic readout. Because of the possibility of air pockets that can bounce an airplane up or down, crew members cannot instantaneously establish a

[4]Actually, the Sides initially used TN 4130, dropped out of the communications link, reentered the link, reacquired the track, and assigned it track number 4131.

trend; digital readouts can jump around. Furthermore, the CRO does not show direction of altitude changes. Crew members must hold prior altitudes in memory to infer changes. Some of the crew members may also have switched their attention from the Large Screen Display to the CRO at their workstation, which would create a distraction each time they looked from one place to the other. The CO of the Vincennes preferred to focus on the Large Screen Displays, and therefore was not seeing any altitude information at all. He has estimated that it would take 5 to 10 seconds of watching the CRO to infer trends. Although this may seem short, it is a long portion of the 90 seconds remaining to be fixated on a single number.

Difficulty #14. The Vincennes misread the target's altitude [T. Impact: High]. The Vincennes crew believed that the Airbus was descending. It is possible that the decreasing range (also inferred from the static CRO display) was confused as a decreasing altitude. It is also possible that there was a misreading due to the changed target number (see #10 and #11). Roberts and Dotterway (1995) found that at the time the A-6 now designated as TN 4474 was sharply descending and increasing in speed. According to Roberts and Dotterway, possibly nine crew members may have entered the wrong data, TN 4474, and reported that the aircraft was descending, while possibly five crew members entered the correct TN 4131 and reported that it was ascending.

Difficulty #15. The FAAWC may have been overloaded by having responsibilities as radio talker along with helping to manage the air picture [O. Impact: Medium]. This was an observation from the Fogarty Report, and bears on the organization of crew members in the CIC. The CO of the Vincennes felt that these problems had not arisen during training or in previous activities, but in retrospect the load on the FAAWC seems heavy, and the recommendation seems fair, a point on which the CO of Vincennes agrees with Admiral Fogarty. Because of his extra responsibilities, FAAWC may have had less opportunity to verify the information from other crew members.

Several conclusions follows from this listing of difficulties:

1. There is minimal evidence for any decision biases. Although 11 of the 15 deficiencies were related to the training of the crew, only one of these, the misreading of altitude, can be linked to a decision bias.

Nisbett (1993), in his testimony to Congress, has suggested that this altitude error shows expectancy bias, because the crew believed the target to be hostile, and expected the target to be descending and that, in fact, was what they believed they saw even though the CRO altitude readings showed no descent. (see "Vincennes: Findings Could Have Helped Avert Tragedy, Scientist Tells Hill Panel" (1988) for description of testimony)

Klein (1989) argued that this "expectancy" bias can also be viewed as a proper use of base rates. The frequency of commercial airliners being warned by USN ships in the Persian Gulf at that time was very low, less than 2%, so base rates led the CO to believe that commercial airliners would monitor IAD. The CO of the Vincennes had the opportunity to warn commercial airliners twice in the past, using IAD, and he said that both times the pilots were monitoring IAD. If the CO had let the unidentified aircraft keep approaching, if it had been an F-14, and if he had been attacking, then decision analysts could have claimed that he failed to use base rates (the 2% data). Therefore, either way, he was open to charges of decision bias if he made a mistake: expectancy bias if he shot the aircraft down, or failure to use base rates if he did not shoot it down. It seems that any wrong choice can be explained as decision bias.

The strongest reason for rejecting Nisbett's claim of expectancy bias is that the misreading of altitude seems best explained as confusion generated by the shifted track numbers. Roberts and Dotterway (1995) showed that the Vincennes' Combat Information Center wasn't a clash of different reports (which would be the case if expectancy bias was operating), but centered around two different sets of values for airspeed and altitude trends. One set of reports bore a strong resemblance to the probable A-6 data, and the other to the Airbus data. The error was most likely due to crew members entering in the erroneous TN 4474 requested by the CO. There is no need to invoke expectancy bias to account for the misreading of altitude trends.

Even without this error the Airbus may have been shot down, because a hostile aircraft might well have been expected to hold a high altitude as long as possible. The visibility was murky, and without using its radar an aircraft would have had difficulty locating the Vincennes unless it first gained altitude, prior to making the attack. Therefore, the reported descent was a definite trigger for self-defense, but a lack of a descent would not have meant much; a hostile aircraft could have delayed its descent until it was already within the Vincennes' defense perimeter. The Fogarty Report suggested that at 9 NM and 13,000 ft. (the coordinates for the Airbus when it was hit), an F-14 would still be able to attack the Vincennes.

2. There were several deficiencies that could be attributed to training and experience. The misinterpretation of the Mode II IFF (Difficulty #3) was due, in part, to the way the range gate operated under unusual circumstances. In this case, a second return associated with military aircraft on the ground appears to have been mistakenly correlated with the signal of the commercial aircraft that was already airborne. However, the crew member involved was judged to be highly experienced.

Failure by the AAWC to operate the Fire Control System (#6) was due to inexperience. The failure of the Sides to share information (#7) was due more to failure to adhere to procedures than to expertise. The misreading of altitude changes (#13) cuts across experience and design. The failure of the crew to catch the erroneous altitude report (#12) was a problem of team coordination and training.

3. The equipment design seems to have been a major contributor to the

problem: lack of capacity for querying the basis for the AEGIS classification, and difficulty in interpreting altitude trends. However, the ambiguity of the situation was such that even with no system design flaws, there is a good chance that the Airbus would have been shot down.

4. The key decision strategy was the use of mental simulation (Klein & Crandall, 1995) to formulate an understanding of the intent of the pilot of the approaching aircraft. The available data presented to the CO as facts were all consistent with a hostile aircraft, and a number of facts were inconsistent with a commercial airliner. It may have been easy for the Vincennes' crew to understand why a hostile pilot would fly directly at the Vincennes during a battle with surface boats, why it would ignore radio warnings, why it would generate, even for a second, Mode II IFF (the signal of a military aircraft), why it would descend when it got close to the Vincennes. It may have been hard to understand how a commercial pilot would generate Mode II IFF, ignore radio warnings, depart from the center of the corridor, and descend as it reached the Vincennes. With hindsight, we are able to see the true interpretation of the data, but given the information available during the incident, the misinterpretation seems reasonable.

5. There is more evidence for difficulty in team decision making, rather than individual decision making. These team problems include the inability to achieve a shared mental model, the failure to catch errors, the difficulty in self-managing so that workload was redistributed during the incident, the failure to at least search for a way to contact Air Traffic Control, and the lack of clarity about how much time was remaining before the Vincennes would fire on the track.

6. There were a large number of problems here, and possibly in any incident there are many things that have gone wrong, regardless of the correctness of the ultimate decision. Let us try a different approach—imagine that we could undo any of these 15 deficiencies. Which one would we choose? Perhaps the easiest error to undo is the attempted communication with the Airbus. The Vincennes crew, following SOP, assumed that the Airbus would be monitoring the International Air Distress (IAD) frequency. However, some pilots feel this was unlikely—that with two available ears, a pilot would be in touch with the airport of origin and the airport of destination. The Fogarty Report concurred with this view. Had the Vincennes crew contacted Air Traffic Control at Bandar Abbas airport, the accident may have been averted. The limited communications capability onboard the Vincennes made this impractical.[5]

7. Time pressure did not appear to affect the way the decision was made. The effect of time pressure was to restrict the amount of information the Vincennes' crew could collect in making its decision and to prevent the crew from verifying

[5]Another possibility is that the pilot of the Airbus did hear the warning but ignored it. The Iranian government had instructed its commercial aircraft not to reply to requests from the U.S. Navy, after a similar request to alter course almost resulted in a midair collision.

key data. With more time the crew might have obtained more information, such as verifying the altitude reading, but time pressure itself did not degrade the decision *process*. If, at any point in the incident, time had been artificially stopped so the crew had an extra 30 minutes to deliberate on the facts that were than available, the outcome would probably have been the same.

8. The Vincennes incident supports the ideas of Woods et al. (1993) and Reason (1990) that human errors are indicators of the latent pathogens that exist in most systems. The review of the incident does not support the thesis that stress results in decision biases.

CONCLUSIONS

This chapter has described the effect of different stressors on naturalistic decision making. The acute stressors I have considered are time pressure, noise, threat, and workload. Stressors such as time pressure and excessive workload are sources of task interference that would result in effects such as narrowing of attention and filtration of cues. Noise could interfere with internal speech. Other stressors such as threat and fear could increase self-monitoring and require attentional resources, thereby serving as a distracting secondary task to monitor and manage the emotional reactions. The impact could be to reduce the capacity in working memory available to perform a task.

Fortunately, naturalistic decision strategies are fairly robust to these effects. In most settings, decision makers can draw on experience to rapidly recognize the dynamics of a situation, and thereby know how to react to events. Cognitive effort seems to be required for some naturalistic decision-making processes like constructing scenarios to explain how a situation might have arisen, or vicariously playing out a course of action to see where it might go wrong. Stressors might diminish this ability to construct scenarios, but they would not affect the recognitional processes.

There seem to be typical reactions to stressors. These reactions do not appear to be biases, but rather may be viewed as effective and adaptive adjustments to constraints such as time pressure and secondary task requirements. They include the selection of simpler and more robust decision strategies, narrowed and focused attention, use of heuristics, increased conservatism, and rapid closure on a course of action.

To help decision makers avoid potential disruptions due to stressors, it may be useful to train them to better manage time pressure, distracting levels of noise, and high workload. Rapid pattern matching may be a particularly useful means of handling time pressure. In addition, interfaces can be designed with greater sensitivity of the need to find pertinent items of information under time pressure, and to highlight critical trends and relationships to allow a rapid situation assessment.

One danger in considering adaptations to stress is to treat them as reactions to be changed, in one way or another. Training programs should not encourage decision makers to adopt more complex, analytical strategies. Display designers should not try to pack more information on displays, or encourage more systematic methods for scanning displays. It is not helpful to discourage the use of heuristics, or to slow down reaction time to make sure all relevant hypotheses have been considered.

If time pressure is part of the task, then decision makers can be trained to handle such pressure. During routine exercises, time management must be learned, but it may be possible to develop more systematic approaches to teaching time sensitivity and time management. This would constitute training in metacognition, because the decision maker would be learning to manage his or her own strategies.

There appear to be a common set of cognitive impacts for different acute stressors, and as we learn more about these we should be able to do a better job of supporting decision makers who must work in operational settings that are challenging and difficult.

ACKNOWLEDGEMENTS

The preparation of this chapter was supported by contracts from Naval Command, Control Ocean Surveillance Center (N66001–90–C–6023), and the U.S. Army Research Institute (MDA903–89–C–0032).

I would like to thank the Technical Advisory Board of the TADMUS project for their stimulating discussions about decision making and stress. In particular, William Howell, William Vaughan, and Martin Tolcott provided me with a number of important insights, especially regarding the dangers of treating stress as a hypothetical construct. I also wish to thank Eduardo Salas, Jan Cannon-Bowers, and James Driskell for their helpful comments and ideas. Michael Hennessy has been an important source of information and suggestions about the psychophysiology of stress.

Alan Stokes provided useful suggestions about the research issues discussed, and Ted Kramer, Pete Edgar, and Allen Hale at Martin Marietta helped clarify the account of the Vincennes shootdown. Captain (retired) Will Rogers III was very generous with his time in being interviewed about this painful experience. Finally, Caroline Zsambok made many helpful comments on a number of drafts of this chapter.

REFERENCES

Bacon, S. J. (1974). Arousal and the range of cue utilization. *Journal of Experimental Psychology, 102,* 81–87.

Baddeley, A. D. (1972). Selective attention and performance in dangerous environments. *British Journal of Psychology, 63,* 537–546.

Baron, J. (1988). *Thinking and deciding.* NY: Cambridge University Press.

Barnett, B. J. (1993). Perspectives on decision analysis for decision support system design. (Symposium Summary). *Proceedings of the Human Factors and Ergonomics Society 37th Annual Meeting.* Santa Monica, CA: Human Factors Society.

Beach, L. R. (1990). *Image theory: Decision making in personal and organizational contexts.* West Sussex, England: Wiley.

Beach, L. R., & Mitchell, T. R. (1978). A contingency model for the selection of decision strategies. *Academy of Management Review, 3,* 439–449.

Calderwood, R., Klein, G. A., & Crandall, B. W. (1988). Time pressure, skill, and move quality in chess. *American Journal of Psychology, 101,* 481–493.

Christensen-Szalanski, J. J. J. (1986). Improving the practical utility of judgment research. In B. Brehmer, H. Jungermann, P. Lourens, & G. Sevón (Eds.), *New directions for research in judgment and decision making* (pp. 383–410). New York: North-Holland.

Combs, A. W., & Taylor, C. (1952). The effect of the perception of mild degrees of threat on performance. *Journal of Abnormal and Social Psychology, 47,* 420–424.

Cohen, M. S. (1993). The naturalistic basis of decision biases. In G. A. Klein, J. Orasanu, R. Calderwood, & C. E. Zsambok (Eds.), *Decision making in action: Models and methods* (pp. 51–99). Norwood, NJ: Ablex.

Driskell, J. E., Hughes, S. C., Guy, W., Willis, R. C., Cannon-Bowers, J., & Salas, E. (1991). *Stress, stressors, and decision-making.* Orlando, FL: Technical Report for the Naval Training Systems Center.

Driskell, J. E., & Salas, E. (1991). Overcoming the effects of stress on military performance: Human factors, training, and selection strategies. In R. Gal and A. D. Mangelsdorff (Eds.), *Handbook of military psychology* (pp. 183–193). New York: Wiley.

Driskell, J. E., Salas, E., & Hall, J. K. (1994). *The effect of vigilant and hypervigilant decision training on performance.* Paper presented at the Annual Meeting of the Society of Industrial and Organizational Psychology, Nashville, TN.

Easterbrook, J. A. (1959). The effect of emotion on cue utilization and the organization of behavior. *Psychological Review, 56,* 183–201.

Edland, A. (1989). On cognitive processes under time stress: A selective review of the literature on time stress and related stress. *Reports from the Department of Psychology.* University of Stockholm, Stockholm.

Edland, A., & Svenson, O. (1993). Judgment and decision making under time pressure: Studies and findings. In O. Svenson and J. Maule (Eds.), *Time pressure and stress in human judgment and decision making* (pp. 27–40). New York: Plenum.

Ferrari, J. R. (1994). Dysfunctional procrastination and its relationship with self-esteem, interpersonal dependency, and self-defeating behaviors. *Personality and Individual Differences, 17,* 673–679.

Ferrari, J. R., & Emmons, R. A. (1994). Procrastination as revenge: Do people report using delays as a strategy for vengeance? *Personality and Individual Differences, 17,* 539–544.

Fishburn, P. (1974). Lexicographic order, utilities and decision rules: A survey. *Management Science, 20,* 1442–1471.

Folkman, S. (1984). Personal control and stress and coping processes: A theoretical analysis. *Journal of Personality and Social Psychology, 46,* 839–852.

Fraser, J. M., Smith, P. J., & Smith, J. W. (1992). A catalog of errors. *International Journal of Man–Machine Studies, 37,* 265–307.

Gigerenzer, G. (1987). Survival of the fittest probabilist: Brunswik, Thurstone, and the two disciplines of psychology. In L. Kruger, G. Gigerenzer, & M. S. Morgan (Eds.), *A probabilistic revolution: Ideas in the sciences, 2* (pp. 49–72). Cambridge, MA: MIT Press.

Hamilton, V. (1982). Cognition and stress: An information processing model. In L. Goldberger & S. Breznitz (Eds.), *Handbook of stress: Theoretical and clinical aspects* (pp. 105–120). New York: The Free Press.

Hammond, K. R., Hamm, R. M., Grassia, J., & Pearson, T. (1987). Direct comparison of the efficacy of intuitive and analytical cognition in expert judgment. *IEEE Transactions on Systems, Man, and Cybernetics, SMC–17*(5), 753–770.

Harvey, L. O., Jr., Hammond, K. R., Lusk, C. M., & Mross, E. F. (1992). The application of signal detection theory to weather forecasting behavior. *Monthly Weather Review, 120*(5), 863–883.

Hintzman, D. L. (1986). Schema abstraction in a multiple-trace memory model. *Psychological Review, 93*, 411–428.

Hockey, G. R. J. (1970). Effect of loud noise on attentional selectivity. *Quarterly Journal of Experimental Psychology, 22*, 28–36.

Hockey, G. R. J. (1984). Varieties of attentional state: The effect of environment. In R. S. Parasuraman & D. R. Davies (Eds.), *Varieties of attention* (pp. 449–483). Orlando, FL: Academic Press.

Hockey, G. R. J., Dornic, S., & Hamilton, P. (1975). *Selective attention during reading: The effect of noise*. Report from the Institute of Applied Psychology, University of Stockholm, No. 66.

Hutchins, S. (in press). Decision-making errors demonstrated by experienced naval officers in a littoral environment. In C. E. Zsambok and G. Klein (Eds.), *Naturalistic decision making*. Hillsdale, NJ: Lawrence Erlbaum Associates.

Isenberg, D. J. (1984, November/December). How senior managers think. *Harvard Business Review*, 80–90.

Janis, I. L., & Mann, L. (1977). *Decision making: A psychological analysis of conflict, choice, and commitment*. New York: The Free Press.

Kaempf, G. L., Klein, G., Thordsen, M. L., & Wolf, S. (in press). Decision making in complex command-and-control environments. *Human Factors*.

Kahneman, D., & Miller, D. T. (1986). Norm theory: Comparing reality to its alternatives. *Psychological Review, 93*, 136–153.

Keinan, G. A. (1987). Decision making under stress: Scanning of alternatives under controllable and uncontrollable threats. *Journal of Personality and Social Psychology, 52*, 639–644.

Klein, G. (in press). The recognition-primed decision (RPD) model: Looking back, looking forward. In C. E. Zsambok and G. Klein (Ed.), *Naturalistic decision making*. Mahwah, NJ: Lawrence Erlbaum Associates.

Klein, G. A. (1989). Recognition-primed decisions. In W. Rouse (Eds.), *Advances in man–machine systems research* (Vol. 5, pp. 47–92). Greenwich, CT: JAI Press.

Klein, G. A., Calderwood, R., & Clinton-Cirocco, A. (1986). Rapid decision making on the fire ground. In *Proceedings of the Human Factors Society 30th Annual Meeting* (pp. 576–580). Dayton, OH: Human Factors Society.

Klein, G., & Crandall, B. W. (1995). The role of mental simulation in problem solving and decision making. In P. Hancock, J. Flach, J. Caird, & K. Vicente (Eds.), *Local applications of the ecological approach to human–machine systems* (pp. 324–358). Hillsdale, NJ: Lawrence Erlbaum Associates.

Klein, G. A., Orasanu, J., Calderwood, R., & Zsambok, C. E. (1993). *Decision making in action: Models and methods*. Norwood, NJ: Ablex.

Klein, G., Wolf, S., Militello, L., & Zsambok, C. (1995). Characteristics of skilled option generation in chess. *Organizational Behavior and Human Decision Processes*, 63–69.

Kuhl, J., & Beckman, J. (1985). *Action control: From cognition to behavior*. Berlin: Springer-Verlag.

Lieblich, A. (1968). Effects of stress on risk taking. *Psychonomic Science, 10*, 303–304.

Lopes, L. L. (1981). Decision making in the shortrun. *Journal of Experimental Psychology: Human Learning and Memory, 1*, 377–385.

Lusk, C. M. (1993). Assessing components of judgments in an operational setting: The effects of time pressure on aviation weather forecasting. In O. Svenson and J. Maule (Eds.), *Time pressure and stress in human judgment and decision making* (pp. 309–321). New York: Plenum.

McGrath, J. E. (Ed.). (1970). *Social and psychological factors in stress*. New York: Holt, Rinehart & Winston.

Minsky, M. (1986). *The society of mind*. New York: Simon and Schuster.

Mintzberg, H. (1975, July/August). The manager's job: Folklore and fact. *Harvard Business Review*, 49–61.

Mross, E. F., & Hammond, K. R. (1990). *The effects of stress on judgment and decision making: An annotated bibliography* (Report No. 295). Boulder: Center for Research on Judgment and Policy, University of Colorado.

Nisbett, R. E. (1993). *Rules for reasoning*. Hillsdale, NJ: Lawrence Erlbaum Associates.

Noble, D. (1993). A model to support development of situation assessment aids. In G. A. Klein, J. Orasanu, R. Calderwood, & C. E. Zsambok (Eds.), *Decision making in action: Models and methods* (pp. 287–305). Norwood, NJ: Ablex.

Orasanu, J., & Connolly, T. (1993). The reinvention of decision making. In G. A. Klein, J. Orasanu, R. Calderwood, & C. E. Zsambok (Eds.), *Decision making in action: Models and methods* (pp. 3–20). Norwood, NJ: Ablex.

Payne, J. W. (1976). Task complexity and contingent processing in decision making: An information search and protocol analysis. *Organizational Behavior and Human Performance, 16*, 366–387.

Payne, J. W., Bettman, J. R., & Johnson, E. J. (1988). Adaptive strategy selection in decision making. *Journal of Experimental Psychology: Learning, Memory, and Cognition, 14*(3), 534–552.

Pennington, N., & Hastie, R. (1993). A theory of explanation-based decision making. In G. A. Klein, J. Orasanu, R. Calderwood, & C. E. Zsambok (Eds.), *Decision making in action: Models and Methods* (pp. 188–201). Norwood, NJ: Ablex.

Poulton, E. C. (1976). Continuous noise interferes with work by masking auditory feedback and inner speech. *Applied Ergonomics, 7*, 79–84.

Rasmussen, J. (1985). The role of hierarchical knowledge representation in decision making and system management. *IEEE Transactions on Systems, Man, and Cybernetics, 2, (SMC–15)* 234–243.

Reason, J. (1987). Generic error-modelling system (GEMS): A cognitive framework for locating common human error forms. In J. Rasmussen, K. Duncan, & J. Leplat (Eds.), *New technology and human error* (pp. 63–83). New York: Wiley.

Reason, J. (1990). *Human error*. Cambridge, England: Cambridge University Press.

Roberts, N. C., & Dotterway, K. A. (1995). The Vincennes incident. Another player on the stage. *Defense Analysis, 11*, 31–45.

Sage, A. P. (1981). Behavioral and organizational considerations in the design of information systems and processes for planning and decision support. *IEEE Transactions on Systems, Man, and Cybernetics, 11*, 640–678.

Schneider, W., & Detweiler, M. (1988). The role of practice in dual-task performance: Toward workload modeling in a connectionist/control architecture. *Human Factors, 30*, 539–566.

Serfaty, D., Entin, E. E., & Volpe, C. (1993). Adaptation to stress in team decision-making and coordination. In *Proceedings of the Human Factors and Ergonomics Society, 37th Annual Meeting* (pp. 1228–1232). Santa Monica, CA: Human Factors and Ergonomics Society.

Shanteau, J. (1989). Cognitive heuristics and biases in behavioral auditing: Review, comments, and observations. *Accounting Organizations and Society, 14*(1–2), 165–177.

Shanteau, J. (1988). Psychological characteristics and strategies of expert decision makers. 11th

Conference on: Subjective probability, utility, and decision making (1987, Cambridge, England). *Acta Psychologica, 68,* 203–215.

Sheridan, T. (1981). Understanding human error and aiding human diagnostic behavior in nuclear power plants. In J. Rasmussen & W. B. Rouse (Eds.), *Human Detection and Diagnosis of System Failures* (pp. 19–35). New York: Plenum.

Smith, J. F., & Kida, T. (1991). Heuristics and biases: Expertise and task realism in auditing. *Psychological Bulletin, 109*(3), 472–489.

Stiensmeier-Pelster, J., & Schurmann, M. (1993). Information processing in decision making under time pressure: The influence of action versus state orientation. In O. Svenson & J. Maule (Eds.), *Time pressure and stress in human judgment and decision making* (pp. 241–253). New York: Plenum.

Stokes, A., Belger, A., & Zhang, K. (1990). *Investigation of factors comprising a model of pilot decision making: Part II. Anxiety and cognitive strategies in expert and novice aviators (ARL–90–8/SCEEE–90–2).* Urbana-Champaign, IL: Institute of Aviation.

Stokes, A. F., Kemper, K. L., & Marsh, R. (1992). *Time-stressed flight decision making: A study of expert and novice aviators* (TR ARL–93–1/INEL–93–1). Urbana-Champaign, IL: Institute of Aviation.

Streufert, S., & Streufert, S. C. (1981). *Stress and information search in complex decision making: Effects of load and time urgency* (Tech. Rep. No. 4). Arlington, VA: Office of Naval Research.

Svenson, O. (1979). Process descriptions in decision making. *Organizational Behavior and Human Performance, 23,* 86–112.

Tversky, A. (1972). Elimination by aspects: A theory of choice. *Psychological Review, 79*(4), 281–299.

Tversky, A., & Kahneman, D. (1974). Judgment under uncertainty: Heuristics and biases. *Science, 185,* 1124–1131.

Vincennes: Findings could have helped avert tragedy, scientist tells Hill panel. (1988, December). *APA Monitor,* pp. 10–11.

von Winterfeldt, D., & Edwards, W. (1986). *Decision analysis and behavioral research.* Cambridge, England: Cambridge University Press.

Wachtel, P. L. (1968). Anxiety, attention, and coping with threat. *Journal of Abnormal Psychology, 73,* 137–143.

Wickens, C. D. (1987). Attention. In P. A. Hancock (Ed.), *Human factors psychology* (pp. 29–80). Amsterdam: Elsevier.

Wickens, C. D., Stokes, A., Barnett, B., & Hyman, F. (1988). *The effects of stress on pilot judgment in a MIDIS simulator* (ARL–88–5/INEL–88–1). Savoy: University of Illinois, Aviation Research Laboratory.

Woods, D. D., Johannesen, L. J., Cook, R. I, & Sarter, N. B. (1993). *Behind human error: Cognitive systems, computers and hindsight.* Columbus: Cognitive Systems Engineering Laboratory, Ohio State University.

Wright, P. (1974). The harassed decision maker: Time pressure, distraction and the use of evidence. *Journal of Applied Psychology, 59,* 555–561.

Yates, J. F. (1990). *Judgment and decision making.* Englewood Cliffs, NJ: Prentice-Hall.

Zakay, D. (1993). The impact of time perception processes on decision making under time stress. In O. Svenson & J. Maule (Eds.), *Time pressure and stress in human judgment and decision making* (pp. 59–72). New York: Plenum.

Zakay, D., & Wooler, S. (1984). Time pressure, training, and decision effectiveness. *Ergonomics, 27,* 273–284.

Zsambok, C. E., Beach, L. R., & Klein, G. (1992). *A literature review of analytical and naturalistic decisionmaking.* Fairborn, OH: Klein Associates Inc. (Prepared under contract N66001–90–C–6023 for the Naval Command, Control and Ocean Surveillance Center, San Diego, CA).

3 Stress and Military Performance

Judith M. Orasanu
NASA–Ames Research Center

Patricia Backer
San Jose State University

Combat with its very real threat of death or mutilation might represent the ultimate in naturally occurring events of stress.

—Bourne (1970, p. 22)

Military training aims to prepare personnel to perform under life-threatening conditions, often in sweltering heat or freezing cold, wearing clumsy protective garments, in situations that are highly ambiguous, and where individual soldiers, sailors, or airmen have little apparent control over their fates. Modern military training incorporates many stressors in an effort to "harden" personnel for the rigors of combat; however, much of that training is nonspecific with respect to stressors and uninformed with respect to predicted performance enhancements. In their review of 35 articles and books on combat stress, Michel and Solick (1983) found that neither the exact amount of performance degradation nor proof of the nature or source of those degradations could be determined from the literature.

The debilitating effects of stress on military performance have been long recognized. Stress effects during the Normandy campaign of World War II were such that "the soldier was slow-witted; he was slow to comprehend orders, directions, and techniques. . . . Memory defects became so extreme that he could not be counted upon to relay a verbal order" (cited in Siegel et al., 1981, p. 13). Marshall (1947) also reported that in World War II only a small percentage of combat troops actually fired their weapons in combat because of the stress they perceived. Labuc (1981) found that especially fearful to soldiers is the prospect of artillery shelling accompanied by loud noise, smoke, earth tremor, and debris. Combat stress has been recognized as a critical area for research for some time by both U.S. researchers and those from other countries, particularly Israel, Great Britain, and Russia. In summarizing the importance of combat-related research, the U.S. Army School of Advanced Military Studies concluded that "combat stress will be one of the most significant causes of loss of manpower" (Coomler, 1985, p. 34).

Just as combat stress is a complex issue, so is the soldier's reaction to the combat situation—battle fatigue. According to some estimates, battle fatigue has accounted for as much as 50% of all casualties in wars (Abraham, 1982; Mareth & Brooker, 1982). Battle fatigue differs from ordinary stress reactions in that battle fatigue is disabling; beyond a certain point, service members are not able to do their jobs. McCaughey (1991) found that battle fatigue played a critical part in determining the outcome of Operation Desert Storm; the lack of effective management of battle risk factors worked severely against the Iraqi Army, possibly causing or hastening its defeat.

Combat stress is different from other types in that it is a combination of other stressors. It includes danger and threat, environmental factors such as noise, heat, cold, and crowding, fatigue and sleep deprivation, uncertainty, lack of control, and workload factors such as time pressure and information overload. Because combat stress includes many stress factors in addition to the actual danger of combat itself, it is difficult to replicate in controlled laboratory or simulation experiments. Combat stress has been studied mainly by psychiatric and psychophysiological researchers. One particular model based on cognitive theories of stress and coping that emphasizes the individual's response to stressful conditions (Arnold, 1960; Gal & Lazarus, 1975; Lazarus, 1980; Lazarus & Folkman, 1984) is derived from Israeli and other combat experiences and is used in the Israeli military (Gal, 1988). This model is interactional in that it contains a number of antecedent variables (individual, unit, and battlefield characteristics) acting through mediating variables (cognitive interpretations of the antecedent variables) to affect the individual's appraisal of the combat situation, which results in the combatant's mode of response in coping with the realities of combat. This model has advantages over classical models of stress in that it is derived from a combat perspective and based on combat stress situations.

In order to design training programs to enhance performance under combat conditions, it is necessary to understand the effects of stressors on various types of performance. Although a large literature exists on stress effects, not all of it is relevant to combat performance and training. Effects of stressors often are measured through self-report or physiological indicators rather than by examining performance itself. Unfortunately, the correlations among the three types of measures are far from perfect. This may reflect relative independence of the three measures or a difference in the time course of the correlated measures. For example, people are likely to report "feeling stressed" before showing evidence of performance impairment, which means that commanders are likely to hear complaints about stress before the evidence is seen under combat conditions (Driskell, Mullen, Johnson, Hughes, & Batchelor, 1992). A review of the literature on the effects of stressors on complex miliary-relevant task performance by Backer and Orasanu (1992) found that different stressors have different effects and that the effects vary depending on the type of skill being measured. Although large holes exist in the literature, certain effects appear to be robust. This chapter summarizes some of those effects that relate specifically to military performance and describes three approaches to training people to perform under stressful conditions.

For this review, stressors are classified into two major categories: physical and environmental factors versus psychological factors. Stressors in the first category (e.g., sleep deprivation/fatigue, noise, temperature, crowding or isolation) impinge on all subjects in a physical space and are assumed to have primary and direct physiological effects. Although there may be individual differences in reactions to these stressors, the effects do not depend primarily on a person's perceptions or interpretations. In contrast, psychological factors are more heavily dependent on interpretation for their effects. Psychological factors can be broken into two subcategories: those that signal danger or threat of physical or psychological harm, and those that represent limitation of cognitive or physical capacity to meet a demand, such as high information load, workload, lack of control over a situation, ambiguity, and time pressure. A third class of stressors includes social factors, which reflect interactions with another person, either in a personal sphere such as with family or friends, or in a job context, such as with platoon or squadron members or with leaders. Although social factors can have powerful effects, this chapter will not deal with them.

THE NATURE OF THE MILITARY ENVIRONMENT

All individuals are exposed to numerous stressors at one time or another, but in the military many stressors are inherent in the working environment. The military operates under different conditions in peacetime and wartime; however, training in peacetime is directed at preparedness for combat situations. Soldier performance in a combat situation is critical, with very little margin for error. The issue of performance under stress becomes more pressing in the high-tech battlefield because of the possibility that equipment may not be operated optimally by soldiers whose cognitive abilities are degraded by stress (Buckalew, 1988). Advances in sensor and communication technologies mean that more people will have access to greater amounts of information than in the past. This will raise the information processing requirements of many military jobs, increasing the need to scan, sort, and interpret ambiguous or conflicting data. Modern weapon systems operate at such great speed that little time may be available for interpreting signals (friend or foe?) and deciding on a course of action (e.g., the Vincennes incident; see Klein, this volume). Therefore, stressors that disrupt cognitive processes will become more of a threat to performance. If training is not adequate, the complexity of modern military systems and the intensity of the high-tech battlefield may increase the overall effect of combat stress on the soldier of the future. The impact of stress may reduce the advantages gained by technological advantages in warfare and jeopardize performance.

The cognitive demands placed on military personnel vary according to their job classification. Under particular stress are soldiers who must interact with complex equipment under heavy workload conditions. One particular instance of this changing environment is for crews of military aircraft (Kanki, ch. 4). According to Hart (1989), mission demands have increased to the point that pilots become overloaded during critical flight phases—a factor that can contribute to increased errors, reduced safety, and mission failure. Modern military action requires coordinated action and sharing of information about timing of actions and interpretation of large amounts of electronically transmitted data. This places a great cognitive burden on present—and future—personnel.

Given the concern of this chapter with stress effects on combat performance, we review several classes of stressors that may all be present in combat situations and that may interact to produce their effects. Of particular concern will be identifying the types of tasks or task components that are most vulnerable to specific stressors. This chapter consists of two main sections: In the first, we review research on the performance effects of several combat-related stressors. In the second we describe three approaches to ameliorating the effects of stressors and assess the appropriateness of each approach for different types of tasks.

The first section will be organized according to the following four classes of combat-related stressors:

- Danger and threat.
- Fatigue, including sleep deprivation and sustained and continuous operations.
- Workload/information load.
- Environmental factors (noise, heat/cold, altitude).

STRESS EFFECTS ON COMBAT PERFORMANCE

Table 3.1 summarizes the major findings from a large body of research on the effects of the above stressors on complex tasks. These findings are restricted to studies that deal with cognitively complex tasks and do not reflect an exhaustive review of the literature.

TABLE 3.1
Summary of Effects of Selected Stressors on Complex Performance

Stressor	*Major effects*
Danger/threat/loss of control	Subjective anxiety Freezing; escape from situation, reduced motivation Performance decrements (sometimes due to haste); memory decrement
Sleep deprivation CONOPS/SUSOPS	Effects seen with 18–24 hours of sleep deprivation Long and boring tasks most vulnerable; Cognitively complex and high workload tasks more vulnerable; Attention lapses, perceptual focus narrows, performance more variable, timing accuracy decreases, complex psychomotor skills degrade, loss of discrimination about task priorities
Workload Time pressure	Speed up processing Limit information scanning; focus on selected information Working memory disrupted; decision strategies shift Restricted number of team members participate in decision process
Information load	Restricted information search; less information used in decisions Greater risk taking
Noise	Depends on whether variable or continuous Continuous: Disrupts continuous attention; increases attention selectivity; disrupts memory component of complex tasks Variable: May increase alertness
Heat	Long duration: Cognitive confusion; impaired attention, memory, judgment, complex calculation; piloting skills disrupted
Cold	Slowing of responses, loss of manipulative ability (hands); little effect on cognitive tasks
Altitude	Impaired concentration and acquisition of new information

Danger, Threat, and Lack of Control

Perhaps the most significant but most difficult stressor to study is danger or threat of physical harm. Few experimental studies exist on this topic. Those that do generally show that threat effects depend on how strongly participants believe in the threat of death or injury. According to Idzikowski and Baddeley (1983), the magnitude of an individual's response to threat depends on a number of factors, including the individual's predisposition toward feeling anxious, the individual's assessment of the dangerousness of the situation, and his or her previous exposure and ability to cope with it.

A series of classic studies was done by Berkun and colleagues on the effect of physical threat on performance (Berkun, 1964; Berkun, Bialek, Kern, & Yagi,

1962; Berkun, Bialek, Yagi, Berry, & Kern, 1959). In a war-simulation study, each trainee was left in an isolated outpost where a series of nearby explosions could be heard and was told that artillery shells were landing outside the designated area, producing a threatening environment. Eight of the 24 recruits left the outpost before they completed their job of repairing a radio transmitter, but the 16 who remained worked almost as efficiently as an unstressed control group. In other studies Army basic trainees were subjected to simulated aircraft emergencies, explosions, and radioactive fallout. Overall, Berkun et al. found that the perception of physical threat and danger influenced memory and performance of job tasks, and that experienced soldiers were less affected by the threat than recruits.

Although few studies have investigated danger or threat by observing individuals in inherently threatening situations, even less research has examined cognitive task performance. In one study with a cognitive focus, Villoldo and Tarno (1984) studied seven explosive-ordnance disposal personnel in a simulated field operation. Stressors included battlefield noise, ordnance detonation noise, operator fatigue, and disorientation. Personnel completed a Render Safe Procedure in which a simulated explosive device in a briefcase was to be disarmed. Performance measures included speed, accuracy, and the care with which subjects worked. Analyses showed that during periods that included battlefield stressors subjects tended to work hastily with a concurrent decrease in performance. Overall there was a loss of accuracy in the Render Safe procedure when stressors were present.

The detrimental effect of danger also has been noted in chemical and biological radiological defense exercises at sea where critical shipboard tasks related to decision making were degraded relative to performance under normal circumstances (Tijerina, Stabb, Eldredge, Herschler, & Mangold, 1988). Similar decrements were found in a study of 185 medical unit personnel participating in a 3-day simulated chemical warfare field training exercise where real casualties occurred within the context of chemical warfare training (Carter & Cammermeyer, 1985).

One reason that threat or danger is so stressful is that it represents the ultimate loss of control over one's fate. Glass and Singer (1972) found that people are more comfortable and under less stress when they believe they have control in a situation. In World War II, studies of military fliers (discussed in Rachman, 1978) indicated that controllability was a major dimension in fear or courage, along with competence and group membership. Later studies of Project Mercury astronauts also found controllability to be very important. An increase of control over one's environment can lead to a reduction in fear reactions, which in turn leads to increased performance. In studying combat air crews, Stouffer et al. (1949) found that the degree of motivation for combat was a function of type of aircraft, with the lowest motivation being in heavy bomber crews, followed by medium and light bomber crews, and then fighter pilots. These differences in feelings were attributed to perceived control over the situation: Fighter pilots felt more "in control" because of the perceived advantages of flying a fighter (superior speed, maneuverability, and power). In a recent study, Gal-or, Tenenbaum, Furst, Shertzer (1985) investigated the effects of self-control and trait anxiety on the performance of 11 novice parachutists. They found that subjects who rated themselves higher on self-control performed better after training.

These studies show that the *anticipation* of dangerous or threatening situations and lack of control over them decreases performance levels. The aforementioned studies also show that individuals differ greatly in their vulnerability to stressors, and suggest that it may be possible to select personnel who are most resilient to the stress effects of dangerous situations.

Fatigue

The threat of fatigue is acute in the military. For example, in the most recent U.S. military undertaking, Operation Desert Storm, the requirement for rapid deployment of a high number of sorties resulted in tremendous pressure to prepare and repair aircraft on a 24-hour basis. Recent examples of catastrophic accidents in related environments highlight the effect of fatigue on performance. For example, the U.S. space shuttle management team had been on duty for periods ranging from 12 to 19 hours when they made the decision to launch the Challenger. Both the Chernobyl and Three Mile Island accidents happened in the early morning hours, indicating an increased level of fatigue of the workers. A recent analysis of causes of commercial air transport accidents found that a disproportional number of procedural and judgment errors occurred when crew members were fatigued (National Transportation Safety Board, 1994).

Two separate types of fatigue can be identified: sleep deprivation per se and continuous (CONOPS) or sustained operations (SUSOPS). Sleep deprivation is distinguished from CONOPS and SUSOPS thus: In sleep deprivation, the subject is not necessarily required to perform a continuous task during the period of sleep deprivation, but may be required to remain alert and vigilant, for example, monitoring a system. In CONOPS and SUSOPS, the subject is both sleep deprived and physically fatigued from performing a task over a long period of time. Continuous operations (CONOPS) refer to continuous combat situations and work tasks with short periods of rest ranging from 3 to 5 hours in a 24-hour period. Sustained operations (SUSOPS) "is used when the same soldiers and small units engage in continuous operations with no opportunity for the unit to stand down and very little opportunity for soldiers to catch more than a few minutes of sleep" (Department of the Army, 1983, p. 1–2).

Mackie, Wylie, and Evans (in press) completed a review of over 500 articles on the effect of fatigue and/or sleep deprivation on the performance of military personnel. Overall, they identified the following general problems with the existing research: Performance tests were too short and infrequently administered, administration of the performance tests was displaced from the fatiguing events, results of performance tests can not be applied to operational tasks in the military, and relatively few studies involved continuous work on complex tasks. Because of the methodological problems with many of the studies that Mackie et al. reviewed, they did not offer any generalized findings from the sleep deprivation literature.

Despite Mackie et al.'s rejection of generalizations from their review, we present here several findings that bear on military performance. Most important is the vulnerability of particular skill components to sleep deprivation or fatigue effects. Effects of sleep deprivation per se appear to depend on the type of task (e.g., more interesting tasks are more resistent to sleep deprivation than are less interesting ones, Wilkinson, 1964, 1965), the length of rest preceding sleep deprivation (Rutenfranz, Aschoff, & Mann, 1972), and the circadian rhythms of subjects (see Winget, DeRoshia, Markley, & Holley, 1984, for a review). Also, research indicates that decrements in performance due to sleep deprivation itself occur sooner than previously reported, as early as 24 hours (Wilkinson, 1971; Williams, Lubin, & Goodnow, 1959).

Several reviews have been made on the effects of continuous and sustained operations on soldier and unit performance (Belenky, Balkin, Krueger, Headley, & Solick, 1986, 1987; Dewulf, 1978; Siegel, Pfeiffer, Kopstein, Wilson, & Ozkaptan, 1979; Siegel, Pfeiffer, Kopstein, Wolf, & Ozkaptan, 1980). Their findings indicate the following effects of fatigue: (a) Tasks that primarily require physical performance are relatively immune to sleep loss; (b) there is a positive correlation between the length of a task and its sensitivity to sleep loss, perhaps reflecting a demand for continuous attention; (c) the more cognitively demanding a task, the greater is its sensitivity to sleep loss; and (d) workload interacts with sleep deprivation, producing more severe disruption of performance. Under sustained operations, degradation in cognitive performance can be seen as early as 18 hours into a SUSOP, and after 48–72 hours without sleep soldiers become militarily ineffective. Under continuous operations, a wide variety exists in the amount of sleep needed by the individual. In CONOPS with minimal sleep (less than 3 hours per 24-hour period), the limit before severe decline in performance is several days to as much as a week; with moderate sleep (4 or more hours per 24-hour period), the limit can be up to 2 weeks.

Few studies in the literature have included performance measures similar to actual military tasks. The following individual studies on sleep deprivation have direct implications for military training and performance under stress. Bartlett (1943) completed one of the first studies on complex task fatigue, the Cambridge Cockpit study. Subjects were exposed to tasks that consisted of responding on aircraft-type controls to changes in a variety of instruments. As alertness declined, progressively larger deviations in the instruments were tolerated before any corrective actions were taken by subjects. Fatigued subjects appeared to shift their standards of performance, their attention lapsed, their perceptual focus narrowed, and their performance became more variable, with the appropriate control responses no longer smoothly sequenced. With respect to helicopter performance, Perry (1974) summarized the changes in performances due to skill fatigue in helicopter crews as follows: deterioration in the accuracy of timing of the components of the skill with subsequent decrease in the level of skill, acceptance of lower standards of accuracy and performance, narrowing of the range of attention so that peripheral ones are forgotten or ignored, and reduction of the field of view so that instruments are no longer integrated into an overall pattern.

Drucker, Canon, and Ware (1969) and Haggard (1970) tested 142 enlisted men under nearly continuous work conditions of 48 hours duration. They found that sleep-deprived subjects performed significantly worse than control subjects on a driving task (continuously tracking a winding road on a driving simulator), and that the decrements in performance were much larger during the second night than during the first. Ainsworth and Bishop (1971) duplicated this study in a field setting with 120 men. They found that the fatigued group exhibited little performance decrement in communication, gunnery, maintenance exercises, or in two driving exercises (ditch and minefield driving). The performance of the fatigued group was significantly worse than that of the rested group only in moving surveillance and in two heavy workload driving exercises (slalom-type driving and log obstacles). They concluded that activities that demanded a high level of alertness or complex perceptual motor activity were the most sensitive to sleep loss. Successful performance of less cognitively demanding skills persisted longer, implying that these tasks were performed with automaticity. In another study, Angus and Heslegrave (1985) studied 20 female students during a 54-hour period of wakefulness. Subjects assumed the role of operations duty officers. They were required to monitor a communication network, which involved accessing, reading, understanding, interpreting, and filing information from an ongoing military conflict. In addition to monitoring and logging messages, subjects had to update tactical maps of the battle area. Embedded in and distributed around their tasks were a variety of cognitive tests that continuously monitored and assessed their performance. Angus and Heslegrave found that sleep loss produced large mood and performance decrements in this cognitively demanding environment.

Experience level may interact with the performance of subjects under conditions of sleep deprivation. Rognum and colleagues (1986) studied 24 Norwegian military cadets during a period of heavy, sustained work lasting for 107 hours, during which time they had less than 2 hours of sleep. After 1 day of sustained activity, all subjects were judged to be ineffective as soldiers and showed severe decrements in performance on simulated military tasks.

Of particular interest in sleep deprivation is the series of studies completed by Haslam and colleagues on soldiers in continuous operations. Haslam (1981) tested three platoons consisting of 68 soldiers over a 9-day period on vigilance, shooting, grouping capacity (which required the subject to fire five rounds of shots into as small an area as possible), weapons handling, and cognitive memory tests. For the majority of the cognitive tests, she found a rapid deterioration in performance over the first 4 days of sleep deprivation. Tasks with a mainly physical content suffered the least, and those with a cognitive and vigilance component suffered most, deteriorating to about 50% of control values. These results were verified in later studies (Haslam, 1982, 1985).

Another series of trials by Rejman & Green (in Allnut, Haslam, Rejman, & Green, 1990) assessed team rather than individual performance. Five three-man crews, consisting of experienced noncommissioned officers, were used as subjects on a 3 day/2 night scenario of 65 hours continuous operation. The simulation facility consisted of a distributed microprocessor system containing a fully interactive Ground Control Station in which the crew could plan and execute realistical target acquisition and intelligence-gathering missions. These missions included such cognitively demanding tasks such as mission allocation, photographic interpretation, map rendering, navigation, and air vehicle control. The tasks required them to work as a team: They had to fly the aircraft tactically from a launch point to the target area while making various observations and reports. Four types of performance were evaluated: system measures embedded in the Ground Control Station, standard cognitive laboratory tasks (e.g., serial reaction time and spatial decision making), subjective evaluations of fatigue, and physiological measures composed of saccadic eye movements, cortisol levels, and oral temperature. Each task the team had to complete was given one of three priorities (P1, P2, P3), and the team was told that the higher priority tasks should take precedence.

Performance analyses showed that the total mean number of tasks accomplished per hour remained relatively constant. However, tasks assigned different priorities differed from each other and from the baseline level. During the 2-day baseline session prior to the period of continuous operation, the crews adhered to the priority scheme—they completed more high priority (P1) than lower priority tasks (P2 and P3). Under sleep deprivation conditions, performance of medium priority (P2) tasks rose above baseline values so that there was no significant difference between high (P1) and medium (P2) priority tasks. This finding suggests that sleep deprivation affected the subjects' ability to discriminate between priorities. With increasing sleep loss, both high (P1) and medium (P2) priority tasks declined together. Overall performance measured by the Ground Control Station system showed relatively little change over the 65 hours of sleep loss, whereas the cognitive, subjective, and physiological measures showed performance decrements consistent with sleep deprivation. In addition, group performance measures embedded in the Ground Control Station also showed less performance deterioration. The researchers attributed this finding to the positive effects of team interaction on the group performance and to the stimulating nature of the task itself.

The body of research indicates that continuous performance has an effect on soldier and unit performance. Complex tasks involving cognitive or vigilance components suffer the most, whereas those involving mainly physical components suffer the least. However, as indicated in the Rejman and Green study (Allnut et al., 1990), detrimental effects of continuous operations can be partially ameliorated by working in teams. At this time the extent to which group interactions can overcome both sleep deprivation and continuous operations or only one of these factors is unknown, as are the classes of tasks that can benefit from group processes.

Cognitive Overload

One definition of *stress* is that it is a demand that exceeds the capacity of the organism to respond; clearly, issues of workload, time pressure, and information overload fall into this category. This class of stressors appears to be most sensitive to individual differences in training and experience and is amenable to coping or management strategies. For these reasons, this class of variables is most difficult to assess as stressors. The problem is to determine how much information is too much, so that processing it becomes stressful. If performance decrements are found, are they due to stress effects or to limitations in the subject's knowledge or skill? Moreover, the criterion measures used to assess stress effects are more problematic for this class of stressors. Much of the literature on time pressure and information load has examined complex cognitive tasks like problem solving and decision making (for an extensive review of stress and judgment and decision making, see Hammond, 1990, and Hammond & Doyle, 1991). For many of these tasks there is no single right or best answer. Rather than using an outcome measure or answer as the criterion, process variables frequently are used instead. These include measures of amount of information selected and used, information integration, amount and type of planning, number and quality of options considered, and so on. Workload effects are somewhat more amenable to assessment because of the availability of self-report instruments (e.g., the NASA TLX or Task-Load Index, Hart & Staveland, 1988) that can be correlated with performance measures.

Workload. Workload is defined as the amount of work or number of tasks that must be performed within a fixed time period. Extensive research has been done on workload as it relates to pilots, both military and commercial. (See Hart & Wickens, 1994, for a review of this extensive literature.) A study using realistic task performance was conducted by Hughes Aircraft (1977). Single and two-seat cockpits were compared as pilots flew simulated air-to-ground strike missions. They found that as threat density increased, the performance difference between the one- and two-person crews became greater, with the latter consistently showing better performance. This difference is attributed to the effects of multiple threats on single-pilot workload. The additional crew member freed the pilot from defensive tasks such as monitoring radar and other displays and allowed him more time to scan visually for ground and air threats. These findings agree with others that have studied pilot workload; overall, as workload increases for single-pilot aircraft, there is a decrease in performance (Moroney, Reising, Biers, & Eggemeier, 1993).

Hart (1989) reported on a series of helicopter simulations that were conducted at NASA/Ames to compare pilot workload and performance with 20 combinations of stability control/augmentation systems and pilot-controlled features (altitude, airspeed, etc.). Single and dual pilot configurations were compared, yielding the following conclusions: Single pilot workload was found to be higher than dual pilot, regardless of the automation and control augmentation provided; a second crew member smoothed workload peaks, reducing the differences between mission segments; and the effects of vehicle augmentation on pilot workload varied by mission segment. Mission management tasks also increased the already high demands of single-pilot flight path control and contributed to higher workload and poorer performance.

In contrast, a recent study by Clothier (1991) illustrates the complexities of pilot workload effects. After analyzing 6,129 cases in 1989 and 3,756 cases in 1990, she found that two-person crews consistently outperformed three-person crews on the line and in full mission flight simulation (LOFT or Line Oriented Flight Training). Clothier stated "While the third person is an extra set of eyes, that extra communication node seems to detract more than aid behavioral operations" (p. 336). However, Orasanu and Fischer (1992) showed that having a third crewmember in the cockpit presents options for load distribution during high-workload, time-critical periods, especially following a system malfunction. Efficient use of the third crewmember in an emergency can benefit mission performance.

Overall, pilot workload is a complicated phenomenon. The nature of crew utilization and the level of automation present seem to affect the amount of perceived workload and its effects on performance.

Time Stress. Time stress is defined as a ratio of time to perform required tasks divided by the time available. Therefore, in principle workload and time stress work together: If more time is available to accomplish tasks, perceived or reported workload is reduced and performance may improve. Much of the recent research on time pressure has examined problem solving and decision making. Smart and Vertinsky (1977) stated that during "crises when individuals are under great stress and important decisions much be made within a short time, certain pathologies may arise in the decision process that reduce its quality" (p. 642). Time pressure forces decision makers to respond quickly, which may induce them to use fewer channels to process information, to omit certain actions, to delay responding, to filter information, and to process incorrect information (also see Miller, 1960; Payne, Bettman, & Johnson, 1988). Other researchers have found that time stress causes people to screen out some essential cues or pieces of information and adopt a restricted view of the decision-making process (Ben Zur & Breznitz, 1981; Easterbrook, 1959; Edland, 1985; Svenson, Edland, & Karlsson, 1985; Wright, 1974). Also, time stress may reduce the number of persons in a team who actually participate in the decision-making process (Mulder, van Eck, & de Jong, 1971) and may also reduce the quality of the resulting decision (Levine, 1971; Robinson, 1972).

Aviation accidents frequently occur when the crew is under schedule pressure (to depart and arrive on time) (McElhatton & Drew, 1993). For example, the KLM/Pan Am accident at Tenerife was due in part to the pressure on the flight crew to take off within their duty time and their disregard of the potential hazards. Similarly, the accident at Washington National Airport that occured during a snow storm when the crew took off without de-icing the plane a second time, despite a buildup of snow and ice on the aircraft (National Transportation Safety Board, 1982). A recent National Transportation Safety Board (1994) analysis of crew-caused air transport accidents shows that 55% of the flights were running late (either in departing or arriving), compared to a delay rate of only 17–23% for other flights.

In a study of pilot decision making under stress, Wickens and his colleagues (Wickens, Barnett, Stokes, Davis, & Hyman, 1988; Wickens, Stokes, Barnett, & Hyman, 1991) examined the performance of instrument-rated pilots flying in a computer-based simulator under stress and nonstress conditions. Stress was induced by time pressure, noise, financial risk, and task loading. They found that stress had different effects on different kinds of decision problems: It degraded performance on problems imposing high demand on working memory, but left unaffected problems that required retrieval from long-term memory.

Edland (1989) reviewed the effect of time pressure on cognitive processes involved in decision making and concluded that systematic changes, including more frequent use of non-compensatory decision rules, use of a smaller number of attributes or data, and more avoidance of negative aspects, occur when decision makers are under time pressure. Specifically, she stated:

> It may be suggested that the changes occurring when people are under time pressure start with an acceleration of the processing. Then, when there is no possibility to process the information faster and still reach the same result, one has to filtrate the information by increasing the selectivity and focusing on a subset on the available information (i.e., the negative cues to avoid the negative consequences) and base the decisions or judgments on that. (p. 26)

The effects of time pressure on decision making have been well documented and include contraction of authority, restriction of information searching, and restriction of the decision-making process.

Information Load. Military personnel often have to make decisions and take actions without having the essential information in hand. In military settings, particularly under combat conditions and with the presence of high-tech sensing and communication devices, the need for rapid planning is obvious. A military decision maker may be flooded with data that need to be integrated and interpreted for them to be useful. Information load differs from workload in that it indexes the amount of data handling that must be completed. With increasing amounts of data there is the possibility of conflicting or uninterpretable data, which increases the decision maker's load.

In an early study, Lanzetta and Thornton (1957) found that the error rate on task performance was correlated positively with the increased volume of information received by decision makers. Clearly, more information is not necessarily good. Streufert, Suedfeld, and Driver (1965) assessed the effects of information load on information search in a complex decision-making task. In their study, teams of four directed the fate of a small developing nation that was threatened by economic problems and a military takeover attempt by a simulated opponent. They found that search activity decreased with increases in information load and that less integrative decision makers were more affected by information load increases than were more integrative decision makers (where "integrative" was a measure of how successful the subject was in utilizing information). In a follow-up study, Streufert and Streufert (1981) found that the numbers of quick decisions and integrative decisions were greatest at intermediate load levels, and declined with increasing load. However, when time urgency and information load were both high, there were fewer search actions and a complete absence of integrative decisions, but subjects showed an increase in quick decisions (decisions that did not utilize all available information). Further studies found that risk-taking behavior also increased as information load increased (Streufert, 1983; Streufert, Streufert, & Denson, 1982).

The problem of defining good or adequate performance levels in problem-solving and decision-making tasks has led to recent controversy (see Klein, Orasanu, Calderwood, & Zsambok, 1993, and Plous, 1993, for contrasting views). On the one side are those who maintain that human judgment and decision making are flawed when compared to normative models. As evidence they cite the fact that decision makers use a small number of heuristics (rules) in making their decisions (Tversky & Kahneman, 1973), fail to consider all possible decision and outcome options (Slovic, Fischhoff, & Lichtenstein, 1977), are inconsistent in dealing with risk (Lopes, 1983), are subject to situational context in which decision are made (Tversky & Kahneman, 1981), and have inappropriate levels of confidence in their own decisions (Einhorn & Hogarth, 1978). As Janis (1983) noted, when stress is high, the decision maker is likely to display premature closure—terminating the decisional dilemma without generating all the alternatives and without seeking all available information about the outcomes.

In contrast, Klein and his colleagues (Klein, 1993), in their studies of highly experienced decision makers in field settings, have found that human decision makers assess a situation and retrieve a solution which is evaluated for adequacy. If it is adequate (not necessarily best), it is accepted and other possibilities are not considered. Klein has tested the adequacy of experts' retrieval of solutions and found that when pressed to retrieve multiple solutions, the first one retrieved is often the best one (Klein, Wolf, Militello, & Zsambok, in press). Although some researchers consider this strategy to be an adaptive response, its impact on decision making in the military setting is unknown. Modern combat scenarios are often characterized by rapidly evolving and changing conditions, time stress, high information load, and a high degree of ambiguity and uncertainty. Due to the complexity of the military environment, it is paramount that a situation be fully understood before action is taken. A distinction must be made between inadequate information use in analyzing the problem and premature closure of options.

One major conclusion can be drawn from the aforementioned studies. In cases of high information load, subjects tend to make decisions quickly without examining all relevant information, thereby increasing the errors in performance.

Environmental Stressors

Noise. One common environmental stressor in both work and combat is noise. It has been reviewed extensively elsewhere, although with contradictory results (Cohen, 1980; Jones, 1983; Koelega & Brinkman, 1986). Jones has concluded that intermittent noise produces both general and local effects, with the general effects extending beyond the period of exposure; infrequent bursts of noise may have localized effects (particularly on the intake of information); continuous noise particularly causes decrements in complex multicomponent tasks; judgments become more extreme in noise with excessive confidence; tasks containing a heavy memory component are more susceptible to disruption at relatively low levels of noise; and variable noise may increase alertness at the end of a long vigil. However, Edland (1989), in her review of noise and performance, found that surprisingly little research has been done on the effects of noise on more complex judgment and decision processes, with most effort focusing on memory and attention tasks. She concluded that noise seems to increase attentional selectivity; that is, if a task requires high attention on *every* cue to give the required results, noise may disrupt performance. However, noise may improve performance on other tasks that require focusing on *selected* parts of the information, perhaps because of its arousal value. Therefore, increasing attentional selectivity does not necessarily lead to a decrement in performance.

Temperature. Excessive heat or cold are often unwanted elements of the combat environment. Both heat and cold stress are sometimes products of geography or may result from operating machinery. Ramsey, Burford, Beshin, and Jensen (1983) reported an observational study of thermal conditions that took place over a 14-month period in two industrial plants. Their results indicate that "temperatures below and above those typically preferred by most people have a significantly detrimental effect on the safety-related behavior of workers" (p. 105).

In a recent study of temperature extremes, Carter (1988; Carter & Cammermeyer, 1988) studied 51 U.S. Armed Forces personnel under evaluation for heat stress at a military field training exercise, Wounded Warrior II, at Camp Roberts, CA. Ambient temperature during this 10-day exercise ranged between 90–102° Fahrenheit. She found that half of the heat-injured subjects demonstrated cognitive confusion and showed impairment in attention, delayed memory, situational judgment, and complex calculation. In addition, she found that the "subjects' inability to assess their own condition and/or ability to convince others of their needs may preclude intervention until subjects become acutely ill" (p. 86).

Brief exposures at very high temperatures have also been investigated. Blockley and Lyman (1951) studied temperatures of 160°, 200°, and 235° Fahrenheit with exposure times of 61, 29, and 21 minutes, respectively. They found that performance decrements on a simulated pilot task became more serious only toward latter stages of exposure (after 15 to 25 minutes). In another study with pilots, Iampietro, Chiles, Higgins, and Gibbons (1969) found decrements in pilots' performance on complex tasks with a temperature of 160° for 30 minutes. Ramsey, Dayal, Ghahramani (1975) measured performance on four different sedentary tasks at temperatures ranging between 85° and 105° and for periods up to 4 hours. Their results indicate that the upper limit for unimpaired mental performance is not represented by a single level but varies across individuals.

The other extreme in temperature—cold—also affects performance. Horvath and Freedman (1957) kept 22 men in a room with a -20° temperature for up to 2 weeks. Subjects showed significant performance decrements on manipulative and writing tasks; however, neither mental performance on a code test nor visual performance appeared to be affected. In contrast, the Poulton, Hitchings, and Brooke (1965) study of the effects of cold on ship watchkeepers in the Arctic showed that body temperature declines were accompanied by increasing delays and inaccuracy in reporting of the watch. In his review of the effects of heat and cold on performance, Ramsey (1983) found that brief exposure to high temperatures causes only a minor decrement or even enhances performance on perceptual–motor tasks; the most significant effect of cold exposure is the loss of manipulative ability of the hands; cognitive or mental tasks are much less affected by the cold than are motor tasks.

Altitude. Altitude effects on problem solving have been studied by Bandaret and colleagues (Bandaret & Lieberman, 1988; Bandaret et al., 1987). They simulated high altitudes (15,000–25,000 feet) for up to 40 days, with the dehydration, cold, and muscle atrophy that are associated with it. They found that cognitive performance (tasks as coding, number comparison, compass tasks, and pattern comparison) decreased in a linear fashion with increasing altitude, with impairments usually due to decreases in the speed of performance rather than increased errors. Based on the American Medical Research Expedition to Mount Everest, Townes, Hornbein, Schoene, Sarnquist, and Grant (1984) found that the acquisition of new information is impaired as altitude increases, and that the disruptive effects of altitude on acquisition persisted even on return to a low-altitude environment. The finding was verified by Oelz and Regard (1988), who reported that climbers who repeatedly ventured up to 8000 meters without supplementary oxygen had impaired concentration when returning to sea level.

In summary, environmental stressors clearly interfere with performance, especially on cognitive tasks. Because of the nature of continuous operations in the military, some environmental stressors exist as a part of the soldier's working environment. Of particular concern to military training are the effects of temperature and noise on performance. It appears that under noise and heat, subjects are able to maintain performance speed with no overall degradation; however, performance accuracy is likely to decrease.

IMPROVING PERFORMANCE UNDER STRESS

In the most general terms, the research discussed earlier has shown (a) that the presence of certain stressors leads to decrements in performance, (b) there seem to be significant variations in the effects of stress on different individuals, particularly with respect to experience, and (c) tasks that vary in their demand characteristics are differentially sensitive to various types of stressors. There is a clear need to develop interventions that target specific tasks or components vulnerable to anticipated military stressors. However, the point at which intervention should occur and the type of intervention to use are far from clear. Three different approaches to reducing stress have been developed. The first focuses mainly on the stress itself, with the assumption being that if individuals can be taught to manage their stress reactions effectively, performance will improve. The second approach assumes that stress is the inevitable result of exposure to stressors and that the focus should be on skill training. If individuals can achieve automaticity or durability of certain skills, stress will impair performance considerably less. A third approach is to train strategies for managing stressors through effective use of team resources. An example is crew resource managements training in which participants are taught effective interpersonal skills in order to deal with potentially stressful circumstances. The remainder of this chapter focuses on how each of these three approaches work to overcome the effects of stress.

Stress Training

The primary focus in the literature has been on stress reduction as a means of improving performance. Stress here refers to the subjective feeling of anxiety in response to a stressor, often in conjunction with physiological indicators such as heart or blood pressure. Research on stress management has resulted in several findings, the most notable being that different techniques for reducing stress succeed when they focus on the reduction of uncertainty about and an increase in control over important events in a person's environment (Druckman & Swets, 1988). Certain stress-reduction techniques are not covered in this review because their applicability in the military is unclear. These include biofeedback (see Beatty & Legewie, 1977, and Schwartz & Beatty, 1977, for reviews), rational emotive therapy (Ellis, 1962), and time management (Lakein, 1973). This review focuses on stress management strategies that have proven to be effective in reducing and managing stress in military populations.

Most studies on stress management focus on physiological indicators to document the effects of stress training. Bruning and Frew (1987) studied the effects of three stress intervention strategies (management skills training, exercise, and meditation) and examined physiological measures to determine the effect of each intervention and the combinations of the different strategies. Each of the strategies led to decreases in pulse rate and blood pressure and dual combination strategies showed even more significant decreases in pulse rate under conditions of stress. Migdal & Paciorek (1989) investigated the effect of relaxation exercises on the performance of Polish cadets on a catapult simulator and found that the cadets trained in relaxation techniques displayed less emotional tension, an absence of fatigue, and a lack of sense of guilt on the State-Trait Anxiety Inventory. For pilots, relaxation strategies such as progressive relaxation (Jacobson, 1938), autogenic training (Cowings, 1993; Shultz, & Luthe, 1959), and the pilot stress relaxation exercise (Thomas, 1988) have been used to reduce stress-related symptoms.

Few studies have directly assessed the effects of stress-reduction techniques on task performance. In one large study of 214 Swedish conscripts, Larsson (1987) found that performance of an experimental group of conscripts was significantly better than a control group on actual task examinations and mental tests after the experimental group followed mental-training techniques including relaxation, meditation, and imagery rehearsal. In evaluating respiration control as a stress-management technique, Burke (1980) found that jumpmasters (personnel trained to conduct landings of men and equipment) trained to use respiration control had significantly lower heart rates during the two night jumps of the course. In addition, he found that the stress-management group scored higher grades than the control group for performance as jumpmasters.

Although a variety of stress-management procedures exist, the one that has received the most attention is Stress Inoculation Training, or SIT. Stress inoculation training stands apart from other stress-management methods in that it does not propose any single technique that is presumed to be applicable to all stressful situations. Rather, SIT is based on the premise that a method must be flexible enough to be adapted to the needs of those receiving training (Wertkin, 1985).

SIT consists of an educational, a rehearsal, and an application stage. In the educational stage, individuals are taught about the different ways in which people respond to different types of stress. In the rehearsal stage, individuals learn one of a number of stress management techniques most applicable to their specific situation. Techniques include cognitive restructuring, systematic desensitization, progressive relaxation, deep breathing, guided imagery, and stretching (Wertkin, 1985). During the application stage, individuals apply the techniques that they have learned. This is first done in a simulated environment and then in the actual stressful environment.

Spettel and Liebert (1986) recommended SIT to help mediate the stress present in man–machine systems operations by focusing on improving abilities to ignore distracting stimuli, thereby allowing the subject to handle high information loads during stressful conditions. In a commentary on Spettel and Liebert, Starr (1987) reported that students who received CPR training enhanced by SIT performed faster and more accurately than traditionally taught students when tested without warning 6 or 12 months after original training. In an attempt to make SIT more efficient, Schuler, Gilner, Austrin, and Davenport (1982) compared the effectiveness of SIT with and without the first stage—education. In examining the stressor to stress reaction link, they found that the full SIT group improved significantly more than the group receiving SIT without the education phase on both the behavioral observations and self-report indices.

Hytten, Jensen, and Skauli (1990) studied the effects of a 1-hour SIT on subjective experience, performance, and physiological activation in two fear-provoking training situations using North Sea oil rig workers: smoke diving and free-fall lifeboat evacuation. The former was part of a basic safety course provided by the National Academy of Safety Education in Norway. Potential oil workers were required to dive alone in a complex three-level labyrinth in total darkness with an instructor monitoring the worker's progress on a video screen. The second training situation consisted of a free-fall lifeboat course for offshore workers. Students were randomly assigned to the SIT training condition or to the control condition. In the free-fall lifeboat situation, students receiving SIT training reported higher acceptance or confidence than the control group, as measured by a self-report questionnaire. In smoke diving, the experimental group received less help from the instructor during their dives and reported increased self-confidence, whereas the control group reported a greater increase in skills. However, subjects in the experimental group reported higher anxiety than the control group during training and tended to use more air during the dive.

As discussed earlier, there have been few studies investigating the effects of stress management strategies on task performance. Two of these studies, by Starr (1987) and Larsson (1987) found that using stress-management strategies during training for complex cognitive tasks increased performance compared to those not trained with stress interventions. Because of these findings, further research should be done with an emphasis on task performance measures instead of physiological ones.

Skill Training

Although many different stress training techniques, including SIT, involve some aspects of skill training, their major emphasis is on relaxation or reduction of the stress response. In skill training the focus is on increasing the durability or automaticity of the skill itself, which may have the side benefit of reducing stress. For effective military performance, it is necessary to provide training and practice under operational conditions similar to those found in the real environment; that is, combat. Normal task training without stressors does not necessarily improve task quality when the task is eventually performed under conditions of stress (Zakay & Wooler, 1984). The need for skill training under realistic conditions has been acknowledged since World War II. As part of the studies conducted during that time, military researchers asked combat veterans, "What kind of training did you lack?". The most frequent response was that training was needed under realistic battle conditions (Janis, 1949, p. 229).

Skill training can ameliorate the effects of stress by producing overlearned behavior (Zajonc, 1965). This practice has several benefits: Well-rehearsed tasks are less prone to degradation under conditions of stress; well-rehearsed tasks become "automatic," thus requiring less of the individual's attention; and well-drilled tasks enhance a person's sense of predictability and control (Driskell & Salas, 1991; Logan, 1985). As Wickens (1984) noted, as behaviors become more practiced, cognitive load is reduced and the speed and accuracy of performance is increased. Schendel and Hagman (1982) found that overlearning was an effective means for enhancing performance of a military procedural task, disassembly and assembly of an M60 machine gun. They found that the overtrained group made 65% fewer errors than a control group when retested after 8 weeks. In their meta-analysis review of overlearning and performance, Driskell, Willis, and Copper (1992) found that both 100% and 150% overlearning produced moderate to strong effects on performance. However, for cognitive tasks the longer the delay between the overlearning and performance, the weaker the overlearning effect, with the benefit in performance reduced by one-half after 19 days.

The Army practices this strategy in the form of basic training when recruits are drilled on tasks crucial to their performance as military personnel. Although one goal of overtraining is to make skills resistant to the effects of stressors, few studies have directly tested this notion. Training studies that address effects on stress tend to focus on the effects of training on participants' self-perceptions and sense of control. Increased levels of confidence and competence are associated with subjects' willingness to participate in dangerous tasks and on their actual performance levels. For example, Smith et al. (1990) analyzed the effect of skill training on job proficiency in handling chemical agents. A treatment group composed of 150 soldiers knew that their training would involve lethal agents in the Chemical Decontamination Training Facility at Fort McClellan, AL. There were two control groups: one of 30 soldiers trained in the same facility and the other of 158 soldiers trained in a different, nonlethal environment. The researchers found that there were no differences between the groups in job proficiency as measured by written examinations; however, soldiers who had undergone training involving lethal agents had the *perception* that they were better able to survive in combat and to perform their mission in the event of a chemical attack. In fact, this perception may influence their motivation to perform and their ability to cope with the actual stressful conditions.

Keinan (1988) studied the quality of soldiers' performance and the intensity of experienced stress in a combat situation using 297 male recruits in the Israeli Defense Forces and found that soldiers who assigned a low probability to being physically injured were found to benefit more from dangerous rather than nondangerous training. In addition, exposure to serious physical threats during training yielded better training results than training that did not involve such threats only when the soldiers concluded their training with a feeling of success. From this study, it appears that individual perceptions and differences are strongly related to performance under stress.

Subjects' expectations were manipulated directly in a study by Novaco, Cook, and Sarason (1983) with Marine recruits in San Diego. Recruits were shown a 35-minute videotape called "Making It," which depicted skills and coping strategies needed for success in boot camp. Their results suggested that recruits viewing this film manifested higher expectations of personal control than did recruits seeing a control film. How those perceptions influenced recruits' actual performance was not reported.

A series of studies has investigated the interaction of experience level and stress, using divers and parachutists as subjects. As in most studies on stress training, the measures for these studies were predominately physiological or self-report indices of stress. In an early study, Epstein and Fenz (1965) compared novice parachute jumpers with experienced jumpers on self-report avoidance ratings. With experienced jumpers, they found that the maximal levels of self-reported avoidance occurred the night before, whereas the maximal avoidance for novice jumpers occurred at the "ready" signal. In follow-up work over the next 10 years, Fenz (1975) found that experienced but incompetent jumpers have similar physiological responses to novices. In a similar study, Beaumaster, Knowles, and MacLean (1978) noted that the execution and, to a lesser extent, the anticipation of a jump was stressful for novice parachutists. These findings have been replicated with undersea divers. Biersner and Larocco (1987) found that more experienced divers showed fewer signs of physiological stress than inexperienced divers. Jorna (1985) found that for novice divers performance on a continuous memory task was correlated with amount of diving training; for experienced but inefficient divers, performance on the memory task was reduced by an increased depth of dive. This pattern of findings shows both the benefits of increased training, which reduces stress effects, and the importance of individual differences in vulnerability to stress effects. The stress effects, both physiological and performance-related, found in experienced but less competent individuals are difficult to interpret. Their stress responses may reflect accurate assessment of their modest skill level, or their lower skill level may result from their greater susceptibility to stress. Clearly, experience per se does not eliminate stress effects in all individuals. Together these studies indicate that, although diving and parachute jumping create stress and avoidance for both novice and experienced participants, experience generally is associated with reduced levels of stress, which may reflect both a greater sense of self-efficacy due to higher skill levels and also the development of strategies such as mentally preparing oneself for a stressful event, thus reducing anxiety when actually performing the task.

In certain types of situations, stress can cause a decrement in performance even after high levels of skill training. Stepanov and Stetanov (1979); see also Khachatur'yants, Grimak, & Khrunov, 1976; Khrunov, Khachatur'yants, Popov, & Ivanov, 1974; Lomov, 1966) reported on a series of experiments carried out in the Soviet Union on long-term spaceflight. In a 30-day simulation of spaceflight, Stepanov and Stetanov (1979) noted that 30% of the errors occurred in the 72-hour continuous watch following liftoff and in the period of adaptation (first 3 to 5 days of flight). During the second and third days of the continuous watch, 85% of the operators made errors on technical maintenance tasks. Repeated attempts at correcting technical maintenance problems after the operators detected their errors led to an increase in time expenditures of from 50% to 100%. In short, these findings indicate that severe stress such as that posed by space flight can severely disrupt even highly trained skills. With the passage of time and increased adaptation to flight conditions, the quality of activity normalized, reaching the values obtained in the course of training sessions.

In addition to specific task practice, general fitness training has also been found to benefit task performance. Pleban, Thomas, and Thompson (1985) investigated the role of physical fitness in moderating both cognitive work capacity and fatigue onset under sustained combat operations. Sixteen male Ranger Officers' Training Corps cadets were followed through a 2.5 day Pre-Ranger Evaluation exercise. Cognitive performance and subjective measures of fatigue were assessed at regular intervals before, during, and 1 day after the exercise. Their results indicated that fitness may reduce the effects of stress on cognitive work capacity for tasks requiring prolonged mental effort, particularly as sleep loss and other stressors mounted.

Some general conclusions can be derived from the aforementioned studies. In contrast with stress training, skill training is concerned with hardening specific behaviors to reduce the effect of potential stressors on those behaviors. Few evaluative studies have been done, although the potential of this approach for the military is great. First, higher levels of performance and lower levels of stress are associated with higher levels of skill training. Second, high levels of general physical conditioning have been found to correlate positively with performance on cognitive tasks, even under stressful conditions. Third, skill training appears to affect the attitudes of individuals facing stressful situations; that is, soldiers feel that they are better equipped to perform under stressful conditions after undergoing skill training. Because nonperformance in battle is a continuing dilemma for the military, these findings can be helpful for future training. The last finding relates to the effect of level of experience on performance under stress: As indicated in the studies on divers and parachutists, experienced competent subjects mentally prepare for the stressful event in advance, thus reducing anxiety when performing the task and increasing their ability to perform. Individual differences in both susceptibility to stress and to the benefits of training appear to be important.

Crew Resource Management Training

Crew resource management training was originally developed to improve the performance of air transport crews in high-risk, high-stress conditions (Helmreich & Foushee, 1993). It includes training in communication, decision making, and resource and task allocation. In principle, it applies to any environment in which coordinated action is required by teams of highly trained professionals who must function under dynamic high workload conditions, as in the military or in nuclear power plant, space, and offshore oil operations. One important aspect of CRM training is learning to recognize stressful situations and reactions of team members and to defuse them before they become debilitating. (For a thorough treatment of CRM training see Kanki, this volume.)

Cooper, White, and Lauber (1980) analyzed jet transport accidents that occurred between 1968 and 1976 and found more than 60 involving problems with decision making, leadership, pilot judgement, communications, and crew behavior. From this analysis and other studies (Foushee & Manos, 1981; Ruffell Smith, 1979), there seemed to be a direct correlation between measurable performance of a crew and various aspects of crew interactional processes, including communications. In 1979, the National Transportation Safety Board (NTSB) recommended operational implementation of cockpit resource management (CRM) training programs by the airlines. In the following decade, these programs came into widespread civilian and military use (Alkov, 1991; Hawkins, 1987; Lauber, 1993; Leedom & Simon, 1993; Prince, Chidester, Cannon-Bowers, & Bowers, 1993).

CRM programs are largely based on social psychology and management theory, with many of the programs being developed with data and expertise from NASA (Foushee & Helmreich, 1988; Helmreich, 1991; Lauber, 1984). Several studies have analyzed behaviors associated with more and less effective crew performance. CRM training has been built on these descriptions of effective crew behavior. Oser, McCallum, Salas, and Morgan (1989) have observed that crews do *teamwork* and *taskwork*. Team building is the cornerstone of crew coordination—members must develop a sense of mutual trust, respect, and responsibility for the crew's performance. Taskwork refers to learning how to utilize team resources to get the work done. Communication has been shown to be central, and includes sharing information about problems that develop, intentions and goals, and who will do various tasks (Kanki & Palmer, 1993). Communication is the basis for coordinated action and allows team members to monitor each other and to contribute efficiently to collective tasks. Orasanu & Fischer (1992) found that more effective crews were more explicit about plans and strategies for dealing with problems than were less effective crews. They interpreted this in the framework of building a "shared mental model" for problems encountered in flight (see Orasanu & Salas, 1993). Strategic behaviors necessary for decision making and task management are also mediated by communication. Pepitone, King, & Murphy (1988) found that accurate decision making under time pressure is enhanced by prior contingency planning and that this planning allows crews to develop strategies for further use. Fischer, Orasanu, and Montalvo (1993) and Laudeman and Palmer (1993) showed that effective crews use low workload periods to prepare for higher workload periods, leading to better overall performance.

That CRM training in fact improves crew performance has been documented in many flight environments. Perhaps most dramatically, CRM training was credited with saving the day in the case of the DC-10 that lost all flight controls at 33,000 feet when an engine explosion severed all hydraulic lines. The NTSB investigation concluded that CRM training prepared the crew to use all available resources to figure out how to control the plane using only engine thrust and to bring it in for a controlled crash landing. Many lives were saved that surely would have been lost had the crew not worked together as well as they did under incredibly stressful and totally unanticipated conditions (NTSB, 1990).

The effects of CRM training in one major domestic airline have been analyzed by Clothier (1991). Day-to-day activities of crews flying on the line and in full-mission simulators (known as LOFT or Line-Oriented Flight Training) were evaluated by expert observers using standard checklists. (Because accident rates are already so low in commercial air transport, commercial airplane companies typically evaluate CRM programs using cockpit observers using line or LOFT operations.) The airline provided CRM training to its entire pilot force in 1989, then continued with recurrent training in 1990. A comparison of line flying by 2,000 untrained crews and approximately 1,000 trained crews showed significant differences resulting from CRM training in 12 out of 14 areas on the observers' checklist. In LOFT, 485 trained crews significantly outperformed 1,625 untrained crews in all performance categories. Overall, longitudinal data from organizations with CRM/LOFT programs show continuing improvement in crew performance over time (Helmreich & Foushee, 1993; Helmreich & Wilhelm, 1991).

One goal of current CRM programs is to change attitudes and to raise crews' awareness of the need for improved communication and coordination, as well as to provide the foundation for behavioral change. Several studies have shown positive changes in attitudes following CRM training (Butler, 1991; Helmreich & Foushee, 1993; Helmreich & Wilhelm, 1991; Irwin, 1991). Perhaps more important, reductions in human errors during flight or in accident rates have been used to evaluate CRM training effectiveness in the military or general aviation, where accident rates are relatively high. Diehl (1991) summarized six government-sponsored independent evaluations of various CRM and aeronautical decision making (ADM) programs developed for commuter and military training. These found that training was followed by reductions in crew error rates ranging from 8% to 46% (for details on the individual studies see Berlin et al., 1982; Buch & Diehl, 1983; Connolly & Blackwell, 1987; Diehl & Lester, 1987; Telfer & Ashman, 1986). Bell Helicopter, Petroleum Helicopter, the United States Air Force Military Airlift Command (MAC), and the United States Army Aviation Center have all found substantial improvements in their safety records (i.e., aircraft lost or serious mishaps), crew attitudes, and mission success following CRM training.

Overall, the results indicate that CRM training improves crew performance and contributes to flight safety. Flying airplanes, whether single-engine Cessnas or jumbo jets, is a hazardous activity, particularly when systems fail, weather is bad, or traffic is encountered. Most accidents occur on takeoffs or landings, when workload is highest, with fatigue often being a factor in poor landings. Although CRM training is not designed to reduce stress per se, it does appear to improve performance by providing behavioral mechanisms for coping with specific task demands and conditions that may lead to poor crew performance. The specific mechanisms by which CRM works to improve performance under stressful conditions vary. Crew members may monitor each others' performance, checking and correcting errors before conditions become dangerous. The redundancy of more than one crew member benefits performance when they communicate about problems they have encountered, and each contributes information or opinions about how to cope with it, creating a shared mental model (Cannon-Bowers, Salas, & Converse, 1990; Orasanu, 1993). Moreover, by communicating, crew members direct attention to cues that may be overlooked due to fatigue or other stressors, thereby contributing to performance.

A Comparison of Training Utility

All three types of training have proven to be useful for reducing effects of stressors in some contexts. An important question is whether any of the approaches is more beneficial for one type of stressor (or type of target performance) than another. Given that each class of stressor operates differently and produces different consequences, we suspect that the training approaches may be selective in their efficacy. In Table 3.2 we suggest the major and minor benefits that might be expected from each type of training approach.

TABLE 3.2
Predicted Effects of Various Training Approaches on Each Type of Stressor

	Training approaches		
Stressor	*Stress reduction*	*Skill training*	*CRM-type team training*
Danger/threat/loss of control	XXX	XX	X
Sleep deprivation CONOPS/SUSOPS	—	X	XX
Workload Time pressure	X	XX	XX
Information load	X	X	XX
Noise	—	X	X
Heat	—	—	X
Cold	—	—	X
Altitude	—	—	X

Stress-reduction techniques like SIT should be most useful in reducing the stress of danger, threat, or loss of control because the focus is on anxiety reduction. SIT has been shown to be beneficial in helping people face situations they consider threatening, so it could reduce the nonperformance problem sometimes reported in combat. SIT would probably work best in conjunction with skill training to build a sense of confidence in military personnel. Stress-reduction training might also be expected to benefit stress resulting from high workload, time pressure, and information load, as these can produce anxiety and interfere with attention and concentration. Little benefit from stress-reduction techniques would be predicted for fatigue or the physical stressors (noise, heat, cold, altitude) because these are not primarily anxiety-inducing stressors.

Skill training should be expected to benefit performance subject to both danger and high loads. It has been shown to benefit performance in high-risk situations such as parachute jumping and undersea diving, both by further developing the required skill so that is more durable, but also by giving the participant a greater sense of confidence and mastery. With respect to workload and time pressure, increased levels of skill lead to greater speed and efficiency on many tasks as well as providing greater flexibility. It may be important to distinguish between cognitively complex tasks and skills that are primarily psychomotor—skill training clearly benefits the latter. How increased levels of skill training would affect the stress resistance of more cognitive tasks is not well understood. However, time stress (embedded in a larger set of stressors) has been shown to disrupt working memory more than retrieval from long-term memory, so it is not clear whether greater levels of training would overcome this effect. Some degree of benefit from increased skill training might also benefit performance under fatigue, high information load, and perhaps noise. Again, these three stressors all affect complex cognitive functions and attention, so the benefits from skill training may be limited to skills that have an automatic component.

Crew-resource management-type training may be a powerful adjunct to the other types of stress-reduction training and its effects may help just where the others fail. A team context has been shown to help overcome the effects of fatigue in continuous operations (Allnutt, et al., 1990). Benefits from team training might be greatest in the cases of fatigue, workload/time pressure, and information load. In the fatigue case, benefits could come from the redundancy inherent in a team—members can monitor and assist, as well as stimulate, each other through conversation and prompting. They can help to keep each other on track when all suffer from fatigue. In the overload situations, team members can share task responsibilities and reallocate work to optimize team performance. Secondary benefits might be expected from team training in the case of physical stressors. Noise, heat, cold, and altitude all produce disruptions in concentration and attention and complex processing. Team members may learn to monitor each other and prompt or otherwise support each other. Team training may also benefit personnel in dangerous situations by providing moral support; cohesion has been found be a powerful factor in reducing battle stress (Stewart & Weaver, 1988), although the findings with respect to cohesion overall are not clear cut (Druckman & Swets, 1988; Siebold & Kelly, 1988).

CONCLUSIONS

In the most general terms, the research discussed on effects of various combat-relevant stressors on cognitively complex performance has shown:

- The presence of certain stressors leads to decrements in performance, in addition to physiological and affective reactions.
- Different stressors have different effects; there is no such thing as a single "stress reaction."
- There are significant variations in the effects of stress on different individuals.
- Various tasks are differentially vulnerable to various stressors.

Three different approaches to reducing stress have been described here. The first focuses primarily on reducing the stress reaction itself, with the assumption being that if individuals can be taught to manage their stress effectively, performance will improve. The second approach assumes that stress is the inevitable result of exposure to stressors and that the focus should be on skill training. If individuals can achieve automaticity or robustness on certain skills, stress will impair performance considerably less. A third approach is crew resource management in which participants are taught effective strategic and interactional skills for dealing with potentially stressful situations. All three techniques can give military personnel a greater sense of control over themselves and threatening situations—a critical factor in maintaining performance under stress.

Although the studies of training effectiveness we reviewed used affective or physiological indicators of stress reduction rather than task performance (CRM training being the exception), we have projected for which types of tasks each approach would be most suitable. Feasibility in administering the various types of training may become the deciding issue in the military. Stress-reduction techniques take time and effort and the military is always seeking ways to reduce training time and costs. Team training based on CRM courses has many advantages, in that technical training can be embedded in team training, offering a double benefit. Stress-reduction techniques, like SIT, are probably the most cumbersome to administer, but may be most beneficial for certain combat situations. The challenge to the military will be to determine how to use technology itself, like simulators of varying degrees of fidelity, or virtual reality techniques, to simulate high stress combat scenarios for training purposes. Stress is not going to go away and may be exacerbated by the high-tech battlefield of the future. Military effectiveness may depend on how well the services prepare their personnel to perform under the stressful conditions that they are certain to face.

ACKNOWLEDGMENTS

The authors wish to thank the U.S. Army Research Institute (ARI) for its support of the literature review that served as the foundation for this chapter. The opinions expressed in this chapter are the authors' and do not reflect official policy or the opinion of any government agency.

REFERENCES

Abraham, P. (1982). Training for battleshock. *Journal of Research Army Medical Corps, 128,* 18–27.

Ainsworth, L. L., & Bishop, H. P. (1971). *The effects of a 48-hour period of sustained field activity on tank crew performance* (HumRRO Tech. Rep. No. 71–16). Alexandria, VA: Human Resources Corporation.

Alkov, R. A. (1991). U.S. Navy Aircrew Coordination Training—A progress report. In R. S. Jensen (Ed.), *Proceedings of the Sixth International Symposium on Aviation Psychology* (Vol. 1, pp. 368–371). Columbus: The Department of Aviation, The Ohio State University.

Allnut, M. F., Haslam, D. R., Rejman, M. H., & Green, S. (1990). *Sustained performance and some effects on the design and operation of complex systems.* London: Army Personnel Research Establishment, Ministry of Defense.

Angus, R. G., & Heslegrave, R. J. (1985). Effects of sleep loss on sustained cognitive performance during a command and control situation. *Behavior Research Methods, Instruments, and Computers, 17,* 55–67.

Arnold, M. B. (1960). *Emotion and personality* (2 vols.). New York: Columbia University Press.

Backer, P. R., & Orasanu, J. M. (1992). *Stress, stressors, and performance in military operations: A review* (Contract No. DAAL03–86–D–001). Alexandria, VA: Army Research Institute.

Banderet, L. E., & Lieberman, H. R. (1988). *Treatment with tyrosine, a neurotransmitter precursor, reduces environmental stress in humans* (DTIC Technical Report AD–A199–199). Natick, MA: Army Research Institute of Environmental Medicine.

Banderet, L. E., Schukitt, B. L., Crohn, E. A., Burse, R. L., Roberts, D. E., & Cymerman, A. (1987). *Effects of various environmental stressors on cognitive performance* (DTIC Technical Report AD–A177–587). Natick, MA: Army Research Institute of Environmental Medicine.

Bartlett, F. C. (1943). Fatigue following highly skilled work. *Proceedings of the Royal Society* (Series B), *131,* 247–257.

Beatty, J., & Legewie, H. (Eds.). (1977). *Biofeedback and behavior.* New York: Plenum.

Beaumaster, E. J., Knowles, J. B., & MacLean, A. W. (1978). The sleep of skydivers: A study of stress. *Psychophysiology, 15*(3), 209–213.

Belenky, G., Balkin, T., Krueger, G., Headley, D., & Solick, R. (1986). *Effects of continuous operations (CONOPS) on soldier and unit performance, Phase I Review of the literature.* Ft. Leavenworth, KS: U.S. Army Combined Arms Combat Developments Activity.

Belenky, G., Balkin, T., Krueger, G., Headley, D., & Solick, R. (1987). *Effects of continuous operations (CONOPS) on soldier and unit performance: Review of the literature and strategies for sustaining the soldier in CONOPS.* Bethesda, MD: Walter Reed Army Institute of Research.

Ben Zur, H., & Breznitz, S. J. (1981). The effect of time pressure on risky choice behavior. *Acta Psychologia, 47,* 89–104.

Berkun, M. M. (1964). Performance decrement under psychological stress. *Human Factors, 6*(1), 21–30.

Berkun, M. M., Bialek, H. M., Kern, R. P., & Yagi, K. (1962). Experimental studies of psychological stress in man. *Psychological Monographs, 79,* (Whole No. 534).

Berkun, M. M., Bialek, H. M, Yagi, K., Berry, J. L., & Kern, R. P. (1959). *Human psychophysiological response to stress: Successful experimental simulation of real-life stresses.* Alexandria, VA: Human Resources Research Office, George Washington University.

Berlin, J. I., Gruber, J. M., Holmes, C. W., Jensen, P. K., Lau, J. R., Mills, J. W., & O'Kane, J. M. (1982). *Pilot judgement training and effectiveness* (DOT/FAA Report CT–82–56). Washington, DC: FAA.

Biersner, R. J., & Larocco, J. M. (1987). Personality and demographic variables related to individual responsiveness to diving stress. *Undersea Biomedical Research, 14,* 67–73.

Blockley, W. V., & Lyman, J. (1951). *Studies of human tolerance for extreme heat: IV, Psychomotor performance of pilots as indicated by a task simulating aircraft instrument flight* (Tech. Rep. No. 6521). Washington, DC: U.S. Air Force.

Bourne, P. G. (1970). *Men, stress, and Vietnam.* Boston: Little, Brown.

Bruning, N. S., & Frew, D. R. (1987). The effects of exercise, relaxation, and management skills training on physiological stress indicators: A field experiment. *Journal of Applied Psychology, 72,* 515–521.

Buch, G. D., & Diehl, A. E. (1983). *Pilot Judgement training manual validation.* Unpublished Transport Canada Report.

Buckalew, L. W. (1988). *Soldier performance as a function of stress and load: A review.* Alexandria, VA: Army Research Institute.

Burke, W. P. (1980). *An experimental evaluation of stress-management training for the airborne soldier* (Tech. Rep. No. 550). Alexandria, VA: Army Research Institute. (ERIC Document Reproduction Services No. ED 242 928)

Butler, R. E. (1991). Lessons from cross-fleet/cross-airline observations: Evaluating the impact of CRM/LOS training. In R. S. Jensen (Ed.), *Proceedings of the Sixth International Symposium on Aviation Psychology* (Vol. 1, pp. 326–331). Columbus: The Department of Aviation, The Ohio State University.

Cannon-Bowers, J. A., Salas, E., & Converse, S. A. (1990). Cognitive psychology and team training: Training shared mental models of complex systems. *Human Factors Society, 32*(12), 1–4.

Carter, B. J. (1988). Prevention of heat stress injury. In A. D. Mangelsdorff (Ed.), *Proceedings Sixth Users Workshop on Combat Stress* (Consultation Report #88–003, pp. 83–88). Ft. Sam Houston, TX: U.S. Army Health Services Command. (DTIC Report AD-A199 422)

Carter, B. J., & Cammermeyer, M. (1985). Emergence of real casualties during simulated chemical warfare training under heat conditions. *Military Medicine, 150*(12), 657–665.

Carter, B. J., & Cammermeyer, M. (1988). A phenomenology of heat injury: The predominance of confusion. *Military Medicine, 153*(3), 118–126.

Clothier, C. C. (1991). Behavioral interactions across various aircraft types: Results of systematic observations of line operations and simulation. In R. S. Jensen (Ed.), *Proceedings of the Sixth International Symposium on Aviation Psychology* (Vol. 1, pp. 332–337). Columbus: The Department of Aviation, The Ohio State University.

Cohen, S. (1980). Aftereffects of stress on human performance and social behavior: A review of research and theory. *Psychological Bulletin, 88*(1), 82–108.

Connolly, T. J., & Blackwell, B. B. (1987). A simulator-based approach to training in aeronautical decision making. In R. S. Jensen (Ed.), *Proceedings of the Fourth International Symposium of Aviation Psychology,* (pp. 251–257). Columbus: Ohio State University.

Coomler, J. (1985). *Causes of combat stress in the artillery firing battery supporting high-intensity conflict in the European theatre.* Ft. Leavenworth, KS: U.S. Army Command and General Staff College. (AD–A167674)

Cooper, G. E., White, M. D., & Lauber, J. K. (Eds.). (1980). *Resource management on the flightdeck: Proceedings of a NASA/Industry workshop* (NASA CP–2120). Moffett Field, CA: NASA-Ames Research Center.

Cowings, P. S. (1993). *Autogenic feedback training improves pilot performance during emergency flying conditions.* (NASA Tech. Mem. No. 104005). Moffett Field, CA: NASA-Ames Research Center.

Department of the Army. (1983). *Soldier performance in continuous operations* (Field Manual 22–9). Washington, DC: Government Printing Office.

Dewulf, G. A. (1978). *Continuous operations study (CONOPS) final report.* Ft Leavenworth, KS: U.S. Army Combined Arms Combat Development Activity.

Diehl, A. E. (1991). The effectiveness of training programs for preventing aircrew "error." In R. S. Jensen (Ed.), *Proceedings of the Sixth International Symposium on Aviation Psychology* (Vol. 2, pp. 640–655). Columbus: The Department of Aviation, The Ohio State University.

Diehl, A. E., & Lester, L. F. (1987). *Private pilot judgement training in flight school settings* (DOT/FAA Report 87/6). Washington, DC: FAA.

Driskell, J. A., Mullen, B., Johnson, C., Hughes, S., & Batchelor, C. L. (1992). *Development of quantitative specifications for simulating the stress environment* (Tech. Rep. No. AL–TR–1991–0109). Dayton, OH: Wright–Patterson Air Force Base, Human Resources Directorate Logistics Research Division.

Driskell, J.A., & Salas, E. (1991). Overcoming the effects of stress on military performance: Human factors, training, and selection strategies. In R. Gal & A. D. Mangelsdorff (Eds.), *Handbook of military psychology* (pp. 183–193). New York: Wiley.

Driskell, J. A., Willis, R. P., & Copper, C. (1992). Effect of overlearning on retention. *Journal of Applied Psychology, 77*(5), 615–622.

Drucker, E. H., Canon, L. D., & Ware, J. R. (1969). *The effects of sleep deprivation on performance over a 48-hour period* (Tech. Rep. No. 69–8). Alexandria, VA: Human Resources Office.

Druckman, D., & Swets, J. A. (Eds.). (1988). *Enhancing human performance.* Washington, DC: National Academy Press.

Easterbrook, J. A. (1959). The effect of emotion on cue utilization and the organization of behavior. *Psychological Review, 66,* 183–201.

Edland, A. (1985). Attractiveness judgements of decision alternatives under time stress. *Reports from the Cognition and Decision Research Unit, Department of Psychology* (University of Stockholm, Sweden), No. 21.

Edland, A. (1989, May). On cognitive processes under time stress. *Reports from the Cognition and Decision Research Unit, Department of Psychology* (University of Stockholm, Sweden), Suppl. 68.

Einhorn, H. J., & Hogarth, R. M. (1978). Confidence in judgement: Persistence in the illusion of validity. *Psychological Review, 85,* 395–416.

Ellis, A. (1962). Reason and emotion in psychology. New York: Lyle Stuart.

Epstein, S., & Fenz, W. D. (1965). Steepness of approach and avoidance gradients in humans as a function of experience: Theory and experiment. *Journal of Experimental Psychology, 70,* 1–13.

Fenz, W. D. (1975). Strategies for coping with stress. In I. Sarason & C. Spielberger (Eds.), *Stress and anxiety* (pp. 305–336). Washington, DC: Hemisphere.

Fischer, U., Orasanu, J., & Montalvo, M. L. (1993). Effective decision strategies on the flight deck. In R. Jensen (Ed.), *Proceedings of the Seventh International Symposium on Aviation Psychology* (pp. 238–247). Columbus: The Ohio State University.

Foushee, H. E., & Helmreich, R. L. (1988). Group interaction and flight crew performance. In E. L. Wiener & D. C. Nagel (Eds.), *Human factors in aviation* (pp. 189–227). San Diego: Academic Press.

Foushee, H. E., & Manos, K. L. (1981). Information transfer within the cockpit: Problems in intracockpit communications. In C. E. Billings & E. S. Cheaney (Eds.), *Information transfer problems in the aviation system* (NASA TP–1875, pp. 63–70). Moffett Field, CA: NASA Ames Research Center (NTIS No. N81–31162).

Gal, R. (1988). Psychological aspects of combat stress: A model derived from Israeli and other combat experiences. In A. D. Mangelsdorff (Ed.), *Proceedings Sixth Users Workshop on Combat Stress* (Consultation Report #88–003, pp. 101–122). Ft. Sam Houston, TX: U.S. Army Health Services Command. (DTIC Report AD–A199–422)

Gal, R., & Lazarus, R. S. (1975). The role of activity in anticipation and confronting stressful situations. *Journal of Human Stress, 1*(4), 4–20.

Gal-or, Y., Tenenbaum, G., Furst, D., & Shertzer, M. (1985). Effect of self-control and anxiety on training performance in young and novice parachuters. *Perceptual and Motor Skills, 60*(3), 743–746.

Glass, D. C., & Singer, J. E. (1972). *Urban stress.* New York: Academic Press.

Haggard, D. F. (1970). *HumRRO studies in continuous operations* (HumRRO professional paper 7–70). Alexandria, VA: Human Resources Research Organization.

Hammond, K. R. (1990). *The effects of stress on judgment and decision making: An overview and arguments for a new approach.* Boulder: University of Colorado, Center for Research on Judgment and Policy.

Hammond, K. R., & Doyle, J. K. (1991). *Effects of stress on judgment and decision making: Part II.* (Prepared for the U. S. Army Research Office). Research Triangle Park, NC: U.S. Army Research Office.

Hart, S. G. (1989, May). *Overview of NASA rotorcraft human factors research.* Paper presented at the American Helicopter Society 45th Annual Forum, Boston, MA.

Hart, S. G., & Staveland, L. E. (1988). Development of NASA-TLX (Task Load Index): Results of empirical and theoretical research. In P. A. Hancock & N. Meshkati (Eds.), *Human mental workload* (pp. 239–250). Amsterdam: North-Holland.

Hart, S. G., & Wickens, C. D. (1994). Workload assessment and prediction. In H. R. Booher (Ed.), *MANPRINT: An emerging technology, Advanced concepts for integrating people, machines and organizations*. New York: Van Nostrand & Reinhold.

Haslam, D. R. (1981). The military performance of soldiers in continuous operations: Exercises 'Early Call' I and II. In *The twenty four hour workday: Proceedings of a symposium on variations in work–sleep schedules*. Cincinnati, OH: U.S. Dept. of Health & Human Services.

Haslam, D. R. (1982). Sleep loss, recovery sleep, and military performance. *Ergonomics, 25*(2), 163–178.

Haslam, D. R. (1985). Sustained operations and military performance. *Behavior Research Methods, Instruments, and Computers, 17*(1), 90–95.

Hawkins, F. H. (1987). *Human factors in flight*. Brookfield, VT: Gower.

Helmreich, R. L. (1991). Strategies for the study of flight crew behavior. In R. S. Jensen (Ed.), *Proceedings of the Sixth International Symposium on Aviation Psychology* (Vol. 1, pp. 338–343). Columbus: The Department of Aviation, The Ohio State University.

Helmreich, R. L., & Foushee, H. C. (1993). Why crew resource management?: The history and status of human factors training progress in aviation. In E. L. Weiner, B. G. Kanki, & R. L. Helmreich (Eds.), *Crew resource management* (pp. 3–45). New York: Academic Press.

Helmreich, R. L., & Wilhelm, J. A. (1991). Outcomes of CRM training. *International Journal of Aviation Psychology, 1*, 287–300.

Horvath, S. M., & Freedman, A. (1957). The influence of cold upon the efficiency of man. *Journal of Aviation Medicine, 18*, 158–164.

Hughes Aircraft Company. (1977). *Crew size evaluation for tactical all-weather strike aircraft* (Tech. Rep. No. AFAL TR–76–79). Wright–Patterson Air Force Base, OH: Air Force Avionics Laboratory.

Hytten, K., Jensen, A., & Skauli, G. (1990). Stress inoculation training for smoke divers and free fall lifeboat passengers. *Aviation, Space, and Environmental Medicine, 61*,(11), 983–988.

Iampietro, P. F., Chiles, W. D., Higgins, E. A., & Gibbons, H. L. (1969). Complex performance during exposure to high temperatures. *Aerospace Medicine, 40*, 1331–1335.

Idzikowski, C., & Baddeley, A. D. (1983). Fear and dangerous environments. In G. R. Hockey (Ed.), *Stress and fatigue in human performance* (pp. 123–144). New York: Wiley.

Irwin, C. (1991). The impact of initial and recurrent CRM training. In R. S. Jensen (Ed.), *Proceedings of the Sixth International Symposium on Aviation Psychology* (Vol. 1, pp. 344–349). Columbus: The Department of Aviation, The Ohio State University.

Jacobson, E. (1938). *Progressive relaxation* (2nd. ed.). Chicago: University of Chicago Press.

Janis, I. L. (1949). Problems related to the control of fear in combat. In S. A. Stouffer, A. A. Lumsdaine et al. (Eds.), *The American soldier: Combat and its aftermath*. Princeton, NJ: Princeton University Press.

Janis, I. L. (1983). The patient as decision maker. In D. Gentry (Ed.), *Handbook of behavioral medicine*. New York: Guilford.

Jones, D. (1983). Noise. In R. Hockey (Ed.), *Stress and fatigue in human performance* (pp. 61–96). New York: Wiley.

Jorna, P. G. A. M. (1985). Heart-rate parameters and the coping process underwater. In J. F. Orlebeke, G. Mulder, & L. J. P. van Doornen (Eds.), *Psychophysiology of cardiovascular control* (pp. 827–839) [Proceedings of a NATO Conference on Cardiovascular Psychophysiology: Theory and Methods, held June 12–17, 1983, Noordwijkerhout, The Netherlands]. New York: Plenum.

Kanki, B., & Palmer, M. T. (1993). Communication and CRM. In E. L. Wiener, B. G. Kanki, & R. L. Helmreich (Eds.), *Cockpit resource management* (pp. 99–134). San Diego: Academic Press.

Keinan, G. (1988). Training for dangerous task performance: The effects of expectations and feedback. *Journal of Applied Social Psychology, 18*(4, pt. 2), 355–373.

Khachatur'yants, L. S., Grimak, L. P., & Khrunov, Y. V. (1976). *Eksperimental'naya psikologiya v kosmicheskikh issledovaniyakh* [Experimental psychology in space investigations]. Moscow: Nauka Press.

Khrunov, Y. V., Khachatur'yants, L. S., Popov, V. A., & Ivanov, Y. I. (1974). *Chelovek-operator v komicheskom polete* [The human operator in spaceflight]. Moscow: Mashinostroyeniye.

Klein, G. A. (1993). A recognition-primed decision (RPD) model of rapid decision making. In G. Klein, J. Orasanu, R. Calderwood, & C. Zsambok (Eds.), *Decision making in action: Models and methods* (pp. 138–147). Norwood, NJ: Ablex.

Klein, G. A., Orasanu, J., Calderwood, R., & Zsambok, C. (Eds.). (1993). *Decision making in action: Models and methods.* Norwood, NJ: Ablex.

Klein, G., Wolf, S., Militello, L., & Zsambok, C. (1995). Characteristics of skilled option generation in chess. *Organizational Behavior and Human Decision Processes, 62,* 63–69.

Koelega, H. S., & Brinkman, J. A. (1986). Noise and vigilance: An evaluative review. *Human Factors, 28*(4), 465–481.

Labuc, S. (1981). *Psychological stress and combat efficiency: A review of the literature* (Report 81R005). Farnborough, UK: Army Personnel Research Establishment.

Lakein, A. (1973). *How to get control of your time and your life.* New York: Peter W. Wyden.

Lanzetta, J. T., & Thornton, B. R. (1957). Effects of work-group structure and certain task variables on group performance. *Journal of Abnormal and Social Psychology, 53,* 307–314.

Larsson, G. (1987). Routinization of mental training in organizations: Effects on performance and well-being. *Journal of Applied Psychology, 72*(1), 88–96.

Lauber, J. K. (1984). Resource management in the cockpit. *Air Line Pilot, 54*(9), 20–31.

Lauber, J. K. (1993). Preface. In E. L. Wiener, B. G. Kanki, & R. L. Helmreich (Eds.), *Cockpit resource management* (pp. xv–xviii). San Diego: Academic Press.

Laudeman, I. V., & Palmer, E. A. (1993). Measurement of taskload in the analysis of aircrew performance. In R. S. Jensen & D. Neumeister (Eds.), *International Symposium on Aviation Psychology* (pp. 854–858). Columbus: The Department of Aviation, The Ohio State University.

Lazarus, R. S. (1980). *Stress and the coping process.* New York: McGraw-Hill.

Lazarus, R. S., & Folkman, S. (1984). *Stress, appraisal, and coping.* New York: Springer.

Leedom, D. K., & Simon, R. (1993). U.S. Army crew coordination training and evaluation. In R. S. Jensen & D. Neumeister (Eds.), *International Symposium on Aviation Psychology* (pp. 527–532). Columbus: The Department of Aviation, The Ohio State University.

Levine, S. (1971). Stress and behavior. *Scientific American, 224,* 26–31.

Logan, G. D. (1985). Skill and automaticity. *Canadian Journal of Psychology, 9,* 283–286.

Lomov, B. F. (1966). *Chelovek i teknika* [man and technology]. Moscow: Sovetskoye Radio.

Lopes, L. (1983). Some thoughts on the psychological concept of risk. *Journal of Experimental Psychology, 9,* 137–144.

Mackie, R. R., Wylie, C. D., & Evans, S. M. (in press). *Fatigue effects on human performance in combat: A literature review* [Draft]. Alexandria, VA: U.S. Army Research Institute.

Mareth, T. R., & Brooker, A. E. (1982). Combat stress reaction: A concept in evolution. *Military Medicine, 150,* 186–190.

Marshall, S. L. A. (1947). *Men against fire.* New York: Morrow.

McCaughey, B. G. (1991). Observations about battle fatigue: Its occurrence and absence. *Military Medicine, 156,* 694–695.

McElhatton, J., & Drew, C. (1993). Time pressure as a causal factor in aviation safety incidents: The "hurry-up" syndrome. In R. S. Jensen & D. Neumeister (Eds.), *International Symposium on Aviation Psychology* (pp. 269–274). Columbus: The Department of Aviation, The Ohio State University.

Michel, R. R., & Solick, R. E. (1983). *Review of literature on the effects of selected human performance variables on combat performance* (Field Unit Working Paper FLV–FU–83–4). Ft. Leavenworth, KS: U.S. Army Research Institute.

Migdal, K., & Paciorek, J. (1989). Relaxation exercises as a stress reducing factor during stimulation training (K. Gebert, Trans.). *Polish Psychological Bulletin, 20*(3), 197–205.

Miller, J. G. (1960). Information input overload and psychopathology. *American Journal of Psychiatry, 16,* 695–704.

Moroney, W. F., Reising, J., Biers, D. W., & Eggemeier, F. T. (1993). The effect of previous level of workload on the NASA Task Load Index (TLX) in a simulated flight task. In R. S. Jensen & D. Neumeister (Eds.), *International Symposium on Aviation Psychology* (pp. 882–885). Columbus: The Department of Aviation, The Ohio State University.

Mulder, M., van Eck, J. R. R., & de Jong, R. D. (1971). An organization in crisis and non-crisis situations. *Human Relations, 24,* 19–41.

Novaco, R. W., Cook, T. M., & Sarason, I. G. (1983). Military recruit training: An arena for stress-coping skills. In D. Meichenbaum & M. E. Jaremko (Eds.), *Stress reduction and prevention.* New York: Plenum.

National Transportation Safety Board (1979). *Aircraft accident report: United Airlines, Inc., McDonnell–Douglas DC-8–61, N8082U, Portland, Oregon, December 28, 1978* (NTSB/AAR-79–7). Washington, DC: Author.

National Transportation Safety Board (1982). *Aircraft accident report: Air Florida, Inc., Boeing 737–222, N62AF, collision with 14th street bridge near Washington National Airport, Washington, DC, January 13, 1982* (NTSB/AAR–82–8). Washington, DC: Author.

National Transportation Safety Board (1990). *Aircraft Accident Report: United Airlines Flight 232, McDonnell–Douglas DC–10–10, Sioux Gateway Airport, Sioux City, IA, July 19, 1989* (NTSB/AAR–91–02). Washington, DC: Author.

National Transportation Safety Board. (1994). *A review of flightcrew-involved, major accidents of U.S. air carriers, 1978 through 1990.* (PB94–917001, NTSB/SS–94/01). Washington, DC: Author.

Oelz, O., & Regard, M. (1988). Physiological and neuropsychological characteristics of world-class extreme-altitude climbers. *American Alpine Journal, 30*(62), 83–86.

Orasanu, J. (1993). Decision making in the cockpit. In E. L. Wiener, B. G. Kanki, & R. L. Helmreich (Eds.), *Cockpit resource management* (pp. 137–168). San Diego: Academic Press.

Orasanu, J., & Fischer, U. (1992). Team cognition in the cockpit: Linguistic control of shared problem solving. In *Proceedings of the 14th Annual Conference of the Cognitive Science Society* (pp. 189–194). Hillsdale, NJ: Lawrence Erlbaum Associates.

Orasanu, J., & Salas, E. (1993). Team decision making in complex environments. In G. Klein, J. Orasanu, R. Calderwood, & C. Zsambok (Eds.), *Decision making in action: Models and methods* (pp. 327–345). Norwood, NJ: Ablex.

Oser, R. L., McCallum, G. A., Salas, E., & Morgan, B. B. (1989). *Toward a definition of teamwork: An analysis of critical team behaviors.* (Tech. Rep. No. 89–0043). Orlando, FL: Naval Training Systems Center.

Payne, J. W., Bettman, J. R., & Johnson, E. J. (1988). Adaptive strategy selection in decision making. *Journal of Experimental Psychology: Learning, Memory and Cognition. 14*(3), 534–552.

Pepitone, D., King, T. A., & Murphy, M. (1988). *The role of flight planning in aircrew decision performance* (SAE Technical Paper Series 881517). Warrendale, PA: The Engineering Society for Advancing Mobility Land Sea Air and Space.

Perry, J. C. (Ed.) (1974). *Helicopter aircrew fatigue* (AGARD Advisory Rep. No. 69). Brussels: Advisory Group for Aerospace Research and Development, NATO.

Pleban, R. J., Thomas, D. A., & Thompson, H. L. (1985). Physical fitness as a moderator of cognitive work capacity and fatigue onset under sustained combat-like operations. *Behavior Research Methods, Instruments, and Computers, 17*(1), 86–89.

Plous, S. (1993). *The psychology of judgement and decision making.* New York: McGraw Hill.

Poulton, E. C., Hitchings, N. B., & Brooke, R. B. (1965). Effect of cold and rain upon the vigilance of lookouts. *Ergonomics, 8,* 163–168.

Prince, C., Chidester, T. R., Cannon-Bowers, J., & Bowers, C. (1993). Aircrew coordination:

Achieving teamwork in the cockpit. In R. Swezey & E. Salas (Eds.), *Teams: Their training and performance* (pp. 329–354). Norwood, NJ: Ablex.

Rachman, S. (1978). *Fear and courage*. San Francisco: Freeman.

Rachman, S. (1988). *Psychological analyses of courageous performance in Military personnel* (ARI Progress Report, Second Year Report), Alexandria, VA: ARI.

Ramsey, J. D. (1983). Heat and cold. In G. R. J. Hockey (Ed.), *Stress and fatigue on human performance* (pp. 33–60). New York: Wiley.

Ramsey, J. D., Burford, C. L., Beshin, M. Y., & Jensen, R. C. (1983). Effects of workplace thermal conditions on safe work behavior. *Journal of Safety Research, 14,* 105–114.

Ramsey, J. D., Dayal, D., & Ghahramani, B. (1975). Heat stress limits for the sedentary worker. *American Industrial Hygiene Association Journal, 36,* 259–265.

Robinson, J. A. (1972). Crisis: An appraisal of concepts and theories. In C. F. Hermann (Ed.), *International crises: Insights from behavioral research* (pp. 20–35). New York: Free Press.

Rognum, T. O., Vartdal, F., Rodahl, K., Opstad, P. K., Knudsen-Baas, O., Kindt, E., & Witney, N. R. (1986). Physical and mental performance of soldiers on high- and low-energy diets during prolonged heavy exercise combined with sleep deprivation. *Ergonomics, 29*(7), 859–867.

Ruffell Smith, H. O. (1979). *A simulator study of the interaction of pilot workload with errors, vigilance, and decisions* (NASA Tech. Memo. 78482). Moffett Field, CA: NASA–Ames Research Center.

Rutenfranz, J., Aschoff, J., & Mann, H. (1972). The effects of a cumulative sleep deficit, duration of preceding sleep period and body-temperature on multiple choice reaction time. In W. P. Colquhuon (Ed.), *Aspects of human efficiency* (pp. 217–229). London: The English Universities Press Limited.

Schendel, J. D., & Hagman, J. D. (1982). On sustaining procedural skills over a prolonged retention interval. *Journal of Applied Psychology, 67,* 605–610.

Schuler, K., Gilner, F., Austrin, H., & Davenport, D. G. (1982). Contribution of the education phase to stress-inoculation training. *Psychological Reports, 51,* 611–617.

Schwartz, G. E., & Beatty, J. (Eds.). (1977). *Biofeedback: Theory and research.* New York: Academic Press.

Shultz, J., & Luthe, W. (1959). *Autogenic training: A psychophysiological approach in psychotherapy.* New York: Grune & Stratton.

Siebold, G. L., & Kelly, D. R. (1988). A measure of cohesion which predicts unit performance and ability to withstand stress. In A. D. Mangelsdorff (Ed.), *Proceedings Sixth Users Workshop on Combat Stress* (Consultation Rep. #88–003, pp. 12–15). Ft. Sam Houston, TX: U.S. Army Health Services Command. (DTIC Report AD–A199–422)

Siegel, A. I., Kopstein, F. F., Federmen, P. J., Ozkaptan, H., Slifer, W. E., Hegge, F. W., & Marlowe, D. H. (1981). *Management of stress in army operations* (Research Product 81–19). Alexandria, VA: U.S. Army Research Institute.

Siegel, A. I., Pfeiffer, M. G., Kopstein, F. F., Wilson, L. G., & Ozkaptan, H. (1979). *Human performance in continuous operations: Volume I, Human performance guidelines* (Tech. Rep. No. 80–4a). Alexandria, VA: U.S. Army Research Institute.

Siegel, A. I., Pfeiffer, M. G., Kopstein, F. F., Wolf, J. J., & Ozkaptan, H. (1980). *Human performance in continuous operations: Volume III, Technical documentation.* Wayne, PA: Applied Psychological Services.

Slovic, P., Fischhoff, B., & Lichtenstein, S. (1977). Behavioral decision theory. *Annual Review of Psychology, 28,* 1–39.

Smart, C. & Vertinsky, I. (1977). Designs for crisis decision units. *Administrative Science Quarterly, 22,* 640–657.

Smith, P., et al. (1990, April). *Effects of training with lethal chemicals on job proficiency and job confidence.* Paper presented at the American Educational Research Association, Boston, MA. (ERIC Documentation Reproduction Service No. ED 319 948).

Spettell, C. M., & Liebert, R. M. (1986). Training for safety in automated person–machine systems. *American Psychologist, 41,* 545–550.

Starr, L. M. (1987). "Training for safety in automated person–machine systems": Comment. *American Psychologist, 42,* 1029.

Stepanov, V. N., & Stetanov, E. N. (1979). Engineering–psychological questions oftechnical support in space. In B. N. Petrov, B. F. Lomov, & N. D. Semsonov, (Eds.), *Psikhologicheskiye problemy kosmicheskikh poletov* [Psychological problems of space flights] (pp. 290–299). Moscow: Nauka Press.

Stewart, N. K., & Weaver, S. (1988). A methodological analysis of the link between cohesion and combat stress and post-traumatic stress syndrome. In A. D. Mangelsdorff (Ed.), *Proceedings Sixth Users Workshop on Combat Stress* (Consultation Report #88–003, pp. 16–20). Ft. Sam Houston, TX: U.S. Army Health Services Command. (DTIC Report AD–A199–422)

Stouffer, S. A., Lumsdaine, A. A., Lumsdaine, M. H., Williams, R. M., Smith, M. B., Janis, I. L., Star, S. A., & Cottrell, L. S. (1949). *The American soldier: Combat and its aftermath* (Vol. II). Princeton, NJ: Princeton University Press.

Streufert, S. (1983). *Load effects on the use of strategy in motivated personnel* (DTIC Technical Report AD–P000–818). Paper presented at the Annual Conference of the Military Testing Association (24th), San Antonio, Texas.

Streufert, S., & Streufert, S. C. (1981). *Stress and information search in complex decision making: Effects of load and time urgency* (DTIC Tech. Rep. No. AD-A104–007). Arlington, VA: Office of Naval Research.

Streufert, S. Streufert, S. C., & Denson, A. L. (1982). *Information load stress, risk taking, and physiological responsivity in a visual-motor task* (DTIC Technical Report AD–A118–079). Arlington, VA: Office of Naval Research.

Streufert, S., Suedfeld, P., & Driver, M. J. (1965). Conceptual structure, information search and information utilization. *Journal of Personality and Social Psychology, 2,* 736–740.

Svenson, O., Edland, A., & Karlsson, G. (1985). The effect of numerical and verbal information and time stress on judgements of the attractiveness of decision alternatives. In L. B. Methlie & R. Sprague (Eds.), *Knowledge representation for decision support systems.* Amsterdam: North-Holland.

Telfer, R. A., & Ashman, A. F. (1986). *Pilot judgement training, an Australian validation study.* Unpublished research report, University of Newcastle, New South Wales, Australia.

Thomas, M. (1988). *Managing pilot stress.* New York: Macmillan.

Tijerina, L., Stabb, J. A., Eldredge, D., Herschler, D. A., Mangold, S. J. (1988). *Improving shipboard decision making in the CBR-D environment: Concepts of use for and functional description of a decision aid/training system* (DTIC Report No. AD–A207–219). Edgewood, MD: Chemical Warfare/Chemical Biological Defense Information Analysis Center.

Townes, B. D., Hornbein, T. F., Schoene, R. B., Sarnquist, F. H., & Grant, I. (1984). Human cerebral function at extreme altitude. In J. B. West & S. Lahiri (Eds.), *High Altitude and Man* (pp. 31–36). Baltimore: Williams and Wilkins.

Tversky, A., & Kahneman, D. (1973). Availability: A heuristic for judgement frequency probability. *Cognitive Psychology, 5,* 207–232.

Tversky, A., & Kahneman, D. (1981). The framing of decisions and the psychology of choice. *Science, 211,* 453–458.

Villoldo, A., & Tarno, R. L. (1984). *Measuring the performance of EOD equipment and operators under stress* (DTIC Technical Report AD–B083–850). Indian Head, MD: Naval Explosive and Ordnance Disposal Technology Center.

Wertkin, R. A. (1985). Stress-inoculation training: Principles and applications. *Social Casework: The Journal of Contemporary Social Work, 66,* 611–616.

Wickens, C. D. (1984). *Engineering psychology and human performance.* Columbus, OH: Merrill.

Wickens, C. D., Barnett, B., Stokes, A., Davis, T., & Hyman, F. (1988, October). *Expertise, stress, and pilot judgement.* Paper presented at the NATO/AGARD Panel Meeting/Symposium of the Aerospace Medical Panel on Human Behavior in High Stress Situations in Aerospace Operations, The Hague, Netherlands.

Wickens, C. D., Stokes, A., Barnett, B., & Hyman, F. (1991). The effects of stress on pilot judgement in a MIDIS Simulator. In O. Svenson & J. Maule (Eds.), *Time pressure and stress in human judgement and decision making* (pp. 271–292). New York: Plenum.

Wilkinson, R. T. (1964). Effects of up to 60 hours of sleep deprivation on different types of work. *Ergonomics, 1,* 175–186.

Wilkinson, R. T. (1965). Sleep deprivation. In O. G. Edholm & A. Bacharach (Eds.), *The physiology of human survival* (pp. 399–430). New York: Academic Press.

Wilkinson, R. T. (1971). Hours of work and the twenty-four cycle of rest and activity. In P. B. Warr (Ed.), *Psychology at work* (pp. 31–54). London: Penguin.

Williams, H. L., Lubin, A., & Goodnow, J. J. (1959). Impaired performance with acute sleep loss. *Psychological Monographs, 73,* 1–26.

Winget, C. M., DeRoshia, C. W., & Holley, D. C. (1985). Circadian rhythms and athletic performance. *Medicine and Science in Sports and Exercise, 17*(5), 498–516.

Winget, C. M., DeRoshia, C. W., Markley, C. L., & Holley, D. C. (1984). A review of human physiological and performance changes associated with desynchronosis of biological rhythms. *Aviation, Space, and Environmental Medicine, 55*(12), 1085–1096.

Wright, P. (1974). The harassed decision maker: Time pressure, distraction and use of evidence. *Journal of Applied Psychology, 59,* 555–561.

Zajonc, R. B. (1965). Social facilitation. *Science, 149,* 269–274.

Zakay, D., & Wooler, S. (1984). Time pressure, training, and decision effectiveness. *Ergonomics, 27,* 273–284.

4 Stress and Aircrew Performance: A Team-Level Perspective

Barbara G. Kanki
NASA–Ames Research Center

> There are three kinds of problems you can have with your car: (1) those that will kill you; (2) those that will leave you stranded; and (3) those that will cost you more money if you let them go too long.
>
> Shimmies and shakes—problems that have to do with the suspension and steering of your car—all come under category one. So we're going to look at them in some detail.
>
> The steering pieces—ball joints, center links, idler arms, pitman arms, drag links, tie-rod ends, and so forth—literally hold the front wheels to the rest of the car. If any one of these things breaks, you're in serious trouble. . . . You might say, "well there's nothing I can do about it. If they're going to break, they're going to break." Wrong! If you know what to look for, you can heed early warning signals.
>
> A funny thing happens. People never know when their car is handling badly, because all the things that wear out generally wear out slowly, or incrementally, as the mathematicians say. Everyday, the car gets a little worse—the ball joints get a little looser, the tie-rod ends get a little sloppier, your tires get a little squarer, and all kinds of things happen to make the car handle worse. Except you don't notice because the changes are so small that you adapt to them.
>
> —Magliozzi & Magliozzi (1991, pp. 15–17)

Complex operations, like automobiles, are made up of many interlocking parts, and the way in which the parts work together vary according to the design and condition of the system of parts as well as the design and condition of its constituent parts. Furthermore, as the Magliozzi brothers advise here, a system of parts may wear down imperceptibly because changes due to wear and tear (e.g., stress effects over time) are small, and people adapt to them. However, when warning signals are unnoticed or unheeded, the system can break down unexpectedly, and the resulting problems can be fatal.

THE AIRCREW ENVIRONMENT

In these days of "aging aircraft," there is a valid analogy to be made between automobiles and airplanes that ends with a plea for greater respect and attention for maintenance operations. However, the analogy I am pursuing is the discussion of the aircrew as a system of parts. On one level, it is a simple machine made up of two or three basic components: the captain, the first officer (or copilot), and, in some aircraft, the second officer (or flight engineer). Although every individual is unique, these roles are standardized through training, company policies, and federal regulations. In addition, many pilots share a common military experience. From this standpoint, it is not unreasonable to consider crew members as interchangeable parts; at least within companies, within aircraft type, and within position. At the same time, pilots also learn traditions; that is, norms, expectations, and "unwritten" rules that become ingrained in the culture of flying. Tom Wolfe (1979) described some of this tradition when he wrote about the "right stuff." The familiar stereotype of the white-scarf-and-goggles pilot who maintains his cool demeanor in the face of adversity still has a place in today's pilot culture. When we hear the captain's voice on the PA system, we can still envision the:

> fearless, self-sufficient, technically qualified, and slightly egotistical individual, whose job description calls for the defiance of death on a regular basis. . . . These characteristics were much admired and became not only essential elements of the informal pilot culture which were passed from generation to generation of pilots but they were also institutionalized in regulations governing pilot performance, as well as in policies guiding pilot selection criteria. They are so much a part of the culture that they are still with us in an age where flight operations are routine and aviation is by far the safest mode of commercial transportation. (Foushee & Helmreich, 1988, p. 191)

In addition to acquiring technical competencies and assimilating the pilot culture, crew members also learn the pragmatic day-to-day workings of the system. This includes the real-world knowledge of how well their material and information resources work with weather, airport, and aircraft conditions that are less than ideal. It also includes knowledge about working with people—dispatchers, ramp agents, flight attendants, air traffic control, and each other. In the flurry of activity that takes place during landing and takeoffs, aircrew members must partition their work efficiently. For example, as the captain works a last-minute flightplan change with dispatch, and communicates with the ramp agents to start pushback procedures, the first officer may resolve passenger issues with the gate agent and lead flight attendant. During taxi to the runway, while the captain controls the aircraft both crew members monitor ground control's instructions regarding taxi and takeoff. On takeoff, the captain or first officer assumes flight control (as "pilot flying") while the other takes responsibility for communications with air traffic control (as "pilot-not-flying"). Typically, the captain and first officer alternate who is the pilot flying and pilot-not-flying on each leg of the trip. In short, every takeoff and landing requires many individuals and teams to closely coordinate with each other in a highly precise manner. In busy airports, these operations can take place hundreds of times in just a few hours. For the aircrew, this is also a highly familiar set of behaviors that may take place five or six times a day depending on how many flights can be scheduled into their 12–14 hour maximum duty day. Clearly, the reliability and safety of the system depends on the fact that many aspects of the pilots' tasks are highly routininized in conformity with standard operating procedures (SOPs). Unfortunately, the small stresses and strains of the system also become familiar in time, and as pilots adapt to these "glitches," they may stop noticing them.

A HIGH-RISK WORK ENVIRONMENT

There is no question that aviation accidents and incidents affect the public deeply. Although a highway accident can be equally devastating, people seldom give up driving when they witness an accident. In contrast, plane crashes make people so nervous about flying that their attitudes can affect airline revenue, in spite of the fact that the proportion of total transportation fatalities attributed to aviation is very small. For example, in 1987, the National Transportation Safety Board (NTSB) reported 49,306 transportation fatalities. Broken down into vehicle types, 94% (or 46,330) occurred on highways. All other modes of transportation constituted the other 6% (2,976). Aviation fatalities made up about 39% (1,160) of "other modes," which also included grade crossing, railroad, marine, and pipeline accidents. Therefore, although aviation fatalities probably accounted for a large proportion of media and press attention, they constituted less that 3% of the total transportation fatality statistic. Furthermore, several types of new aircraft that have been introduced within the last 10 years are still without a single accident. Without question, improvements in aircraft reliability have greatly increased flight safety.

This trend is mirrored in statistics reported by Boeing Commercial Airplane Company (Lautman & Gallimore, 1987), in which primary causes are attributed to hull loss accidents (i.e., accidents in which aircraft damage is beyond economic repair). Over the last 30 years the proportion of accidents related to flightcrew-caused factors has been approximately 65–70%, whereas the remaining 30–35% has been distributed over other factors such as the aircraft, maintenance, weather, airport, and air traffic control. Thus, human error has been implicated in a large proportion of accidents for many years, although these errors have taken many forms and have been committed in a variety of conditions. A recent study published by the National Transportation Safety Board (1994) analyzed 37 major accidents that occurred between 1978 and 1990 in terms of the errors identified in the accidents and the contexts in which they occurred:

> Of the 302 specific errors identified in the 37 accidents, the most common were related to procedures, tactical decisions, and failure to monitor or challenge another crew member's error. Monitoring/challenging failures were pervasive, occurring in 31 of the 37 accidents. . . . The type of error most frequently unchallenged was a captain's tactical decision error that was an error of omission. (p. vi.)

In light of these error analyses, we are provided better explanations for how particular crew factors contribute to an accident. However, we also know that accidents are very rare events, and that the errors identified are not causal in a predictive sense; that is, accidents do not occur every time a captain makes a tactical error of omission, or when the first officer fails to challenge a captain's decision.

Nevertheless, we have the classic example of flight-crew caused accident in United Airlines Flight 173, which crashed near Portland on December 28, 1978 (NTSB, 1979). The plane crashed because it ran out of fuel while the flightcrew worked on resolving a landing gear problem and preparing for an emergency landing. Probable cause was determined to be "the failure of the captain to monitor properly the aircraft's fuel state and to properly respond to the low fuel state and the crew members's advisories regarding fuel state. This resulted in fuel exhaustion to all engines" (p. 29). It also resulted in 10 fatalities.

Another monitoring problem was exemplified in Eastern Airlines Flight 401, which crashed in the Florida Everglades in December, 1972. In this case, the entire flightcrew was preoccupied with changing a burned-out landing gear light, and had not noticed that the autopilot had become disengaged. As the aircraft drifted slowly down, the air traffic controller (ATC) noticed and asked, "How are things comin' along out there." The crew, paying attention only to the landing gear problem, said everything was alright and continued their unintended descent. All the critical information was accessible and the aircraft had been fully operational, yet 99 passengers and crew members were killed in this crash (NTSB, 1972).

Systems vulnerability and human error can surface in any work environment that is complex and involves the coordination and management of many team members. In air transport operations (as evidenced by the safety records), small deviations from standards and procedures, and minor miscommunications and control errors, remain small and minor most of the time. However, when human errors do escalate into an unrecoverable situation, the work environment is unforgiving and the consequences are disastrous. The delicate balance among coordinated actions must be carefully protected from its own weaknesses, as well as from external stressors that enter the workplace in unexpected ways.

STRESSORS OF THE WORK ENVIRONMENT

Everyday life is filled with stresses and working in the increasingly complex, crowded, and economically unstable world of aviation transport is no exception. On any given day, the headlines will announce to us a myriad of concerns, ranging from the irritating to the life-threatening. For instance, weather delays, stranded passengers, layoffs, labor–management disputes, grounded aircraft, hijackings, drug testing, maintenance-related accidents, air traffic controller-related accidents, pilot error.

In short, everyday life in transport operations is filled with stresses and they are of many types and definitions. Although I do not advocate a single view as the "best" view, a team-level perspective on the topic of stress and aircrew performance may provide some new, or at least less frequently considered aspects and consequences of stress. Therefore, the first part of this chapter consists of an explanation of what is meant by a team-level perspective in the aviation system. This explanation will include (a) a conceptual framework for distinguishing a team perspective from a traditional individual-based perspective, (b) a general discussion of sources and types of stressors, and (c) a brief discussion of stress management with respect to both prevention and intervention strategies.

The main body of the chapter focuses on three types of stress: stress originating among individuals of a team; stress generated by the relationship of a team to the larger organization and to other teams in the aviation system; and stress generated by the relationship between a team and aspects of the natural and operational environment, as well as technological features of the workplace (e.g., equipment and instruments used in performing one's job). The effects of each type of stressor on aircrew performance is discussed, and research that addresses these issues are described when possible. The assumptions made from the outset are the following:

- Team performance is qualitatively different from individual performance; it is not a simple sum of competencies, skills and attitudes but additionally reflects the way in which the individuals interact with each other as a whole.
- Team performance can be influenced by multiple stressors that do not occur in a vacuum. Rather, these factors may be highly conditional on other factors and the context (both social and environmental) in which they occur. Stressors may occur in combinations that are unanticipated; they may trigger sequences of stressful events, or they may be relatively independent.

Effective interventions must be tailored to such conditions, or be flexible enough to adapt to different situations.

- The effects of stressors on team performance may be dynamically altered by the team's reactions. Depending on the way a team confronts potential stressors and problems, a relatively benign situation may escalate into a big problem and, conversely, a highly stressful situation may be defused quickly.

The final section of the chapter offers a "lesson learned" summary of research in this area, and their relationship to stress management interventions.

A TEAM-LEVEL PERSPECTIVE

Individual Versus Social Stressors

One interpretation of stress and aircrew performance may be described as follows:

> Captain A works for an airline that is struggling financially, and he is worried about his job. For the last two weeks, he has been fatigued because weather is making his commute to the airport longer than usual. His teenage son wants to drop out of school, and his wife's parents are ailing. His plight is no worse than millions of others, but the stress is wearing him down, and may in time, impair his performance. As a part of an aircrew, he could become the weak link in the team that contributes to overall impaired performance; a sort of trickle-up theory.

This story focuses on individual-level stress; stress that comes from an individual and whose effects are assumed to be largely confined to his own individual performance. However, many other stressors are co-present in this scenario; they may come from the other members of the aircrew, other teams in the organization, aspects of the work environment, and flight conditions. How these stresses combine and compound in a complex operation consisting of many teams and many individuals is what makes "stress" a concept with many faces. What I would like to suggest is that although the story is not incorrect, it is incomplete, and only one of the perspectives making up the scenario. In a sense, all individuals bring their own level of "stress" with them to the workplace—sometimes the level is high and sometimes the level is well within control. When individuals come to the workplace, the job itself contains stressful elements; some affect all team members and some affect only the individuals that must deal with them directly. Environmental conditions also affect individuals differently. Bad weather, for example, has different implications for pilots, controllers, dispatchers, and other ground personnel although there are some implications that are shared and recognized by all.

All individuals working in the system must deal with stressful conditions to some extent, and they are selected and trained to be able to handle "normal" amounts. Training, procedures, and system redundancies are provided for "abnormal" conditions that are known to occur. In short, the system prepares and designs its operations to handle stresses that are known and detectable (e.g., weather, equipment malfunctions, runway closures, etc.). It provides prevention and intervention strategies that are meant to help the operators control the situation. For example, wind-shear training is an intervention intended to help pilots deal with environmental conditions known to be difficult. Another type of intervention would be counseling and treatment programs that help individuals with personal or medical problems.

However, it is difficult to prepare and design for configurations of stressful conditions that occur infrequently or are less predictable in their known effects. Similar to the way redundancy is built into the aircraft system in order to avoid total equipment failure by providing backups, a team itself can sometimes compensate and redistribute its energy, attention, and resources in order to overcome difficult conditions. The following excerpt is Ginnett's (1993) description of the United Flight 232 accident (NTSB, 1990) as an example of exemplary team behavior and leadership in the face of insurmountable odds. It is a dramatic illustration of how one team extended and redistributed its resources in order to compensate for extreme stressors on individual crew members:

> Captain Al Haynes and the crew of United 232 enroute from Denver to Chicago suddenly found themselves in a situation that was never supposed to happen. After a catastrophic failure of the DC-10's number 2 engine fan disabled all three hydraulic systems, this crew was left with little or no flight controls. Captain Haynes enlisted the assistance of another captain traveling in the passenger cabin and, with his newly expanded crew, literally developed their own emergency procedures on line. In the midst of crisis the crew of United 232 managed to get the crippled airliner within a few feet of the Sioux City airport before impact. (Ginnett, 1993, p. 84)

To return to an example from everyday life, Captain A starts the workday feeling fatigued and a little upset by his family problems. The first officer is aware of the problem and is extra attentive in monitoring because the captain appears to be a little preoccupied. Performance impairment is avoided because the captain's team member has made a small accommodation that maintains smooth and safe operations. By adapting their behaviors in small ways, team members are themselves interventions for handling many kinds of stresses.

A full mission simulation study conducted by Foushee, Lauber, Baetge, & Acomb (1986) provided evidence that crews that had flown together recently (as opposed to crews that had just met) performed better in spite of more stressful conditions. The crews that had flown together were in the high-fatigue condition because they flew the simulation post-duty, whereas the crews that had just come together were in the low-fatigue condition having flown the simulation pre-duty. Kanki and Palmer (1993) discussed these results as follows:

> crews that flew the simulation immediately after completing a trip together (post-duty condition) performed better than crews that had the benefit of rest before the simulation but had not flown together (pre-duty condition). . . . It has been suggested (Foushee et al., 1986; Kanki & Foushee, 1989) that the time spent flying together before the simulation increases the ability of crewmembers to anticipate each other's actions and interpret the style and content of their communication. Post-duty crews therefore could adopt a more informed or "familiar" style in which first officers might be more willing to initiate directives or question captains' decisions. Furthermore, while there may have been a stronger flow of "bottom-up" communication (i.e., from first officer to captain), the authority structure was not impaired (captains still issued more commands than first officers), and overall, recent experience enhanced performance. (p. 120)

The results suggest that crews may be able to adapt to each other (coordinate more smoothly) after some working experience together and that their joint efforts are synergistic. Obviously, some conditions such as mechanical failures may be irreversible in flight, and the only completely effective intervention is prevention. However, when teams are required to solve in-flight emergencies, every air crew member becomes the potential key to the solution and intervention for managing stress. The opposite to team synergy can also occur; that is, team members may be stressors themselves, and exacerbate problems. For instance, consider that Captain A's first officer misinterprets the captain's fatigue and preoccupation as arrogance or hostility. Reacting thus, the copilot would not only fail to intervene for the captain but could escalate the stress level in the cockpit, and increase the potential threat to overall performance.

One might be tempted to think that complications such as these occur only when humans are involved. However, complex systems also lend themselves to the situation where a seemingly isolated problem cascades into a waterfall of effects. Although increasingly linked computer systems are helping to resolve some problems such as passenger rescheduling, bad weather in one part of the United States can still cause tremendous backups in all parts of the country because of the way flights are connected. Furthermore, even when primary problems are solved, secondary ones may persist. For instance, the blizzard that struck the Northeast in March, 1993 created the following residual problems as reported in *Aviation Week & Space Technology:*

> Airlines are tallying the cost of a winter storm that grounded most of their operations in the eastern U.S. on a major travel weekend and left key airports struggling to resume normal operations for days afterward. . . . Losses likely will run into the millions of dollars when such factors as the crews and aircraft left out of their scheduled positions and the special handling required for thousands of stranded passengers are considered, airline officials said.

Judging by the forecasts, airline and FAA air traffic control officials expected the storm to be bad. Late in the day on Mar. 12, the FAA's Central Flow Control Facility in Washington began a series of telephone conferences with airline planners.

[*By 5:30 p.m., March 13, numerous airports were closed, including Baltimore Washington International, Kennedy International, Newark International, LaGuardia, Philadelphia International, Washington National, Baltimore, Dulles International, Atlantic City International, Logan International, Rochester, and Pittsburgh.*]

While coordination of airline and air traffic operations went smoothly, removal of snow from airports did not. Many facilities saw the temperature rise and precipitation change to rain after getting substantial accumulation of snow, only to have the temperatures drop dramatically again, creating a thick crust of ice over the snow. As a result, crews could not plow the precipitation for more than a few minutes before the ice blocks had to be carted off from in front of the plows. Plow crews had to crack the ice crust, which often set off shear-overload devices, disabling the plows, or wait for chemical deicers or the sun to soften the mixture.

On Mar. 15, operations at Logan were limited to Runway 33 while Massachusetts Port Authority crews continued to work on clearing Runway 9/27.

By Mar. 16, two days after the storm had passed through the New York area, Newark crews were still struggling to clear Runway 4L and the facility's taxiways of snow and ice.

Officials at Philadelphia had Runway 9R/27L back in use. But they were struggling to restore Runway 17/35 so controllers could separate commuter and regional flights from the jet transport traffic stream using 9R/27L.

(McKenna, 1993, March 22, p. 39)

It is not for lack of thought, planning, or good intentions that problems compound with unanticipated consequences. The aviation industry has always appreciated the confluence of factors that can build into problem situations. For instance, numerous accident investigations by the National Transportation Safety Board in the United States provide prime examples of how multiple factors and combinations of events (at times, inconsequential in themselves) are carefully reconstructed in order to better understand how the flow of events reached its conclusion. During the field phase of an investigation, groups focus on numerous factors; operations, survival factors, air traffic control, weather, etc. When human performance investigators are called into an investigation, there are six distinct interest areas that closely parallel the types of factors we have been considering potential sources of stress. Kayten (1993) described these categories of interests as behavioral, medical, operational, task-related, equipment design-related, and environmental. Although a "probable cause statement" is made at the conclusion of an investigation by the NTSB staff, this statement is accompanied and supported by reports, conclusions, findings, and recommendations that reveal a deep understanding of how many factors converged in a particular way. Fortunately, accidents occur infrequently, but is it difficult, unfortunately, to form scientific generalizations from case studies.

An alternative to considering stress factors (and their interventions) as isolated factors, is to investigate these topics within a model that assumes a systems perspective. McGrath (1984) provided a simple model of group process that accommodates multiple input factors, complex interactions, group processes, and multiple outcomes. In addition, it locates the system of factors and processes in real time, with a feedback loop from processes and outcomes that dynamically change the inputs into the system.

Conceptual Model

Simply stated, the McGrath model is a three-part conceptual framework linking input variables to output variables by means of process variables. Adapting it only slightly, input variables may be stressors that potentially affect the output (i.e., team performance outcomes). This is not a direct causal model of input to output, but one that relies on group processes to mediate the effects of input factors on team performance (see Helmreich & Foushee, 1993, pp. 8–23).

If this were a traditional, individual-based model, input variables might include a variety of potential stressors at the individual, team, organizational, and environmental levels. However, distinguishing it from a team-level perspective, the output variables would focus on individual performance outcomes. Individual processes such as perception, attention, cognition, would constitute the mediating process variables. (see Fig. 4.1). From a team perspective, the input variables can be the same—but team performance outcomes reflect more than a simple sum or average of individual performance measures. Furthermore, group processes are categorically different from individual processes and include social activities as communication, information transfer, management processes, team problem solving and decision making.

Input Variables. For purposes of this chapter, input variables will be considered in three main categories: (a) the team, which is made up of individuals; (b) the organization, which is made up of teams, policies and support; and (c) the environment, which is made up of elements of the external environment, such as weather conditions, and elements of the operational environment, such as aircraft type, cockpit displays, etc. Task elements, such as standard operating procedures, checklists, etc. are closely linked with aspects of the operational environment because they define the way in which the equipment and instruments are to be used. Task elements such as rules about who does what are linked to organizational variables because they represent company policies and culture. Individual stressors such as fatigue, health, and so on are important factors but from a team-level perspective, they will be considered as random variation.

As shown in Fig. 4.2, a team-level perspective focuses on (a) stress generated among individuals of a team (e.g., within T1, the relationship of I1 with I2 and I3, etc). (b) Stress may also be generated by the relationship of a team to other

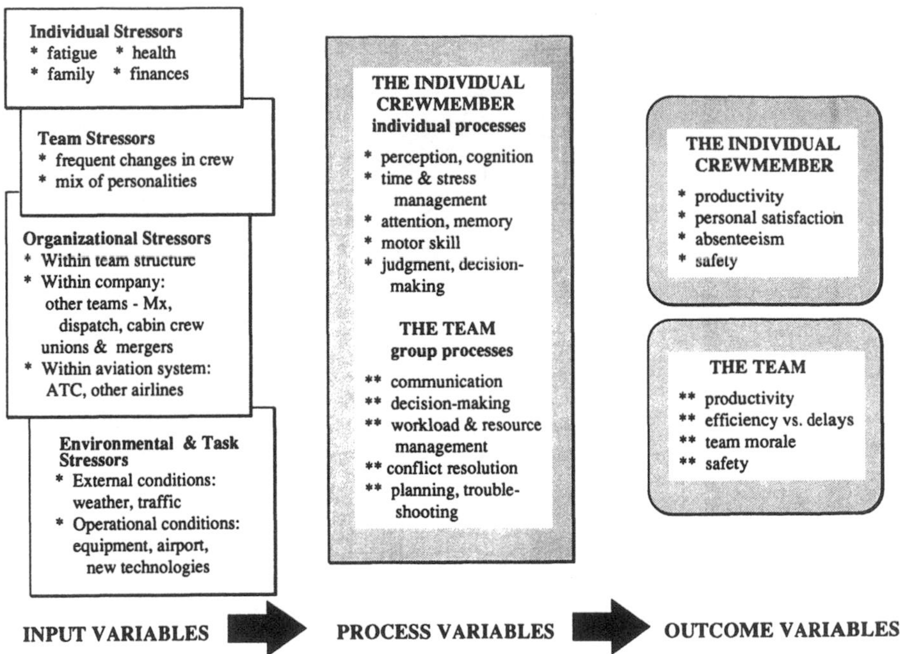

FIG. 4.1. Conceptual model for identifying multiple types of stressors, depicting individual and team-level perspectives.

teams in the parent organization and to other organizations in the aviation system (T1 with T2, T1 with T4, etc). Finally, (c) stress may be generated by the relationship between a team and aspects of the environment; external conditions as well as operational features of the workplace (T1 with E1, E2, E3, etc.)

Figure 4.2 represents a frozen moment in time, and we must keep in mind the dynamic nature of the model where behaviors and outcomes produced during one time frame will influence conditions at a future moment. At the same time, environmental factors may also change simply because the actions taken move them into another sequence of the task and a different physical location (e.g., a different flight phase and weather front). For example, if Captain A failed to hear an ATC clearance, the recognition of this outcome may feed back in the following ways: (a) it may alert the first officer (I2) to assist in monitoring the captain (I1) more closely, thus changing the dynamics between team members. (b) It may alert the air traffic controllers (T4) to coordinate more closely with this flightcrew (T1), thus changing aspects of the team-to-team relationship. (c) It may generate a set of flight conditions (E2) that necessitates a change in flight plans. At the same time, the weather may deteriorate and require a reevaluation of crew actions and review of available options.

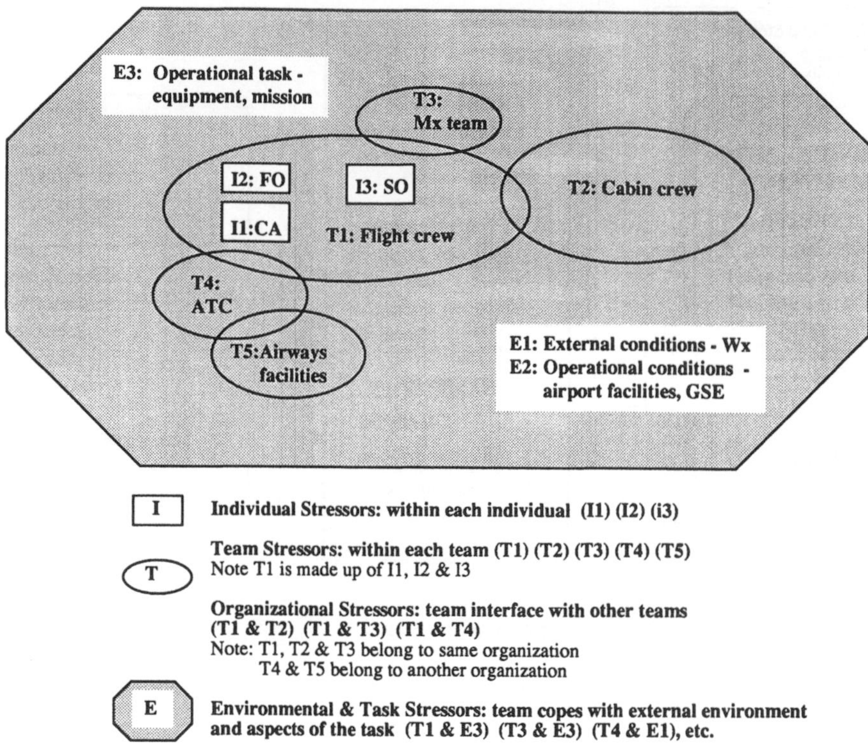

FIG. 4.2. Team-level perspective of individual, team, organizational, and environmental stressors that converge at a single moment in time.

Process Variables. Process variables represent mediators between inputs and outcomes. They refer to the dynamics of the team performance itself, and the means by which crews achieve specific performance outcomes. In contrast, individual processes are the mediators between inputs (e.g., sensory, perceptual, cognitive) and individual performance. Although conceptually distinct, it should be mentioned that team processes in no way substitutes for individual processes. Rather, we assume that the output from individuals processes (perceptions, judgments, decisions, attention, etc.) constitutes much of what individuals contribute to the team process. Obviously the individual uses his or her knowledge, skills, and attitudes to carry out the prescribed task and role in the team. However, at the team level, we must also be concerned with how individual resources and performance are integrated into the total mission. The ways in which team members coordinate and manage these separate contributions are what describe team process; whether formally through procedures, SOPs, and command authority, or informally through active crew member participation, good communication, and management practices.

Outcome Variables. Outcome variables refer to aspects of the group performance, such as productivity and efficiency, but they also refer to regulatory and organizational factors such as whether the crew performed safely and according to flight standards. Additionally, group performance outcomes include how the team felt about their performance, such as whether they felt satisfied, frustrated, compromised, and so on. Certainly the most salient performance outcome is the relative success or failure of a group in achieving formal objectives of their job (in accordance with regulations and company goals). However, from the standpoint of the pilot who is checked and evaluated on an individual basis, the most salient outcome might be his or her proficiency as defined by flight standards. When we consider the viewpoints of the organization, the customers, and other support personnel in the system, performance outcomes may take on different and sometimes conflicting priorities. For instance, we may decide that on-time departure is one good indicator of successful performance. Pilots, maintenance, dispatch, gate agents, flight attendants, ATC, passengers, and the business office of the airline may all agree with this. However, how successful is on-time departure if maintenance feels time pressured and conflicted about causing a delay? How successful is on-time departure if dispatch feels pressured about conflicting weather information? Finally, how successful is it when many passengers from a slightly late flight miss their connecting flight because of a few minutes? Judgements are made everyday and because players in the system have different jobs and goals, success for one may not always imply success for another.

The point is simply that team performance definitions depend on one's particular team perspective. Specifically, they are tied to the particular team's goals. Individual goals are tied to individual performance outcomes, team goals to team outcomes, and so on. Complications arise when we are interested in more than one perspective at a time; for instance, both individual goals and team goals. There are times when an individual performances may be just average but due to good teamwork, the overall performance is excellent. Conversely, we know from accident and incident reports as well as documented research (e.g., Ruffell Smith, 1979) that individual, technical skills do not in themselves guarantee effective team performance.

The following sections deal in turn with (a) team stressors, (b) organizational stressors and (c) environmental stressors. Each section presents research that has addressed these problem areas, and the conclusions and intervention recommendations made on the basis of these studies.

TEAM STRESSORS: THE RELATIONSHIP AMONG TEAM MEMBERS

> It can be inspiring to watch a superb team in action. . . . The first officer, who is flying the leg, calls for retraction of the flaps even as the captain's hand is moving toward the lever. The coordination is smooth and seamless.

> Watching other crews can make you wish you were somewhere else. The captain's hand moves toward the lever and then stops halfway, waiting, while the first officer pointedly keeps his eyes outside and his mouth shut. After some seconds of awkward silence, the first officer announces, "When I'm ready for the flaps, I'll call for them." And things go downhill from there.
> (Hackman, 1993, p. 48)

Although team performance can be impacted by the individuals making up the team and the stresses they bring with them, a different type of stress is attributable to the team as a whole; that is, the particular configuration of individuals together. Because many teams are organized in support of a central figure (leader, captain, manager, principal investigator, etc.), the way a team works together is greatly influenced by the way in which that central person relates to his or her group. In spite of individual competence and skill levels of each member, the way in which a leader coordinates the team as a whole and the way in which the team members follow the lead can facilitate or inhibit smooth, effective teamwork.

Leader Personality Study

A full-mission simulation research study (Chidester, Kanki, Foushee, Dickinson, & Bowles, 1990) was conducted to assess the potential for captain selection along dimensions of personality. Using a selection algorithm described by Chidester (1987), captains were classified as fitting one of three profiles using a battery of personality assessment scales, and the performances of 23 crews led by captains fitting each profile were contrasted over a 2-day simulated trip.

Profile Classification. The personality profiles were based on several dimensions felt to be particularly relevant to aircrew performance. The first was a motivational component or "instrumentality", which was operationally defined as a person's goal orientation, independence, and overall "achievement striving" with respect to task performance. The second dimension making up the profiles was related to communication and interpersonal exchange, or "expressivity," including attributes of interpersonal warmth and sensitivity. In theory, positive expressivity should facilitate communication among team members. The third dimension making up the profiles was related to negative expressivity (frequent complaining, etc.) and negative instrumental traits (verbal aggressiveness, impatience, irritability, etc.), which were expected to inhibit free and open communication among crew members. It was hypothesized that effective leaders would be more often characterized by relatively high levels of both positive instrumentality and positive expressivity (IE+; high levels of both concern for people and concern for performance), and that this type of leadership style would facilitate crew performance. In contrast, higher levels of negative expressive (Ec−) and negative instrumental traits (I−) were expected to lead less effective crew communication, coordination, and performance. Earlier research conducted by Chidester (1987) also found that pilots responded differentially to training in crew coordination as a function of these profiles. Specifically, IE+ pilots benefited the most from training as assessed by cockpit management attitude change.

In summary, three types of crews were engaged in this study:

1. Crews led by captains who were characterized by high levels of instrumentality, expressivity, and achievement striving (designated the Positive Instrumental-Expressive or "IE+" cluster).
2. Crews led by captains who were characterized by high levels of negative expressivity and low levels of instrumentality and achievement striving (designated the Negative Expressive or "Ec−" cluster). This cluster is characterized by traits associated with tendencies to express oneself in a negative fashion (e.g., complaining) and lower than average goal orientation.
3. Crews led by captains who were characterized by higher than average levels of verbal aggressiveness, negative instrumentality, and competitiveness (designated Negative Instrumental or "I−" cluster). This cluster comprises a more "authoritarian" orientation and may well be associated with elements of a profile popularly known as "the right stuff."

Experimental Procedure. Twenty-three three-person crews completed the 2-day full-mission simulation of airline operations in the NASA–Ames Crew–Vehicle Systems Research Facility (CVSRF) B-727 simulator. All crew members were employed by the same major U.S. air carrier, and were currently qualified in the B-727 crew position they occupied in the simulation (i.e., captain, first officer, or second officer).

Crews flew five flight segments under varying conditions of workload in a scenario designed to conform as closely as possible to real operations, including normal flight documentation, cockpit preparations, and communication with ground support personnel normally available to them (dispatch, ground crew, air traffic control, ATIS, and maintenance). Segments 1, 2, and 4 contained routine levels of workload, but segments 3 and 5 involved abnormal conditions that involved complex problem diagnosis, assessment of options, and coordination of procedures required to land an aircraft with mechanical problems. Crew performance data were collected from three sources, expert observations, video coding of crew errors, and computer recording of aircraft handling parameters. An expert observer was present in the simulator cab with every flight crew and evaluated crew performance following every flight segment and individual performance during specific phases of the high-workload segments.

Error analyses were derived from two independent sources of data to assure the reliability of performance assessment. The first source was the expert observer who kept an outline record of all errors he observed. The second source was derived from the videotape records by two observers who reviewed each flight for operational errors. When an error was recognized by one or both observers, the alleged error was reviewed until consensus was reached. The expert observer then reviewed these errors, and held the option of eliminating an error on the basis of his online notes. This was a very conservative error tabulation process and assured that every error data point was reviewed at least twice.

Because some performance errors were more operationally significant than others, errors were categorized according to level of severity. This process was accomplished by the expert observer and both of the observers involved in the videotape error analysis. A three-level classification was utilized. Type 1 errors were defined as minor, with a low probability of serious flight safety consequences. Type 2 errors were defined as of moderate severity, with a stronger potential for flight safety consequences. Type 3 errors were defined as major, operationally significant errors having a direct negative impact on flight safety.

Results and Discussion. Analyses of the observer's ratings of crew performance during the full-mission segments revealed a significant difference among the crews led by different leader personality types ($F(8,80) = 2.80$, $p<.01$). Specifically, crews led by IE+ captains were rated as consistently effective, and these ratings were higher than the other crew types for the segments overall. Crews led by Ec− captains were rated as consistently less effective over all segments than those led by IE+ captains. Crews led by I− captains received ratings that varied across segments. On day 1, crews led by I− captains were similar to Ec− crews; they were rated as less effective than IE+ crews. However, by segment 5 on day 2, I− led crews were rated as performing as well as the IE+ led crews, and significantly more effectively than Ec− led crews.

Consensus was achieved by the expert observer and video observers on identifying a total of 913 errors over all 23 crews, including their severity classifications. Analyses of Type 1 errors revealed no significant main effects or interactions. However, analyses of Type 2 and Type 3 errors revealed differences in performance across the crew types; namely, crews led by Ec− captains tended to make more errors than IE+ or I− led crews ($F(2,19) = 4.53$, $p<.05$).

Although the error analyses were consistent with the online expert ratings in differentiating the performance of Ec− led crews from IE+ or I− led crews, they did not reflect the change in I− led crew performance over the 2 days of the simulation. One possible explanation for this discrepancy is that the process of rating crew performance may take more than errors into account. For instance, because the observer has access not only to errors, but also to the crew communication process and other aspects of crew coordination, these behaviors may be relevant to ratings but not necessarily directed linked to the production of errors.

Crews led by captains fitting the IE+ profile (high achievement motivation and interpersonal skill) were consistently effective and made fewer errors. Crews led by captains fitting a the Ec− profile (low achievement motivation, negative expressive style) were consistently less effective and made more errors. Crews led by captains fitting the I− profile (high levels of competitiveness, verbal aggressiveness, impatience, and irritability) were rated less effective on the first day but equal to the best on the second day. However, in terms of errors, crews led by I− captains made fewer errors than crews led by Ec− captains.

These results validate the impact of leader personality on crew performance. Analyses of the crew effectiveness ratings by an expert observer and of the crew error frequencies revealed three patterns of crew performance as a function of leader personality. As expected, performance was consistently effective for crews led by IE+ captains. Also as expected, less effective performance was shown by crews led by Ec− captains. Unexpectedly, I− led crews performed comparably with IE+ crews in general and, on the basis of ratings, appeared to improve over time. As mentioned earlier, one possible explanation is based on methodological differences; that is, ratings and error codings are not based on exactly the same behavioral criteria.

However, an alternative possibility is that input factors such as leader personality may show a direct effect on performance, but only an indirect (i.e., not causal) link to output variables. In other words, leader personality may affect group processes that may or may not culminate in differences in team performance. Therefore, these two sets of results may be noncontradictory after all. The Ec− and I− captains may both have a negative impact on crew processes (i.e., the smoothness with which the crew coordinates its activities). However, the I− led crews may find a means of circumventing an adverse effect of the captain's personality on overall performance. Perhaps the first officer compensates by adapting to the captain's style, perhaps the captain himself compensates by taking on more of the workload by himself.

Summary. The results of a highly realistic full-mission simulation highlighted the importance of the personality characteristics of pilots during routine and high workload flight segments. The personality profiles of leaders predicted how well crews dealt with significant operational events. IE+ led crews were consistently more effective than Ec− led crews, but I− led crews exhibited a familiarity effect; that is, an improvement in performance over time. Although there are unanswered questions with respect to interpreting the I− results, it seems clear that the impact of personality on leadership behavior is significant, and both selection and training approaches to optimizing crew effectiveness in aerospace operations could be better developed.

ORGANIZATIONAL STRESSORS: THE TEAM, THE ORGANIZATION, AND OTHER TEAMS IN THE AVIATION SYSTEM

Recent *Aviation Week & Space Technology* stories:

"Massive Airline Losses Force Draconian Cuts"—With no upturn in sight, airlines are cutting payrolls, schedules following losses of at least $1.8 billion. (Ott, 1993, p. 30)

"Delta, Pilots in Standoff"—Tension between Delta Air Lines management and the carrier's 9,400-plus pilots is flaring again, fueled by . . . pressure for flight deck crews to accept voluntary pay cuts by Feb. 1. (Editor, 1993, p. 54)

"Cuts, Layoffs Affirm Transport Boom's End"—Decade-long cycle repeats itself as oversupply and poor airline earnings cut back production of commercial transports to level of mid-1980s (Dornheim, 1993, p. 22)

"Northwest, Union Leaders Haggle Over Equity Stake"—Northwest Airlines executives and the leaders of six labor groups are wrestling over how much control union employees will gain over the carrier in exchange for $900 million in concessions that all parties agree are key to the airline's survival (McKenna, 1993, February 1, p. 30)

"Airlines' Choice: Adapt or Perish"—Carriers must streamline their own high-cost structures while fending off the challenge of low-cost entrants. (Phillips, 1994, p. 61)

In this section, we begin by assuming that stressors at levels lower than the team level are under control; that is, the individuals and team-as-a-whole are in good operating condition. Consider that the team is your car; it has been well maintained; it is not rundown, and it has not encountered any abnormal stresses recently (e.g., flat tire, collision, rough roads). Although it may not be perfect, you are confident that you can rely on it for your normal transportation needs. However, your car may fail you for other reasons. There are other drivers on the road causing traffic tieups and accidents; there are natural hazards like storms and floods; there are man-made obstructions like road closures; there are even law enforcers and bill collectors that restrict your car's performance and mobility. In short, you do not operate your car in a social or environmental vacuum. Similarly, the well-prepared, competent, and "unstressed" aircrew does not perform in social and environmental isolation in the aviation system.

The performance of even your best-qualified aircrews can be inhibited or undermined if management and other teams within the company adhere to contradictory policies, hold nonsupportive viewpoints or simply have conflicting goals. Following the framework established by Hackman (1985) and Ginnett (1987), there are two levels of organization factors within a company that promote or hinder effective crew coordination: the way in which the team itself is organized and carries out its joint task, and the way in which the team is supported by management. A third organizational factor is the way the team coordinates with other teams within the aviation system.

The remainder of this section focuses on how organizational factors impact the "health" of a crew and contribute to their flight readiness. Just as fatigue, personal stresses, and substance abuse negatively impact an individual's fitness for duty, organizational factors can affect "crew" fitness. Three organizational issues include: (a) the way the team is structured, (b) the way in which the organizational system supports the crew, and (c) the way in which the crew interfaces with other teams in the aviation system. Research addressing each issue is described later.

Team Structure

Above and beyond the simple summation of knowledge, skills, attitudes, personalities, and experiences of all individual team members, teams have a structure and design, roles and responsibilities that exist quite apart from the individuals that make them up. What I would like to define as organizational stress at this level is a lack of fit between individuals and roles; not simply the case when individuals do not fit their roles adequately, but the situations when there is a lack of role clarity (i.e., the clear delineation of individual task and roles and how individuals interface with each other). On the flightdeck, one might think that role clarity is not an issue because many aspects of the task and roles on the flightdeck are predefined, standardized, and regulated long before any crew member steps into the cockpit. Furthermore, crew members must achieve a standard level of proficiency corresponding to his or her crew position. Thus, crew performance problems rarely surface during "normal" conditions. However, when there is a confluence of "stressful" factors, an unusual sequence or combination of malfunctions, difficult airport or weather conditions, ambiguous or conflicting instrument readings, etc., situations may be created that require a novel and dynamic solution. In such cases, team building and good leadership and followership skills may make the critical difference between staying ahead of the aircraft and scrambling to keep up.

Team Formation Study. As a part of a large-scale cross-organizational field investigation conducted by Hackman (1985, 1987) and Ginnett (1987), an observational study of leadership and the team formation process of aircrews within one organization was undertaken. The research goal was to identify behavioral variations in the team formation process that corresponds with variations in leadership effectiveness; in short, to be able to behaviorally define the differences between effective leaders and less effective leaders. Observations made during actual operations constituted the data gathering mechanism for exploring these differences.

Ten (B-727) crews were chosen, 5 of which were independently determined to have highly effective leaders and 5 having less effective leaders. The field observer was blind to these conditions. Crews were selected on the basis of its leader, although each crew consisted of its captain, first officer (copilot), second officer (flight engineer), and cabin crew members.

Each crew was observed for approximately 30 hours during actual flight operations. This usually broke down into two full trips or about 10 hours of inflight time. Phases of team observation began from the first time the crew met (team creation), continued during the task execution itself (inflight time) and team termination until the team finally split up. From these observations, Ginnett was able to clearly differentiate leadership effectiveness, and using an inductive method, he integrated these observations and constructed "behavioral profiles" corresponding to each type.

The results showed that effective leaders explicitly affirmed or elaborated upon the rules, norms, and task boundaries that constituted the "normative" model (or "shell") of the organizational task environment. Specifically, they briefed both flightdeck and cabin crews about interface tasks, physical and task boundaries, and other norms for performing their task (regarding safety, communication, and cooperation). They established clear authority dynamics, as well as their own technical, social and managerial competence. Each effective leader covered the aforementioned areas in the process of team creation prior to flight (e.g., crew briefings), and behaved consistently with this model during task execution.

In contrast, ineffective leaders were not similar to the effective leaders, nor were they similar to each other. In various ways, they tended to abdicate their leadership responsibilities, and in some cases, actively undermine their crew's normal expectations. In one way or another, these captains tended to leave their crew members guessing. Although it is not clear whether a performance decrement would be detectable under completely routine operations, a confluence of multiple stressors that demand active crew coordination and creative problem solving may require a greater degree of predictability and team preparedness.

The way in which stressors are handled may be highly dependent on how clearly the leader establishes valid expectations for the crew and the kind of atmosphere he or she promotes. Effective leaders promote this kind of predictability by helping to set up expectations for behavior prior to the task AND then behaving in a consistent manner. Less effective leaders tend to abdicate their responsibility for norm setting or actively undermine crew expectations by behaving inconsistently, thus promoting greater unpredictability.

Organizational Support

> Two factors that significantly affect how, and how well, a crew develops are the supportiveness of the organizational context within which the crew works and how well designed the crew itself is (Hackman, 1987). Organizationally, crews need a context that supports and reinforces competent task work during line operations. . . .In addition, crews must be well designed as performing units. Among the design features that appear to be associated with team effectiveness are the group task. . .and group composition. A third class of factors, group norms, should actively support continuous situation scanning and strategy planning. Some of these factors are, for better or worse, built into crews by the very nature of the work itself or the cockpit technology. . . .Even though design and contextual factors, such as those mentioned above, constrain much of the potential variation in

> crew performance, large and significant differences in crew behavior still exist. This is where the leadership provided by the captain (and other crew members) comes into play. (Ginnett, 1990, pp. 443–444)

From an intervention standpoint, there is great payoff in identifying trainable techniques and behaviors for enhancing the team development process. However, these benefits can still be undermined by a system that is nonsupportive or contradictory in its philosophy. Hackman (1985) and Ginnett (1990) discussed at least three types of organizational features that are required for full organizational support. These include (a) a reward system that provides positive reinforcement for excellent team work; (b) an education system that provides the relevant training and consultation resources required by team members; and (c) an information system that provides (in a timely fashion) the data and resources necessary to assess, evaluate and formulate effective crew coordination strategies on the line.

Cross-Organizational Study. On the basis of observations and analyses from a large-scale cross-organizational study, Hackman and his colleagues attempted to directly relate both pilot self-assessment data and field observations of team performance with the broad categories of organizational features mentioned earlier. Seeking to maximize variability across organizations, data collection was completed across three U.S. commercial air carriers, two military transport units and four overseas carriers. Although data revealed differences in what were the organizations' strongest versus weakest attributes, the relationship between attributes and pilots' perceptions of their own teams and team performance were remarkably similar. For example, airline A might be "healthier" than airline B with respect to job security and opportunity for advancement, but the relationship between these variables and pilot attitudes were extremely similar. This similarity held across all organizations sampled including the international carriers. For instance, team performance was perceived to be positively related to the timely availability of information resources by all organizations in spite of the fact that some organizations may be rated "better" on this dimension than others.

Availability of resources (equipment, supplies, personnel, information, etc.) is a major problem in times of economic cutbacks, but organizational stress of this type is not always solved by dollars alone. Hackman (1993) described a further complication:

> Providing crews with timely information and ample material resources is complicated by the pattern of inter-group relationship that develops in some airline organizations. When airlines are structured along strictly functional lines, for example, different groups can easily fall into conflict with one another—especially if resources are scarce. In one airline we studied, it seemed as if the last thing members of one department wanted to do was anything that might make life easier for members of another. (p. 65)

In this scenario, poor organizational design may encourage conflict among teams. Economic stress may exacerbate these conflicts, and if the management is unable or unwilling to solve these conflicts, it may be in danger of violating the first type of organizational support mentioned earlier; that of providing a positive reward system for excellent teamwork.

Other Teams in the Aviation System

System Stressors. A third organizational stressor comes from the interface between aircrew and other teams in the system. Some teams, such as the cabin crew, gate agents, technical operations staff, dispatchers, etc. may be within one's own company; however, the Federal Aviation Administration is also a key participant, as air traffic controllers and airways facilities personnel play critical roles in every flight. One needs only to read a few NTSB accident reports to realize that the interface between teams is not always as clearly defined as it should be. The Avianca crash in January 1990 (NTSB, 1991) is a classic case of complex operations, with multiple teams never successfully communicating with one another. Avianca flight 052 (B-707B) from Bogota to Medellin, Columbia to Kennedy International Airport in New York, ran out of fuel over Long Island and crashed, resulting in 73 fatalities and 85 injuries.

Poor weather conditions led to the flight being held three times by air traffic control for a total of about 1 hour and 17 minutes, and not until the third period of holding did the flightcrew report that (a) the airplane could not remain in holding longer than 5 minutes, (b) it was running out of fuel, and (c) it could not reach its alternate airport, Boston Logan International. Following the execution of a missed approach to JFK, the crew experienced a loss of power to all four engines and crashed approximately 16 miles from the airport. The NTSB attributed probable cause of the accident to the failure of the flightcrew to manage adequately the airplane's fuel load, and their failure to communicate an emergency fuel situation to air traffic control before fuel exhaustion occurred. Contributing to the accident, in the NTSB's view, was the flightcrew's failure to use an airline operational control dispatch system for assistance under these difficult conditions. Traffic flow management by the Federal Aviation Administration (FAA) and the use of standard terminology for pilots and controllers during emergency were also called into question.

Obviously, stress comes into play in a number of ways beginning with the environmental stress of poor weather and all the other problems that compound each other under these circumstances. Traffic accumulates and workload increases for flow control and air traffic controllers in all sites. Pilots have their own particular task stressors such as delayed schedules, anxious and complaining passengers, and fuel considerations when put into holding. With constantly changing conditions, online problem solving and complete vigilance with respect to systems and fuel monitoring as well as weather and airport updates is required.

In short, the organization (one's own company) and the other teams in the aviation system are all critical players in a large, complex cooperative endeavor.

Pilot–Controller Communication Study. Most aviation research confines itself to a single work domain, and the critical relationships among teams in the system, particularly under stressful conditions, have not been studied. One exception, however, is the work conducted by Morrow, Lee, and Rodvold (1991), in which the collaborative nature of pilot–controller communication was studied. Field data was obtained from four of the busiest terminal radar approach control (TRACON) facilities in the United States. Among the results was found that controllers economized workload by composing longer messages including those that require more than one kind of readback, such as readback plus answer request. Associated with this controller strategy, pilots made more procedural deviations such as partial readbacks. Therefore, a decreased workload for controllers appeared to increase the "memory load" for pilots. In addition, procedural deviations were associated with nonroutine transactions (e.g., clarifications, interruptions/repeats, corrections, etc.). Therefore, while on the one hand, a correction of a readback error improved accuracy, it also reduced communication efficiency by lengthening the transaction.

The results point to potential trade-offs inherent in the interactive process between controllers and pilots and the importance of bearing in mind both the speaker and addressee perspective. "ATC communication depends not only on individual skills and capacities (e.g., processing speed and working memory capacity), but on how smoothly controllers and pilots collaborate during routine communication (Morrow et al., 1991, p. 17). An important lesson to learn in this study is that a complete intervention cannot simply address one half of a problem. Just as one crew member may be able to compensate for the weaknesses of another crew member from time to time, the pilot may sometimes compensate for the strategies the controller adopts to ease his or her workload. Likewise, the controller may occasionally compensate for the pilot's extra requests and clarifications by adding to his or her own workload. However, compensation is no substitute for smooth collaboration, and under stressful conditions neither pilot nor controller may have the time or resources to make the accommodations needed to avoid miscommunications.

ENVIRONMENTAL AND TASK STRESSORS: THE TEAM AND ASPECTS OF THE OPERATIONAL ENVIRONMENT

Some typical *Aviation Week & Space Technology* stories:

"NTSB Urges Studies on Mountain Winds"—The NTSB's final report on a 1991 crash of a United Airlines Boeing 737 in Colorado Springs highlights the need for additional knowledge about complex meteorological phenomena downwind of mountain ranges. (Scott, 1993, p. 34)

"Winter Storms Test New Anti-Ice Tactics"—Plans appear successful in assuring pilots that aircraft are free of ice, which has contributed to eight airline accidents since 1982. (McKenna, 1993, January 11, p. 38)

"Parallel Runway Spacing Problem"—Denver Intl Airport, slated to open in October, has already poured concrete for two of three parallel runways even though the FAA has yet to establish national standards for minimum spacing in that configuration. The fast pace of the new airport development forced designers to make their own assumptions and hope they will meet the new FAA standards. . . . Further complicating the issue is the altitude of the Denver airport. In the summer the combination of a mile-high airport and hot temperatures raises the density altitude, which in turn decreases aircraft performance, requiring aircraft to fly faster approaches. . . . As a results of the higher speed, the controllers and pilots would have less reaction time if an aircraft deviated from its final approach course. The aircraft is less maneuverable when operating in higher density altitudes. To have a wider safety margin, more runway separation might be needed for high-altitude triples. (Nordwell, 1993, January 25, p. 56)

"Software Problems Delay ATC Redesign 14 Months"—The FAA's efforts to modernize its air traffic control system have slipped again. (Nordwell, 1993, March 8, p. 30)

"Procedural Issues Slow Benefits of Data Links"—Data links promise benefits for air travel comparable to those provided by radar, but human factor issues involving controller procedures are delaying implementation and must be resolved. (Nordwell, 1993, March 8, p. 32)

In keeping with the assumptions described earlier in this chapter, environmental stressors may be of various types: (a) those associated with flight conditions like weather, traffic, airport changes that are reported in NOTAMs;[1] (b) those associated with the task such as excessive workload (or its opposite, monotony and boredom), time and schedule pressures, lack of resources, or information to complete tasks; and (c) those associated with aspects of the aircraft, inoperative parts, particular aspects of cockpit displays and warning systems, etc. It should be clear that these conditions and tasks are not necessarily problematic to crew members. Nevertheless, they can become sources of stress when they exceed some normal level or occur in combinations that either exceed the normal or generate unexpected conditions.

One area notable for the strain it introduces into a system is the transition to new technology. In the aviation industry, new technology nearly always implies

[1]NOTAMs (Notices to Airmen) are notices containing information concerning conditions, changes, or hazards in national airspace system essential for safe flight operations. This information is not sufficiently in advance to publish by other means.

increased levels of automation, and decreased pilot involvement or control (see Billings, 1991). Although intended to reduce workload and human error, the reviews have been mixed. Many benefits are indisputable; however, other unexpected consequences have been less well received. For example, the results of a field study conducted by Wiener (1989) and summarized in the followup simulation study (Wiener et al., 1991) is the following:

1. Workload is changed, not reduced by the new equipment, and may simply be relocated in time, sometimes to the benefit of safety, sometimes not.
2. Human errors are not eliminated, but their nature may be changed. In many cases, the errors may be more critical; that is, automation may eliminate small errors and create the opportunity for large ones.
3. There are wide differences of opinion about the usefulness and benefits versus risks of automation in the minds of line pilots, and therefore wide differences in patterns of utilization. (p. 4)

Taken together, these points indicate that increased automation may generate unintended and perhaps "stressful" consequences. Furthermore, there does not seem to be a clear understanding of how best to respond to these technology changes. The third point reflects the pilot's attempt to cope. Undoubtedly, a repertoire of practical adaptive strategies are being developed by resourceful pilots everyday. However, until an understanding of the processes underlying these strategies is incorporated into training and standards, wide differences in usage will persist. Whether such differences in practices actually result in systematic differences in team performance, however, is a separate issue, and is the subject of the full-mission simulation described next.

Automation Simulation Study

Level of automation was varied by contrasting crew performance in two aircraft of the same family; namely, the DC-9–30 (low-level automation) and its derivative, MD-88 (high-level automation). A LOFT[2] scenario was designed that satisfied several conditions: (a) to reflect a high degree of operational realism, (b) to favor neither aircraft and (c) to introduce flight conditions that would require crew coordination, communication, and team decision making:

> It was designed to include periods of high workload, generated by system failures, deteriorating weather at the destination, and complex ATC clearances. The flight

[2]LOFT (Line Oriented Flight Training) is the use of full-mission or line-operational simulation in pilot training in order to augment the teaching of technical skills (e.g., systems knowledge, aircraft handling skill), with crew-coordination skills such as resource management, decision making, and communication. Such training involves the complete crew and incorporates both normal and abnormal procedures.

> was based on a regularly scheduled flight for DC-9s and MD-88s from Atlanta to Columbia. Due to low ceilings, A Category II approach was required at Columbia, and this was acceptable for both DC-9 and MD-88 aircraft at this company. (Wiener et al., 1991, p. 28)

All crew members participating in this simulation came from the same company and flew in the aircraft and position they normally held in actual operations. The simulator facilities of their own company were used for test flights. Data collection included three main types of performance measurement: (a) online ratings of performance by observers present in the simulator, (b) self-assessment of subjective workload made by the crew members themselves, and (c) analysis of crew errors coded from videotapes by two expert observers. Performances of 22 crews (12 DC-9 crews and 10 MD-88 crews) were analyzed.

Online ratings of crew performance were made by a company LOFT instructor using a CRM Evaluation sheet. However, because there was no way to insure reliability of ratings across the several instructors who participated, these measures could not be combined with the NASA observer data. On the other hand, the NASA observer was able to make ratings of both overall crew performance and detailed individual performance. These ratings differentiated several CRM topics: crew communication and decision making, interpersonal (management) styles and actions, workload and planning, and crew atmosphere and coordination. In addition, separate ratings were made for four flight phases, two representing normal conditions and two representing abnormal conditions.

Results and Discussion. In spite of all the distinctions mentioned earlier, the only significant difference in ratings across aircraft (level of automation) was for overall crew performance. DC-9 crews performed significantly better ($F(1,20) = 4.75$, $p < .05$), but the differences were slight. A self-assessment of subject workload was obtained by having all participating crew members fill out a questionnaire derived from the NASA Task Load Index (TLX) (Hart & Staveland, 1988) following the flight. Looking first at the composite scores, there was a marginally significant difference between aircraft with the DC-9 crews reporting slightly lower workload than MD-88 crews. Considering captains and first officers separately, this slight difference appeared to be mostly attributable to the first officers. Although all these differences are very small, a closer look at the individual items indicated that differences found came from the items called Physical Demand (first officers only) and Frustration Level (captains and first officers).

The analysis of crew errors did little to illuminate the marginally significant differences in crew performance due to automation level. The methodology employed was similar to that described earlier in the Leader Personality simulation study; namely, two expert observers used a consensus method for arriving at a conservative judgement of errors committed. The errors were categorized according to severity and were correlated (in the expected direction) with the online observer ratings. Although many contrasts were explored, neither frequency nor severity of errors differentiated crew performances in the two aircraft.

These results only highlight the point made earlier that input factors seldom have a simple and direct causal effect on crew performance. This is largely because single factors do not occur in isolation but in combination with many other conditions. Some input factors and aspects of the situation may serve to mitigate some negative effects while others may serve as a catalyst. Furthermore, the group processes mediating the effects of potential stressors on crew performance may hold the explanatory key, because the team itself can dynamically alter the situation that in turn feeds back dynamically into the group process.

Summary. This study investigated the potential negative stressor effect of new technology in general, and the impact of cockpit automation in particular. On the basis of the analyses conducted thus far, it appears that there may be an impact (suggested by the ratings and workload assessments), but the crews are sufficiently adaptive to maintain "no significant effect" on overall crew performance in terms of errors committed. If we are to believe the field study results (i.e., that crews are troubled by increased automation even if they like the new aircraft in time), it is reassuring that crews are also able to compensate for their misgivings. Nevertheless, it is important for the industry to empirically determine how crews successfully adapt to new technology so that these strategies can be trained in a standard fashion and not by chance on the job. As Billings (1991) stated, "Information concerning the effects of automation, and particularly its unwanted effects, does not usually differentiate between 'good' automation, poorly implemented, and 'bad' automation, in terms of roles and functions" (p. 17). In this brief discussion, the differences between implementation issues and design issues should be a prime consideration when interventions are recommended.

INTERVENTIONS

The primary purpose of the conceptual model described earlier was to maintain a systems perspective in designing, conducting, and interpreting research. However, it is also useful in the applications phase of research, as we focus on what interventions can be made on the basis of research findings. Specifically, what recommendations can we make to individuals, teams, airlines, manufacturers, ground support, systems designers, and so on regarding effective management of stressors. Because there are multiple categories of stress, we assume that stress management will probably not fall into one all-purpose formula. More likely, both prevention and intervention strategies will be most effective when tailored to specific types and sources of stress. Returning to one of our initial assumptions, namely:

> Team performance can be influenced by multiple stressors that do not occur in a vacuum. Rather, these factors may be highly conditional on other factors and the context (both social and environmental) in which they occur. Stressors may occur in combinations are unanticipated; they may trigger sequences of stressful events, or they may be relatively independent. *Effective interventions must be tailored to such conditions, or be flexible enough to adapt to different situations.*

In general, interventions fall into three categories: crew selection, crew training, and design recommendations. Although design recommendations typically refer to hardware—namely, the development of new technology or redesign of current instruments and displays—other alternatives may be relevant to a team-level perspective. Recommendations for the more effective design of teams, work schedules, procedures, and information management systems are a few nonhardware fixes that can held reduce the effects of stress on performance.

Interventions for Team Stressors

Considering each of the categories of stress depicted in the model, the stress generated among individuals of a team is probably best addressed by selection and training interventions as well as the design of teams. Crew resource management (CRM) training is an example of one particular training intervention that has been widely implemented in the aviation community. In addition to fairly standard topics such as communication, situation awareness, decision making, leadership, stress management, critique, and interpersonal skills (see Orlady & Foushee, 1987), companies tailor their programs to address the problems and stresses that affect their particular type of operations. Whether flights are domestic or international, night or day, long haul or short haul, passenger or cargo makes a big difference in the types of stressors likely to emerge. In addition, an organization may be undergoing a merger with another organization, recovering from a lengthy labor–management dispute, or transitioning to a new aircraft added to the fleet. CRM programs often develop training modules that address such topical issues.

The research described in the area of team stressors focused on the input variable—leader personality. Although the expected results were only partly obtained, evidence supported the use of selection criteria for leaders as an effective intervention. On the other hand, the data also suggested that the crew as a whole actively contributed to the results. They way in which the ratings of the I–led crews improved over time suggested a familiarity effect. Whether this familiarity effect was driven by the efforts of the captain alone or by the adaptive behavior of the first officers is an empirical question.

Follow-up research of the group processes (e.g., communication, decision making, management strategies) have been conducted (Kanki & Palmer, 1993; Kanki, Palmer, & Veinott, 1991; Orasanu, 1993; Orasanu & Fischer, 1992) in order to identify the trainable behavior patterns associated with high-performing crews. For example, most of the high-performing crews in the leader personality study were led by IE+ and I− Captains, whereas the Ec− led crews did not perform as well. The communication patterns indicate that in the Ec− led crews, the Captain initiated communications less often than his crew members, particularly in categories of questions, and observations. First officers in these same crews asked more questions than other first officers, possibly compensating for the Captain's lack of initiation. In contrast, the high-performing crews generated a more even distribution of speech initiations, with the Captain leading slightly (Kanki et al., 1991). Orasanu and Fischer's (1992) study of the same crews provides some additional information about their decision-making patterns. In their analysis, Captains of the high-performing crews stated more plans or strategies and made more explicit task assignments than Captains of the low-performing crews, and these types of speech were more concentrated in the abnormal phase.

Because group process research focuses on real-time communication and behavioral sequences, it is possible to resolve issues at a fine level of detail. For instance, the studies described earlier could continue their investigations on types of questions asked, response latencies to initiations, patterns of speech sequences and so on. Although exploratory in nature, group-process research is well-suited to identifying specific effective and ineffective behaviors associated with crew performance. When such patterns are translated into specific communication and management skills, they can be useful training interventions.

Interventions for Organizational Stressors

Stress generated by the relationship of a team to the larger organization and aviation system has not been extensively researched, although the stresses in this area are widely recognized and reported (e.g., mergers, pilot–ATC conflict, union vs. nonunion pilots, gender issues, etc.). Problems in this area have sometimes fallen to the chief pilot, and handled on an individual basis, but proactive training interventions are also being developed. For instance, some airlines and military organizations have conducted training in which more than one type of team trains together (e.g., pilots and dispatchers, pilots and cabin crew). However, these innovations are rare, and many other members of the aviation system have yet to be included in such training efforts.

On the research front, we have seen from Ginnett's study of leadership and team formation that the behavioral results identify trainable team-formation skills. These skills can and have been incorporated in many CRM programs. The new directions in research involving the collaborations of teams with other teams (such as the pilot–ATC communications study described earlier) also holds training potential. Results that illuminate the process by which teams accomplish their work in both normal and stressful conditions can have a direct training relevance, but may also point out policy and procedure changes that could facilitate cooperative teamwork. In this sense, design recommendations are more likely to refer to "software" rather than hardware changes. Certainly there are new technologies that are intended to facilitate more effective teamwork across teams (e.g., better communication links), but system-wide organizational support for the effective coordination of their activities is prerequisite.

Interventions for Environmental and Task Stressors

Interventions tailored to alleviate stress generated by the relationship between a team and aspects of the workplace come in many varieties, but traditional design interventions naturally fit in this category (e.g., new technology intended to reduce workload, provide more information, warn operators of dangerous conditions, etc.). What has been typically lacking in this area, however, has been a systems perspective in testing how these interventions would be used during actual operations in relation to other teams in the aviation system. Therefore, the real need in this area is to accompany design interventions with (a) the appropriate task and procedures redesign, (b) training that focuses on how the operators can best use the technology (rather than simply how the technology works), and (c) organizational philosophy and policies that are consistent with the best implementation of the new technology.

The research of Wiener and colleagues described earlier supports this notion. Essentially, the overall results contradicted both real world opinion and the researcher's hypotheses in finding no significant differences in performance across levels of aircraft automation. However, on reflection there emerge many reasons why these findings could arise. First, as Wiener discussed in the results, the DC-9 crews were more experienced in their aircraft than those in the MD-88 (This same imbalance was not true for total flying time), and this could have given them the performance advantage. Experience, as well as many other factors such as familiarity, training, scheduling, etc., are known to systematically affect performance.

However, I have also raised the issue that overall performance measurement may obscure the various group processes that underlie them. In fact, DC-9 and MD-88 crews may be invoking different kinds of strategies for solving the problems, handling stressful conditions, and coordinating their activities. In addition, captains and first officers may find that their communication and monitoring roles may be quite different. In studying the actual group processes that underlie their performances, we hope to discover the behavioral differences that account for the "no difference" in performance.

Summary

In this chapter, I have expanded the meaning of stress and aircrew performance by extending the traditional individual-level interpretation of stress to a team-level perspective. In addition, I have described three broad categories of stressors: team stressors, organizational stressors, and environmental stressors. It has probably become evident that at any given moment in time, any and all of these types of stressors can be copresent, and in some unusual cases, they combine in potentially dangerous ways. The research described has, in some cases, provided supporting evidence of the negative impact of specific stressors; however, in almost all cases, the results have been complicated by the effects of other factors within the situational context or by the nature of the group processes themselves. Although it is overwhelming to think of how so many uncontrollable factors can converge in relatively unpredictable ways, we have a system that provides backups, warnings, increasingly accessible information, and highly reliable equipment. Most impressive of all, however, is the adaptability and creativity shown by team members themselves in complex, stressful operations. Although it is important to eliminate known stressors whenever possible, it is equally important to prepare aircrews to become their own best resource, equipped with team management skills and strategies to deal with a stressful work environment. Throughout team selection, training, and design, these skills must be recognized, taught, and valued.

FUTURE TRENDS

Future Stressors

The prior discussions have described interventions for team, organization and environment, and task stressors. Although they cover a wide range of problems and solutions, they are not intended to be comprehensive; rather, they are simply examples from several stress categories. For instance, the section on team stressors describes how leader personality and management style can affect team performance. There are other potential team stressors, however, and the anticipated change in the workforce is already an issue. It was once highly probable that a new pilot was of the same gender and culture as the group he was joining. In addition, he shared a similar military background as a part of his initial flight training. The new workforce will be considerably more diverse, however, and many new pilots will have received ab initio flight training through an aviation school. As long as standards and proficiency levels are maintained, we do not presume that any of these individual attributes will affect team performance directly. However, the informal, unspoken traditions and understandings that were previously shared by an extremely homogeneous group, may no longer be recognized or accepted by the more diverse workforce. Under normal and standard procedures, such differences are probably irrelevant, but the possibility of misunderstandings arising under nonstandard, pressured situations is a concern.

Organizational stressors will continue to affect the aviation system in numerous ways. Airline mergers and international partnerships that create changes in company policies, philosophies and priorities are common, and serious labor disputes are perennial problems for one company or another. In the reinvention of the federal government, massive changes in the FAA's management of the air traffic control system also may be forthcoming. Even within companies, changes are occurring in which some ground-support services formerly handled by company maintenance or customer service are shifting to outside contractors. If organizational restructuring and shifts in workforce were simple substitutions, it would not be so troubling, but changes in organization often imply drastic changes in ways of doing business, as well as subtle differences in how tasks and procedures are interpreted. Ambiguities in task elements, boundaries, and responsibilities cannot be allowed to remain unresolved.

One of the most pressing environmental and task stressors will be the combined demands for increased efficiency and traffic flow under all-weather conditions, and the introduction of new technology associated with increased productivity goals. Earlier, we discussed how technology changes on the flightdeck have influenced the pilots' tasks. Based on anticipated changes in the air traffic management system, there are likely to be numerous changes in information technology that will affect all communicators in the system. As researchers and developers create these new information systems, users will test the systems and evaluate them on the basis of how well they can accomplish their job. Less obvious will be the indirect impact on other interactants of the system; that is, as each operator accommodates to the new technology, they will change their behavior patterns in some ways. Others in the system will have to adjust to these changes without the benefit of knowing why they are occurring, and how they will vary under different conditions. Because the unwritten, informal aspects of communication may be providing more information than is immediately obvious, the change or absence of such information may have unexpected impact on overall system performance.

Future Interventions

Fortunately, the future holds the promise of innovative intervention strategies to counteract some of problems described throughout this chapter. Over the last decade, airlines have made great progress in the development of crew resource management training programs. They have responded quickly to problems in the system, and have incorporated related training materials and practical skills into their training programs. With the goal of integrating technical and team training (Advanced Qualification Program), the FAA is collaborating with airlines in developing proficiency-based requirements that will allow programs to incorporate many innovations in training curriculum, implementation, and media. Of obvious benefit will be the more efficient and focused use of simulators of various levels of fidelity. With computer-based training and new multimedia possibilities, this is an opportunity for companies to upgrade their programs in both substance and efficiency. Because companies must justify the way in which they collect evaluation data, great emphasis is placed on the qualifications of instructors and evaluators and the development of reliable and standard methods for evaluating crew performance.

Airlines have also recognized that team training offers benefits beyond the cockpit, and other organizations (maintenance, cabin crew, customer service, air traffic controllers) have initiated team training in their own domains. In addition, there is acknowledged need for greater cross-operational education both within and outside of the company, and joint training where possible. These initiatives are still young in their development, but the trend reflects more of a systems perspective. Finally, interventions related to developing new technology have been notoriously late in incorporating human factors feedback. There is some indication from designers and manufacturers that they now take the philosophy of human-centered automation seriously, and that human considerations need to be incorporated at the very beginning of the development process. Hopefully these are not empty promises.

EPILOGUE

This chapter began with a car analogy on the topic of avoiding costly and fatal problems in operating this complex piece of machinery. Although people tend to adapt to everyday stresses and strains that wear a car down, this, in the long run, destroys the car. On the other hand, paying attention to the warning signals and practicing good preventive maintenance can save the day. Along these lines, Pirsig (1974) described good motorcycle maintenance in the following way:

> A "mechanic's feel" implies not only an understanding for the elasticity of metal but for its softness. The insides of a motorcycle contain surfaces that are precise in some cases to as little as one ten-thousandth of an inch. If you drop them or get dirt on them or scratch them or bang them with a hammer they'll lose that precision. It's important to understand that the metal behind the surfaces can normally take great shock and stress but that the surfaces themselves cannot. When handling precision parts that are stuck or difficult to manipulate, a person with mechanics's feel will avoid damaging the surfaces and work with his tools on the nonprecision surfaces of the same part whenever possible. . . . Handle precision parts gently. You'll

never be sorry. If you have a tendency to bang things around, take more time and try to develop a little more respect for the accomplishment that a precision part represents. (p. 293)

It is important to heed the warning signals of the stresses that aircrews (and others in the aviation system) face everyday and in unusual circumstances. Like the aforementioned precision parts, human operators can be difficult, complex, and fragile, but their ability to absorb great shock and stress is enormous. The abilities and skills crew members exhibit in adapting to, and intervening for, their fellow crew members is an accomplishment that should be respected.

REFERENCES

Billings, C. E. (1991). *Human-centered aircraft automation: A concept and guidelines.* (NASA Tech. Memorandum 103885). Moffett Field, CA: NASA–Ames Research Center.

Chidester, T. R. (1987). Selection for optimal crew performance: Relative impact of selection and training. In R. S. Jensen (Ed.), *Proceedings of the Fourth International Symposium on Aviation Psychology* (pp. 473–479). Columbus: Ohio State University.

Chidester, T. R., Kanki, B. G., Foushee, H. C., Dickinson, C. L., & Bowles, S. V. (1990). *Personality factors in flight operations: I. Leader characteristics and crew performance in full-mission air transport simulation* (NASA Tech. Memorandum 102259). Moffett Field, CA: NASA–Ames Research Center.

Delta, pilots in standoff. (1993, January 25). *Aviation Week & Space Technology,* p. 54.

Dornheim, M. A. (1993, February 1). Cuts, layoffs affirm transport boom's end. *Aviation Week & Space Technology,* pp. 22–24.

Foushee, H. C., & Helmreich, R. L. (1988). Group interaction and flightcrew performance. In E. L. Wiener & D. C. Nagel (Eds.), *Human factors in aviation* (pp. 189–227). New York: Academic Press.

Foushee, H. C., Lauber, J. K., Baetge, M. M., & Acomb, D. B. (1986). *Crew factors in flight operations III: The operational significance of exposure to short-haul air transport operations.* (NASA Tech. Memorandum 88322). Moffett Field, CA: NASA–Ames Research Center.

Ginnett, R. G. (1987). The formation of airline flight crews. In R. S. Jensen (Ed.), *Proceedings of the Fourth International Symposium on Aviation Psychology* (pp. 399–405). Columbus: Ohio State University.

Ginnett, R. C. (1990). Airline cockpit crew. In J. R. Hackman (Ed.), *Groups that work (and those that don't): Creating conditions for effective teamwork* (pp. 427–448). San Francisco: Jossey-Bass.

Ginnett, R. C. (1993). Crews as groups: Their formation and their leadership. In E. L. Wiener, B. G. Kanki, & R. L. Helmreich (Eds.), *Cockpit resource management* (pp. 71–98). San Diego: Academic Press.

Hackman, J. R. (1985). The design of work teams. In J. W. Lorsch (Ed.), *Handbook of organizational behavior* (pp. 315–342). Englewood Cliffs, NJ: Prentice-Hall.

Hackman, J. R. (1987). Group level issues in the design and training of cockpit crews. In H. W. Orlady & H. C. Foushee (Eds.), *Proceedings of the NASA/MAC Workshop on Cockpit Resource Management.* (NASA Conference Publication 2455, pp. 23–39). Moffett Field, CA: NASA–Ames Research Center.

Hackman, J. R. (1993). Teams, leaders, and organizations: New directions for crew-oriented flight training. In E. L. Wiener, B. G. Kanki, & R. L. Helmreich (Eds.), *Cockpit Resource Management* (pp. 47–69). San Diego: Academic Press.

Hart, S. G., & Staveland, L. E. (1988). Development of a multi-dimensional workload rating scale: Results of empirical and theoretical research. In P. A. Hancock & N. Meshkati (Eds.), *Human Mental Workload* (pp. 139–183). Amsterdam: Elsevier.

Helmreich, R. L., & Foushee, H. C. (1993). Why crew resource management: Empirical and theoretical bases of human factors training in aviation. In E. L. Wiener, B. G. Kanki, & R. L. Helmreich (Eds.), *Cockpit resource management* (pp. 3–45). San Diego: Academic Press.

Kanki, B. G., & Foushee, H. C. (1989). Communication as group process mediator of aircrew performance. *Aviation, Space and Environmental Medicine, 60*(5), 402–410.

Kanki, B. G., & Palmer, M. T. (1993). Communication and crew resource management. In E. L. Wiener, B. G. Kanki, & R. L. Helmreich (Eds.), *Cockpit resource management* (pp. 99–136). San Diego: Academic Press.

Kanki, B. G., Palmer, M. T., & Veinott, E. (1991). Communication variations related to leader personality. In R. S. Jensen (Ed.), *Proceedings of the Sixth International Symposium on Aviation Psychology* (pp. 253–259). Columbus: Ohio State University.

Kayten, P. J. (1993). The accident investigator's perspective. In E. L. Wiener, B. G. Kanki, & R. L. Helmreich (Eds.), *Cockpit resource management* (pp. 283–314), San Diego: Academic Press.

Lautman, L. G., & Gallimore, P. L. (1987, April–June). Control of the crew caused accident: Results of a 12-operator survey. *Airliner: Magazine of the Boeing Commercial Airplane Co.*, pp. 1–6.

Magliozzi, T., & Magliozzi, R. (1991). *Car Talk.* New York: Bantam.

McGrath, J. E. (1984). *Groups: Interaction and performance.* Englewood Cliffs, NJ: Prentice-Hall.

McKenna, J. T. (1993, January 11). Winter storms test new anti-ice tactics. *Aviation Week & Space Technology,* pp. 38–40.

McKenna, J. T. (1993, February 1). Northwest, union leaders haggle over equity stake. *Aviation Week & Space Technology,* p. 30.

McKenna, J. T. (1993, March 22). U.S. airlines see red after storm havoc. *Aviation Week & Space Technology,* p. 39.

Morrow, D. G., Lee, A. T., & Rodvold, M. (1991). Collaboration in pilot–controller communication. In R. S. Jensen (Ed.), *Proceedings of the Sixth International Symposium on Aviation Psychology.* (pp. 278–283). Columbus: Ohio State University.

National Transportation Safety Board. (1972). *Aircraft accident report—Eastern Air Lines, Inc. L-1011, N310EA, Miami, Florida, December 29, 1972.* (Rep. No. NTSB–AAR–73–14). Washington, DC: Author.

National Transportation Safety Board. (1979). *Aircraft Accident Report—United Airlines, Inc., McDonnell–Douglas DC-8–61, N8082U, Portland, Oregon, December 28, 1978.* (Rep. No. NTSB–AAR–79–7). Washington, DC: Author.

National Transportation Safety Board. (1990). *Aircraft Accident Report—United Airlines Flight 232, McDonnell–Douglas DC-10–10, Sioux City Gateway Airport, Sioux City, Iowa, July 19, 1989.* (Rep. No. NTSB–AAR–90–06). Washington DC: Author.

National Transportation Safety Board (1991). *Aircraft Accident Report—Avianca, The Airline of Columbia, Boeing 707–321B, HK2016, Fuel exhaustion, Cove Neck, New York, January 25, 1990.* (NTSB–AAR–91–04). Washington, DC: Author.

National Transportation Safety Board. (1994). *A review of flightcrew-involved, major accidents of U.S. air carriers, 1978 through 1990.* (Rep. No. NTSB/SS–94/01). Washington, DC: Author.

Nordwell, B. D. (1993, January 25). Parallel runway spacing problem. *Aviation Week & Space Technology,* p. 56.

Nordwell, B. D. (1993a, March 8). Software problems delay ATC redesign 14 months. *Aviation Week & Space Technology,* p. 30.

Nordwell, B. D. (1993b, March 8). Procedural issues slow benefits of data links. *Aviation Week & Space Technology,* pp. 32–33.

Orasanu, J. (1993). Decision-making in the cockpit. In E. L. Wiener, B. G. Kanki, & R. L. Helmreich (Eds.), *Cockpit resource management* (pp. 137–172). San Diego: Academic Press.

Orasanu, J., & Fischer, U. (1992). Team cognition in the cockpit: Linguistic control of share problem solving. In *Proceedings of the 14th Annual Conference of the Cognitive Science Society* (pp. 189–194). Hillsdale, NJ: Lawrence Erlbaum Associates.

Orlady, H. W., & Foushee, H. C. (Eds.). (1987). *Proceedings of the NASA/MAC workshop on cockpit resource management* (NASA Conference Publication 2455). Moffett Field, CA: NASA–Ames Research Center.

Ott, J. A. (1993, January 11). Massive airline losses force draconian cuts. *Aviation Week & Space Technology,* pp. 30–33.

Phillips, E. H. (1994, March 14). Airlines' choice: Adapt or perish. *Aviation Week & Space Technology,* pp. 61–63.

Pirsig, R. M. (1974). *Zen and the art of motorcycle maintenance.* New York: Bantam.

Ruffell Smith, H. P. (1979). *A simulator study of the interaction of pilot workload with errors, vigilance, and decisions* (NASA Tech. Memorandum 78482). Moffett Field, CA: NASA–Ames Research Center.

Scott, W. B. (1993, January 11). NTSB urges studies on mountain winds. *Aviation Week & Space Technology,* pp. 34–35.

Wiener, E. L. (1989). *Human factors of advanced technology ("glass cockpit") transport aircraft* (NASA Contractor Rep. 177528). Moffett Field, CA: NASA–Ames Research Center.

Wiener, E. L., Chidester, T. R., Kanki, B. G., Palmer, E. A., Curry, R. E., & Gregorich, S. E. (1991). *The impact of cockpit automation on crew coordination and communication: I Overview, LOFT evaluations, error severity, and questionnaire data.* (NASA Contractor Rep. No. 177587). Moffett Field, CA: NASA–Ames Research Center.

Wolfe, T. (1979). *The right stuff.* New York: Farrar, Straus, & Giroux.

5 Moderating the Performance Effects of Stressors

Clint A. Bowers
Jeanne L. Weaver
Ben B. Morgan, Jr.
University of Central Florida

BACKGROUND

As illustrated by several of the chapters in this volume, there is an increasing pressure on human operators to perform complex tasks effectively under a variety of stressful conditions. Consequently, there is also an increasing pressure on scientists to understand the performance effects of stress, to predict the effects of novel stressors, and to intervene to reduce potential negative stress effects. Unfortunately, the literature regarding stress and human performance frequently provides little guidance for responding to these demands. Although there are literally hundreds of published manuscripts in this area, the accumulated findings are somewhat contradictory, and it is often difficult to draw unequivocal conclusions about the effects of a given stressor.

For example, one can imagine several situations in which it would be useful to understand how performance is affected by noise. A recent literature review considered 58 such investigations and found that 29 studies reported a performance decrement due to noise exposure, 22 reported no effects, and 7 reported improved performance (Gawron, 1982). It is not difficult to understand why system designers, policy makers, and researchers might be frustrated by this mixed pattern of results.

In the case of laboratory research, this variability of results is frequently attributed to methodological problems such as errors in experimental design and the use of assessment methods with questionable validity (Weaver, Morgan, Adkins-Holmes, & Hall, 1992). Although applied research is vulnerable to these same sources of error, field researchers must also consider additional error variance that might be introduced by the frequent inability to employ random sampling procedures. In addition, it has been suggested that stress research is particularly susceptible to the effects of "moderator" variables which are often unevenly distributed within treatment conditions. The presence of these moderators can alter an individual's perception of, or reaction to, stressors, resulting in confounded results that make it increasingly difficult to understand the human response to stress (Holt, 1982; Jex & Beehr, 1991).

Although the effects and importance of moderator variables have been considered as part of several theoretical models of stress (i.e., French & Kahn, 1962; House, 1974), there has not been a systematic line of research designed to assess the degree to which moderators influence the results of stress research. Research regarding the effects of stressors typically focuses on the direct impact (i.e., main effects) of specific variables. In addition, moderator effects are often discovered serendipitously, precluding the generalization of results beyond a narrow set of conditions. The existing literature includes only a few reviews of the effects of moderator variables and their impact on stress research. In one of these reviews, Holt (1982) identified five categories of moderator variables that have been found to impact the effect of occupational stress. Specifically, he discussed the moderating effects of physiological, individual characteristic, situational, organizational, and sociological variables. Although his chapter provides a rather extensive review, it focuses on variables that might alter the effects of stress on health outcomes. It adds very little to our understanding of the performance effects of stress in the work place.

Of greater interest to applied researchers, Jex and Beehr (1991) discussed the effects of moderator variables in their recent review of methodological and theoretical issues in work stress research. However, although their chapter highlights the importance of moderator variables and provides several useful examples of their effects, its scope is limited by the authors' focus on the effects of stress in industrial work places. In another recent review, Weaver and her colleagues (Weaver et al., 1992) reviewed the literature concerning the impact of moderators on the relationship between stress and decision-making performance. Although this review provided extensive discussion of moderator effects, its emphasis on application to decision making might limit its generalizability to other areas of human performance.

The current chapter attempts to build on these recent discussions of moderator effects by broadening the review of the literature and identifying general trends concerning the effects of moderator variables on the stress–performance relationship. In so doing, the goal is to assist applied researchers in identifying moderator effects that are most likely to affect the relationship between specific stressors and performance outcomes of interest. The achievement of this goal is made more difficult by the fact that the stress literature includes a large number of variables that might be considered as potential moderator variables. Although it might be impossible to control (or even measure) the effects of so many variables, the arbitrary selection or dismissal of these effects might result in inappropriate conclusions about the effects of stress. Therefore, we believe that it will be helpful to provide an organizational framework within which the available findings can be interpreted. Thus, in order to highlight the types of variables which appear to present the greatest risk for confounding the stress–performance relationship, the current review examines the impact of different classes of moderator variables on the effects of various types of stressors. The categorization schema to be used in this review is described in the following section.

Categories of Stressors

The organization of the current discussion is based, in part, on the stress model presented by Ivancevich and Matteson (1980). This model, which is depicted in Fig. 5.1, is most useful because of its thorough coverage of stressor variables. However, for purposes of this review, only the intraorganizational factors contained within the antecedent component of the model will be considered. Specifically, the Ivancevich and Matteson model describes four classes of stressors that operate within organizations; namely, those that derive from physical, individual, group, and organizational conditions. As indicated in the following paragraphs, conditions are categorized into these classes primarily on the basis of the original source of the stress in question.

Stressors included in the *physical environment* meet two criteria: First, these stressors "refer to physical conditions in the environment which require that an employee adapt in order to maintain homeostasis" (Ivancevich & Matteson, 1980, p. 105). The second characteristic of a physical stressor is that it results in a direct physical impact on the operator. These stressors are derived from physical features of the environment. They include factors such as heat, cold, noise, vibration, and so forth, which are typically experienced directly through one of the five senses.

Individual-level stressors occur as a result of functions required in the process of individual performance, or those "stressors that are directly associated with the role we play or the tasks we have to accomplish within the organization" (p. 110). Ivancevich and Matteson suggest that stressors within this category account for the majority of the stress experienced within organizations. In addition, they indicate that this category of stressors has received more attention by researchers than any other. Examples of stressors that might be considered at the individual level are role ambiguity, role conflict, and work overload.

Ivancevich and Matteson also suggested that an organization's effectiveness is influenced by the relationships among its members. Therefore, stressors can also develop at the group level of organizations. *Group-level* stressors are defined as conditions that create stress within the individual due to some group influence. Examples of group-level stressors to be discussed within this chapter are *crowding* and *competition*.

According to the Ivancevich and Matteson model, the final category of

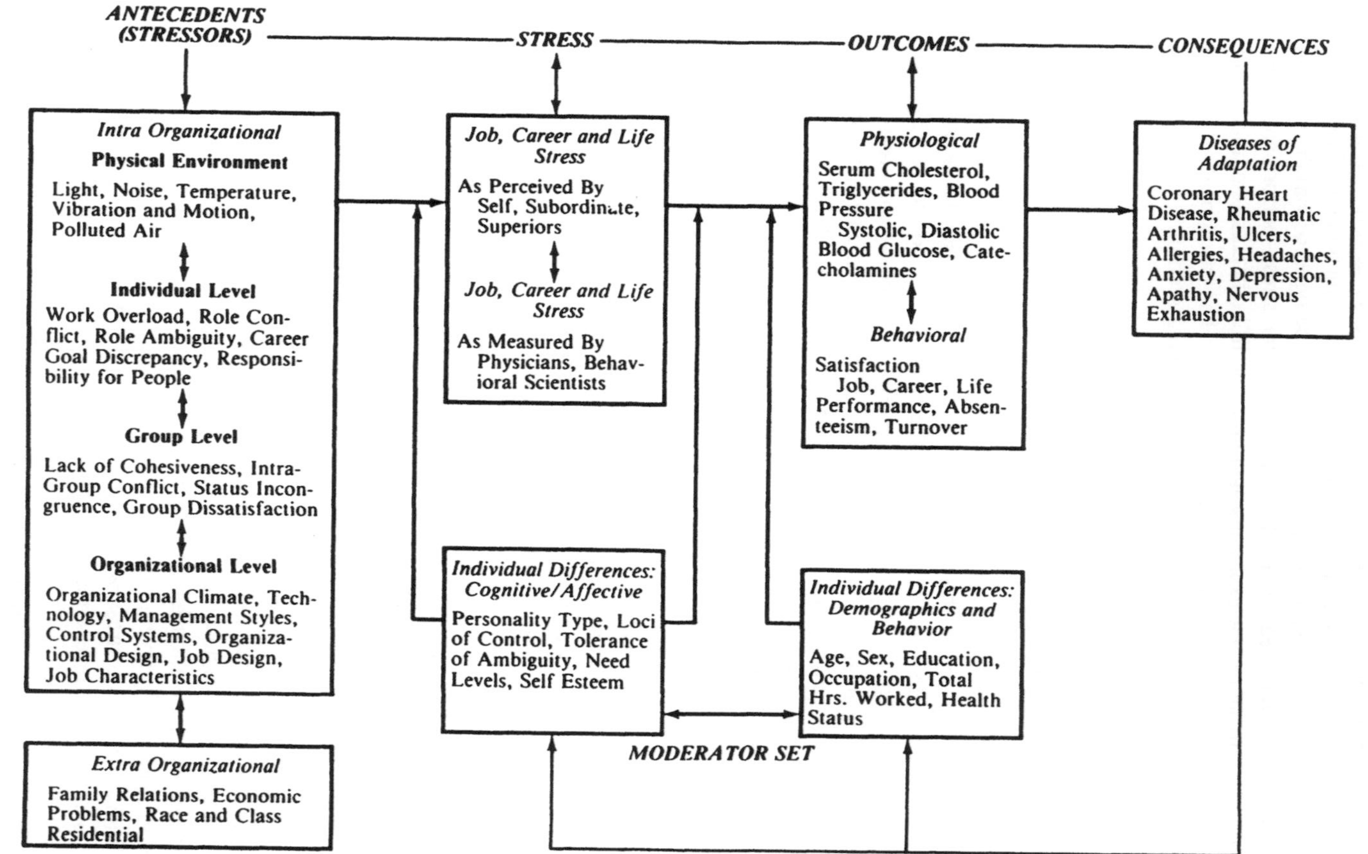

FIG. 5.1 Organizational stress research model. From Ivancevich and Matteson (1980). Reprinted by permission.

stressors is *organization-level* stressors. These derive from the general climate and working conditions of the organization. Variables that might be examined within this category include occupational stress (e.g., organizational structure, organizational climate, managerial conditions, etc.), shift work, and continuous work. A degree of overlap is apparent between this category and individual-level stressors in relation to job characteristics. For example, although work overload was mentioned earlier as an individual-level stressor, it might also be considered as a characteristic of the overall organizational setting. In this review, variables that relate to an individuals' role in a specific job will be considered as an individual-level variable. Those that derive from general organizational conditions or organizational design factors that impact a variety of jobs will be classified as organizational-level variables. We will tend to consider such stresses as individual-level stressors unless it is clear that they derive from general occupational characteristics. For example, occupational differences in workload that might be observed between jobs such as air traffic controllers and sales people will be classified as organizational variables. Specific job conditions that impact a specific individual (e.g., the traffic conditions experienced by a given air traffic controller) will be classified as an individual variable.

Categories of Moderators

As noted earlier, the systematic examination of moderators and their impact on stress and performance is relatively new; thus, there is little precedent for categorizing variables into types of moderators. However, for purposes of simplicity and logical consistency, we have chosen to categorize moderators based on the source of their effects into three of the stress categories identified by Ivancevich and Matteson (1980). That is, we will categorize moderator variable in terms of their origination at the level of the individual, group, or organization. Moderators at the *individual level* include variables such as personality characteristics, demographic variables (i.e., age and nationality), experience, and so forth. Moderator variables at the *group level* derive their effects from some group influence or interaction. For example, social support, team size, and group cohesion are all moderators that might be considered to be within this category. Finally, moderators that originate within the *environment/organization level* include physical characteristics of the environment that require responses in order to maintain an homeostatic condition (i.e., temperature, noise) and characteristics of the organization of the workplace (i.e., decision-making participation, or organizational level).

One factor that makes it difficult to establish relationships between stressors and their potential moderators is the fact that variables sometimes serve as stressors in one case and as moderators in others. For instance, some studies treat role ambiguity as a stressor (i.e., Beehr, 1976), but role ambiguity has also been identified as a moderator between occupational stress, satisfaction, and job involvement (Abdel-Halim, 1981). For the most part, this distinction is made by the manner in which effects are evaluated in reporting the results. Moderator effects are most often observed as interactions, covariates, or elements of sampling, whereas primary stress effects are typically inferred from the main effects of analysis. We employ this guideline in the current analysis in order to interpret the results of studies in which the distinction between stressors and moderators is unclear.

Performance Effects

Because the performance effects of stressors are of interest to researchers and practitioners alike, eventually it would be very useful to produce a three-dimensional model to illustrate the relationships among categories of stressors, moderators, and specific types of performance outcomes such as productivity, errors, and so forth. However, the current state of the literature does not support the development of a model of moderator effects at that level of detail. That is, there are too few studies to allow a categorization of results in terms of classes of performance effects. Therefore, the current discussion summarizes performance effects as concisely as possible in relation to the stressor and moderator categories that are considered. In so doing, an attempt is made to identify areas in which necessary research is clearly lacking. Hopefully, this discussion will provide direction for future research concerning the impact of moderators on specific stress–performance relationships.

MODERATOR EFFECTS

Stressors from the Physical Environment

Main Effects. The influence of the physical environment on workers' performance has been a very popular area of study in industrial and organizational psychology. Some of the most frequently investigated stressors within this category are noise, temperature, and motion. These stressors are discussed here as illustrative of the typical effects of stressors from the physical environment.

Although heavily researched, the effects of noise stress on performance remains equivocal. Early research indicated that the presence of noise causes decreases in frustration tolerance (Glass, Singer, & Friedman, 1969) and performance (Hockey, 1978; Weinstein, 1977). However, others have demonstrated that the effects of noise are more complex (Gawron, 1982). As previously mentioned, Gawron's review of 58 experiments found results that indicated improved performance under some conditions of noise, decreased performance in others, and no noise effects in others. The author attributed the equivocality of these results to a variety of factors. Of the experiments included in the review, conditions differed on such factors as duration of the noise used, method of noise generation, and type of task. The discrepant nature of research regarding noise and performance on various tasks suggests that noise stress is especially susceptible to the influence of moderator variables.

Investigations regarding the effects of temperature typically explore the effects of extreme heat or cold on task performance. Studies of the effects of heat on performance have generally found that temperatures of sufficient intensity and duration to cause fatigue will cause decrements in perceptual–motor performance, but will have a negligible effect on more cognitively based tasks (e.g., Ramsey, 1983). For example, Mackworth (1961) reported that increased temperatures caused performance decrements on a physical pull task. Similarly, Azer, McNall, and Leung (1972) reported significantly longer reaction times and poorer tracking performance for subjects exposed to 95° temperatures and high humidity.

Investigations of the effects of cold on performance have emphasized its effects on the hands. Results indicate that the hands become less flexible (Ramsey, 1983) and manual dexterity is reduced (Gaydos & Dusek, 1958) under cold conditions. Clark and Cohen (1960) investigated the performance of subjects who were required to place their hands in a cold box and tie knots in rope. Decreased temperature resulted in greater performance decrements under slow cooling rate conditions.

The physiological and performance effects of motion have likewise been investigated. Whiteside (1965) reported physical symptoms including pallor, increased heart rate, and sweating in response to motion-induced sickness. Brand, Colquhoun, Gould, and Perry (1967) found that mathematical problem solving was significantly poorer in subjects with induced motion sickness. However, the authors noted that these results might have been partially attributable to other factors (e.g., crowding).

Although some discrepancy exists regarding the effects of environmental stressors, it can be concluded generally that excessive levels of this type of stress are associated with diminished performance. Typically, the more extreme the stressor the greater the performance decrement. However, there are also indications that deleterious effects of environmental stress on task performance can be altered by the presence of some moderator variables. The following section summarizes these moderating effects.

Individual-Level Moderators. One variable that has been found to moderate the effects of environmental stressors is *perceived control.* Perceived control can be defined as the extent to which individuals believe that they regulate the events within their environment (Weaver, et al., 1992). It refers to the individual's belief that he or she is able to respond in a way that will influence the aversiveness of the event. Investigations regarding perceived control focus on the individual's beliefs regarding the situation and not necessarily their actual control.

Glass and Singer (1972) investigated the relationship between exposure to uncontrollable versus controllable noise and subsequent performance on two cognitive tasks. Subjects who were exposed to uncontrollable noise performed significantly poorer on problem solving (as measured by the frequency of their attempts to solve puzzles) and proofreading tasks. It was concluded that these subjects performed less well because of feelings of helplessness that were induced by the uncontrollable noise. The moderating effects of controllability has also been supported in a recent meta-analytic review (Driskell, Mullen, Johnson, Hughes, & Batchelor, 1991). This review concluded that inducing a perception of control by the subjects reduced stress in cases where shock was threatened but never actually delivered. Although subjects with control who did receive shocks perceived more stress, Driskell and his colleagues concluded that this might have been due to the feedback component of the shock. That is, because subjects were told that poor performance would result in shock, they might have perceived the shocks as feedback of failure and, therefore, experienced more stress.

The detrimental effects of environmental stress have also been shown to be moderated by experience or *prior exposure* to a stressor. For example, Klonowicz (1989) investigated the relationship between reactivity, experience, and noise in relation to proofreading performance. The author defined experience in this context as previous frequent or long exposure to street noise. It was hypothesized that noise experience would interact with reactivity to determine proofreading performance under high-noise conditions. The hypothesis was supported in that an interaction was found between experience and reactivity. Students with high levels of experience and high reactivity displayed fewer errors than did high-reactivity students with less experience. Given these results, the authors concluded that the performance of low-experience persons worsens under increased noise conditions, and that the effects of stress are reduced for individuals with increased experience. Experience seems to be especially beneficial for high-reactivity individuals.

Stressor experience has also been investigated in relation to stress-inoculation training (Meichenbaum, 1985). It has been argued that prior stress experiences allow individuals to acquire coping strategies for dealing with subsequent stress exposures (Eysenck, 1983). These strategies subsequently serve as a buffer against stress effects.

Norris and Murell (1988) investigated the relationship between stressor experience and anxiety in response to a natural disaster. Their results indicated that individuals who had been exposed to a previous flood experienced significantly less anxiety than people with no prior exposure. Therefore, experience with a stressor appears to reduce the deleterious effects of subsequent stress exposures and, in general, raise the threshold at which stress affects individuals.

The performance effects of environmental stress has also been found to be moderated by the effects of *incentive*. In a study of the relationship between heat and tracking performance, Pepler (1958) found that better accuracy was obtained under both of two target speed conditions when an incentive was provided. His results also indicated that the adverse climatic effects on performance under low-incentive conditions were completely offset by the introduction of the higher incentives. Pepler suggested that the performance of individuals exposed to an unfavorable climate for several hours can be maintained at a level equivalent to those in favorable conditions, provided that incentives are increased accordingly.

Self-control has also been investigated in relation to the effects of physical stressors. For example, Kanfer and Seidner (1973) tested the effectiveness of self-control as a moderator between cold pressor exposure and tolerance to cold. Their results indicated that high-self-control individuals were able to tolerate ice water immersion of the hand significantly longer than low-self-control persons. Interestingly, although high-self-control individuals were able to tolerate the cold longer than low-self-control subjects, they reported equivalent levels of pain intensity.

Self-control has also been found to moderate the adverse effects of motion on performance (Rosenbaum & Rolnick, 1983). This study investigated the relationship between motion sickness and the work performance of Navy crewmen. Consistent with the results of Kanfer and Seidner (1973), individuals with high levels of self-control performed better than those with low levels of self-control. Given that high-self-control persons also reported using more specific self-control methods to cope with their sickness, it might be hypothesized that these coping procedures allowed the individuals to maintain their performance.

Group and Environmental/Organizational Moderators. The literature concerning environmental stressors and related moderators has concentrated on individual-level moderators. We were unable to identify any studies in which group or organizational variables were reported to moderate the effects of physical environmental stressors. This lack of research emphasis is understandable in light of the fact that research regarding the impact of moderators on environmental stressors is relatively rare in and of itself. A likely reason for this situation is the lack of an intuitive relationship between environmental stressors and group or organizational-level moderators. For example, although social support might be an effective moderator of some individual stresses, its utility is not necessarily self-evident in relation to noise or extreme temperature. Whatever the reasons for this lack of research, useful information might be gained from empirical investigations of those relationships.

Summary. In general, the available research indicates that the negative effects of environmental stressors can be attenuated by individual-level moderators such as perceived control, experience, incentive, and self-control. These results are summarized in Table 5.1. Of particular interest among these results is the finding that prior exposure to stress can reduce the subsequent impact of exposure to the same stress. This finding can have important practical applications in

TABLE 5.1
Physical Environment Stressors

Stressor	*Moderator*	*Effect*	*Authors*
Noise	Perceived control	Subjects exposed to controllable noise perform better than subjects with no control.	Glass & Singer, 1972
	Experience	Detrimental effects of noise on performance were moderated by noise experience.	Klonowicz, 1989
Heat	Incentive	High incentive resulted in improved tracking performance.	Pepler, 1958
Cold	Self-control	High-self-control subjects endured cold longer than low-self-control subjects.	Kanfer & Seidner, 1973
	Self-control	High-self-control subjects endured cold longer than low-self-control subjects.	Rosenbaum, 1980
Motion	Self-control	High-self-control seamen maintained performance superior to those low in self-control.	Rosenbaum & Rolnick, 1983

the form of stress-inoculation training. The effectiveness of such training, perhaps in combination with other efforts to increase perceived control (of self or other events), needs further investigation in applied settings.

It appears that moderators at the individual level are effective in raising the threshold at which environmental stressors impact performance. This is illustrated by Folkman's (1984) conceptualization of stress as "a relationship between the person and the environment that is appraised by the person as taxing or exceeding his or her resources and as endangering well-being" (p. 840). Based on this notion, it is hypothesized that these variables are effective moderators because they contribute to the individual's perceived coping resources. That is, the moderators shown in Table 5.1 increase the individual's sense of control, thereby raising the stress response threshold to environmental stressors.

Individual Stressors

Main Effects. As mentioned previously, stressors at the individual level are related to the role that one fills within the organization, to the tasks one must accomplish, or to both. Individual stressors such as workload, threat, time pressure, and a variety of role stressors have received the greatest empirical attention. *Workload* refers to the mental or physical demands of a task and can be generally defined as the availability of resources that an individual possesses in order to perform a task relative to the resources required (Sanders & McCormick, 1987). Generally speaking, performance decrements are typically caused by increased task demands, the depletion of available resources, or both (Tole, Stephens, Harris, & Ephrath, 1982). Physical threat, often manipulated by the possibility of receiving electric shock, has also been investigated. The effects of threat are often investigated in terms of the performance differences that are caused by changes in an individual's perception of control of the threatening stimulus (Houston, 1972). Additional individual-level stressors are role ambiguity, role conflict, role strain and role overload. Generally, increased levels of these stressors result in decreased productivity and job satisfaction (Caplan & Jones, 1975).

Individual-Level Moderators. A number of *personality* variables have been found to moderate the effects of individual stressors. For example, Caplan and Jones (1975) investigated the relationships among *Type A* personality (a hard-driving, persistent, and highly work-involved behavior pattern), role ambiguity, and workload. These authors found that role ambiguity failed to increase in relation to increased workload; however, subjective workload measures indicated a positive relationship between workload and anxiety. This relationship was greatest for individuals with Type A personalities. This finding is consistent with other such research indicating that Type A persons experience more anxiety when faced with potential failure (Gurin, Veroff, & Feld, 1960).

The moderating effects of the Type A behavior pattern have also been demonstrated in relation to the effects of threat. Evans and Moran (1987) investigated the relationships between the Type A personality, gender, vigilant coping, and heart rate reactivity under the threat of possible electric shock. Their results indicated that female Type As exhibited higher reactivity than both male and female Type Bs. In addition, female Type As also exhibited greater reactivity than male Type As. In addition, female Type As adopted a vigilant coping strategy in comparison to males.

Anxiety is another important moderator, and can be conceptualized in two distinct ways (Spielberger, 1972). Anxiety is an emotional state that results when a stressor is interpreted as requiring avoidance, and some sort of harm is anticipated. State anxiety is transitory in nature and may vary in intensity. On the other hand, trait anxiety refers to relatively stable individual predispositions toward anxiety. Thus, trait anxiety can be described as a susceptibility to respond to situations perceived as threatening with an increased level of state anxiety.

Archer (1979) investigated the relationships between locus of control, situational control, state anxiety, and trait anxiety under conditions of physical threat manipulated by the possibility of electric shock. It was hypothesized that individuals' ratings of situational control would interact with locus of control. Although this hypothesis was not supported, other interactions were detected. Specifically, a significant interaction was discovered between situational control and trait anxiety, such that high trait anxiety subjects in the ambiguous control condition reported a significantly lower expectancy of avoiding shocks than did subjects with low trait anxiety. The authors concluded that trait anxiety predicts an individual's behavior in an ambiguous situation, but not in those conditions with clear and explicit situational cues regarding reinforcement contingencies.

The moderating effects of anxiety have also been demonstrated in relation to both cognitive and motor performance (Katchmar, Ross, & Andrews, 1958; Weinberg & Ragan, 1978). Katchmar and his colleagues investigated the relationship between trait anxiety and performance on a complex verbal coding task under stress induced by failure. Results indicated that "time to task completion" increased under stress for low- and high-anxiety groups. However, the high-anxiety group exhibited the greatest performance change and the longest times to task completion. In addition, high-anxiety subjects also committed the highest number of errors.

Weinberg and Ragan (1978) also found that motor performance (throwing balls at a target) under stress differed for high- and low-trait-anxiety individuals. The predicted inverted U hypothesis (Evans, 1979) was tested by manipulating three levels of trait anxiety and psychological stress induced by providing false performance feedback. Results indicated that the best performance was exhibited by subjects in the moderate stress condition. In addition, high-trait-anxiety subjects performed best in the low stress condition and low-trait-anxiety subjects performed best in the high-stress condition.

Closely related to trait anxiety is the concept of *reactivity.* Reactivity has been defined as an individual's comparatively stable and characteristic strength of reaction, or sensitivity, to stimuli (Strelau & Maciejczyk, 1977). Individual reactivity has been investigated in relation to time pressure and the decision-making performance of pilots. The authors hypothesized that high-reactivity pilots would exhibit inferior performance as compared to that of low-reactivity pilots in the performance of decision making and motor tasks. Results indicated that low-reactivity pilots made quicker decisions than did the high-reactivity pilots under stressful conditions. In addition, the decision-making quality of low-reactive pilots was superior to that of the high-reactive pilots in the stressful condition. There was no performance difference on the motor task.

The moderating effects of several aspects of *control,* including locus of control, self-control, and perceived control, have also been investigated. In an investigation of the relationships between workload, locus of control, and job satisfaction, Perrewe (1986) found that individuals who were less external in their locus of control experienced higher levels of satisfaction under stressful job demands than did persons with a higher external locus of control orientation.

In a more performance-oriented study, Gal-or, Tenenbaum, Furst, and Shertzer (1985) investigated the effects of self-control and anxiety on training performance under conditions of physical threat in novice parachutists. Threat in this investigation was the physical threat of parachuting. It was hypothesized that self-control would be a better predictor of performance than anxiety, and that high-anxiety subjects would not perform as well as low-anxiety subjects. Results supported this hypothesis, such that subjects high in self-control performed better than low-self-control subjects regardless of their anxiety level. Contrary to the author's hypothesis, however, the best performance was exhibited by persons high in both anxiety and self-control. The poorest performance was exhibited by individuals with high anxiety and low self-control. A similar pattern of moderator effects regarding self-control and threat has been reported by Kanfer and Seidner (1973).

Actual control has been investigated by Solomon, Holmes, and McCaul (1980) in a paradigm where subjects were required to perform tasks that varied in terms of difficulty level and controllability of threat. Electric shocks were presented as either being beyond or within the subjects' control. Results indicated that control over the threat reduced anxiety, but only when the control required little effort. When the control was difficult to exercise, individuals reported as much anxiety as those under uncontrollable threat.

It has also been found that *participation in decision making* can enhance the level of perceived control so as to moderate the stress–performance relationship. That is, when individuals are given opportunities to make input into decisions that are made by the organization, their level of perceived control is often increased. This tends to increase the threshold for stress effects and leads to enhanced performance. This relationship is illustrated by the results of Jackson (1983), who found that individuals who were allowed more opportunities to participate in organizational decision making were less likely to experience role strain. Similarly, Miller, Ellis, Zook, and Lyles (1990) found that increased participation in decision making moderated the relationship between stress and burnout in hospital employees; that is, participation in decision making was related to lower levels of burnout. The authors hypothesized that this was due to a reduction in uncertainty regarding the requirements of their roles.

The effectiveness of experience as a moderator of workload stress has also been investigated. Brictson, McHugh, and Naitoh (1974) investigated the relationships between workload, experience, and landing performance of pilots. The results of this study were mixed, indicating that total flying experience was a significant predictor of landing performance during low-cumulative-workload (low-stress) conditions, whereas specific aircraft experience predicted performance during moderate-cumulative-workload conditions. In high-workload (high-stress) conditions, experience failed to predict pilot landing performance.

Group-Level Moderators. One group-level moderator that seems important in moderating the effects of workload is *social support*. Social support has been defined as:

> attachments among individuals or between individuals and groups that serve to improve adaptive competence in dealing with short-term crises and life transitions

> as well as long-term challenges, stresses, and privations through (a) promoting emotional mastery, (b) offering guidance regarding the field of relevant forces involved in expectable problems and methods of dealing with them, and (c) providing feedback about an individual's behavior that validates his conception of his own identity and fosters improved performance based on adequate self-evaluation. (Caplan & Killilea, 1976, p. 41)

One problem with the research regarding the moderating effects of social support is that social support is often studied in relation to outcomes other than performance (e.g., satisfaction, psychological symptoms, etc.). However, social support is such an effective moderator of these outcomes that it appears reasonable to hypothesize that it might also moderate the effects of stress on performance. For example, Kirmeyer and Dougherty (1988) investigated the relationship between supervisor support, workload, tension, and the coping actions in police radio dispatchers. The dispatchers were observed and rated on the dimensions of objective load, tension/anxiety, and coping actions. Coping actions were defined as problem-focused strategies for managing or altering work load. Results indicated that supervisor support moderated the effects of perceived and objective load. Dispatchers with high social support, under high perceived load, engaged in more coping actions and experienced less tension or anxiety than dispatchers with low social support. The authors hypothesize that high supervisor support might increase employee willingness to take actions which reduce overload.

An investigation by Koeske and Koeske (1989) explored the relationship between workload and burnout in social workers. The results indicated that social support, especially from coworkers, buffered the negative impact of workload on burnout. Only under conditions of low social support and accomplishment did workload produce significant amounts of work stress. These results appear to indicate that workload stress is effectively moderated by coworker social support in some occupations.

Another variable that functions much like social support is group cohesion. Griffith (1989) noted that the "classic" definition of cohesion suggests that groups which are very cohesive provide a buffer for group members against potentially debilitating stressor effects. Griffith found that individual morale and well-being were increased under cohesive conditions. The author argued that cohesion acts as a buffer against stress, allowing individuals to perform their duties more effectively under stressful conditions.

Group cohesion has also been investigated in relation to command style by Tziner and Vardi (1982). The results of this study revealed a significant interaction between cohesion and command style. The best performance was exhibited by high-cohesive tank crews with leaders who were high in "task and people orientation." Similar performance effects were found for low-cohesive tank crews with "people-emphasizing" leaders.

Steiner and Neumann (1978) investigated the effects of cohesion in a study conducted in the Israel Defense Forces. The study indicated that soldiers from highly cohesive units incurred less stress casualties from combat than low-level cohesion units.

Environment/Organizational-Level Moderators. Schuler (1975) investigated the relationships among role ambiguity, conflict, job satisfaction, and job performance at three levels of an organization (upper-level managers, midlevel managers, and entry-level professionals, and lower level jobs such as clerical workers, tradesmen, and maintenance personnel). Role conflict and role ambiguity were both negatively related to job satisfaction at all three levels of the organization; however, in midlevel employees, both role ambiguity and role conflict were related to poorer task performance. This relationship was not present at the higher organizational level. Similarly, Ivancevich and Donnelly (1975) explored the relationship between job stress and organizational structure. The results of this study indicated that salesmen in a relatively "flat" organization (with fewer management levels) experienced less anxiety and stress and more satisfaction than salesmen in medium and tall organizations (with more management levels).

Summary. As summarized in Table 5.2, the review of the performance effects of individual stressors indicates that the deleterious effects of these stresses can be effectively moderated by variables at the individual, group, and environmental/organizational level. However, it appears that individual moderators have received considerably more research attention than group or environmental/organizational moderators. As is the case with moderators of environmental stressors, it appears that the determining factor in moderating the effects of individual stresses on performance is the degree to which these variables increase the individual's coping capacity or perceived situational control. For example, it is fairly evident that organizations of increasingly complex structure can lead to increased role ambiguity in its employees. This increased role ambiguity can often lead to increased levels of stress, reduced productivity, and so forth. It is hypothesized that individuals working in highly complex organizations might perceive less control due to increased isolation; therefore, any variable that acts to decrease the ambiguity of the individual's role (or increase their perceived control) can serve to raise the threshold at which stress is perceived, which might result in improved performance.

As the earlier review indicates, these variables include personality characteristics, situational variables, and work-group and organizational conditions. Research designed to specifically test the relationships between perceived control and other variables thought to have moderating capabilities might shed additional light on the underlying mechanism of variables that moderate stress effects.

TABLE 5.2
Individual Stressors

Stressor	*Moderator*	*Effect*	*Authors*
Threat	Type A personality	Female Type A subjects engaged in vigilant coping strategies.	Evans & Moran, 1987
	Anxiety	Subjects high in anxiety under ambiguous threat conditions had a lower expectancy of avoiding shock.	Archer, 1979
	Self-control	High-self-control subjects performed better than low-self-control subjects.	Gal-or, Tenenbaum, Furst, & Shertzer, 1985
	Control	Control that required little effort reduced anxiety.	Solomon, Holmes, & McCaul, 1980
Time pressure	Reactivity	The performance of low-reactivity pilots was better than that of high-reactivity pilots under stress.	Strelau & Maciejczyk, 1977
High job demands	Locus of control	External locus of control persons experience less satisfaction in high-workload conditions.	Perrewe, 1986
Role strain	Participation in decision making	Increased decision making was associated with less experienced role strain.	Jackson, 1983
Workload	Experience	Under low-cumulative-workload conditions, total flying experience predicted landing performance, whereas specific aircraft experience predicted landing performance under moderate workload.	Brictson, McHugh, & Naitoh, 1974
Workload and role ambiguity	Type A personality	The relationship between role ambiguity and workload was greatest for Type A individuals.	Caplan & Jones, 1975
Workload and burnout	Social support	Coworker social support buffered the negative impact of workload on burnout.	Koeske & Koeske, 1989
Combat stress	Group cohesion	Cohesion acts as a buffer against stress and increases morale and well-being.	Griffith, 1989
Combat stress	Group cohesion × command style	Performance effectiveness was highest under high cohesiveness with leaders	Tziner & Vardi, 1982

(Continued)

TABLE 5.2
(Continued)

Stressor	*Moderator*	*Effect*	*Authors*
		with high in task and people orientation and low cohesiveness with leaders high in people orientation.	
Role stress	Organizational level	Role ambiguity and role conflict were negatively related to job satisfaction at all three levels of the organization; however, their relationship to performance varied at all three levels.	Schuler, 1975
Occupational stress	Supervisor support	Under high perceived load, subjects with high supervisor support experienced less tension/anxiety and engaged in more coping actions.	Kirmeyer & Dougherty, 1988
	Social support	Work stress was moderated by supportive work environments for men and supportive life sources for women. The effects of life stress on burnout were not moderated by social support.	Etzion, 1984
Occupational stress	Organizational structure	Salesmen in a flat organization perceive more satisfaction and less anxiety than salesmen in medium and tall organizations.	Ivancevich & Donnelly, 1975
	Participation in decision making	Subjects with increased decision making participation experienced less burnout.	Miller, Ellis, Zook, & Lyles, 1990
Occupational stress	Job complexity, role ambiguity, leader consideration, and locus of control	Subjects experiencing high role ambiguity under low leader consideration were less satisfied as were subjects with simple, structured jobs under low leader consideration. An internal locus of control was associated with high job involvement under leaders with high initiating structure.	Abdel-Halim, 1981

Group-Level Stressors

Main Effects. The effects of group-level stressors have received much less empirical attention than individual stressor effects. As noted previously, group-level stressors are factors that result in individual stress perceptions due to some group influence. Among the group-level stressors that have been explored in relation to moderator variables are *crowding* and *competition*. It has been noted (Evans, 1979) that in contrast to other stressors such as heat and noise, crowding is by necessity a group phenomenon. Therefore, the consideration of crowding as a group stressor appears to be especially appropriate. In regard to the performance effects of crowding, it has been found that "performance on simple, well-learned tasks is facilitated by the presence of others; whereas performance on complex or not-well-learned tasks is interfered with" (Freedman, Klevansky, & Ehrlich, 1971, p. 12). Langer and Saegert (1977) also found complex task decrements under conditions of crowding.

Another variable that can be considered a group-level stressor is competition. Competition is considered to be related to apprehension due to the evaluation of others; that is, "Being bested in competition is probably regarded by most people as a failure, so that realization of the possibility of losing in competition may serve as a cue for evaluation anxiety" (Geen, 1989, p. 27). Geen also stated that being in the presence of coworkers increases the possibility of being bested in performance. Because of this potential for increased anxiety under competitive circumstances, researchers have devoted increasing attention to investigations of competition and performance. That is, because training is not always a good predictor of performance in competition, researchers have sought to investigate factors that might influence performance in competitive circumstances (Lee, 1982). Generally, increased levels of competition and crowding are associated with increased perceptions of stress.

Individual-Level Moderators. Johnston and Briggs (1968) investigated the relationship of two team compensation levels to task load and performance on a simulated air traffic control task. Team compensation referred to the degree to which individuals could coordinate with their partner in order to compensate for an unexpected approach. Results indicated that better performance was exhibited by subjects in the compensatory condition under high-task-load conditions. Thus, the ability to coordinate among crewmembers helped to moderate (reduce) the effects of workload stress.

Evans (1979) investigated the physiological and behavioral effects of crowding on two types of task performance. One task was an information processing task that could be varied in complexity by varying the rate of response. The second used a dual task paradigm in which a shape-identification task was presented concurrently with a secondary task that required subjects to memorize details of a high-information-content story. The authors hypothesized that arousal would moderate the effects of stress in response to crowding such that complex, but not simple, task performance would be impacted. Results indicated that crowding had little effect on the performance of the simple information processing task. However, crowded individuals did commit more errors on the complex task. In the dual task condition, no difference was seen on the primary task, but secondary task performance was poorer under the crowded condition. These results support the author's hypothesis that crowding acts as a stressor which is mediated by overarousal.

The moderating effect of perceived control in crowded conditions was investigated by Sherrod (1974). This study investigated the performance of simple and complex tasks under three conditions of crowding. Subjects were exposed to crowded, uncrowded, or crowded with perceived control conditions. The simple tasks were paper-and-pencil tasks such as number comparison, addition, and "finding As." The complex task was a paper-and-pencil version of the Stroop Color–Word Test. Additional postcrowding measures of frustration tolerance were also obtained. Results indicated that crowding had no immediate effect on simple or complex task performance. However, crowding did significantly affect subsequent performance on a frustrating task (insoluble puzzles) given following the crowded condition. These effects were ameliorated by the perception of control over the crowded situation; that is, the performance aftereffects of crowding are less for individuals with a perception of control over the crowded conditions. Therefore, the existence of perceived control over the crowded conditions prevented the appearance of adverse aftereffects.

Self-efficacy is one factor that has received a great deal of research attention as a moderator of competition stress. It can be defined as "the strength of one's conviction that he or she can successfully execute a behavior required to produce a certain outcome" (Bandura, 1977). Bandura argued that an individual's expectations of mastery affect both the initiation and continuance of coping behaviors. Bandura (1988) also proposed that it is the person's perception of control over potential threats which determines their level of anxiety. That is, threat is experienced when a discrepancy exists between the individual's perceived coping abilities and potentially harmful characteristics of the environment. Control over stress is accomplished when both behavioral and cognitive coping efficacy are present. This relationship has been tested empirically by researchers who have investigated the relationship between competition induced stress and self-efficacy beliefs.

The effect of private and public efficacy expectations on the competitive performance of a muscular leg-endurance task were investigated by Weinberg, Yukelson, and Jackson (1980). Self-efficacy levels were manipulated by varying the descriptions of the person against which individuals were to compete. Subjects were told that they were to compete against a varsity track athlete or an individual with a knee injury. The results of the study indicated that individual's with high self-efficacy performed better than those with low self-efficacy. There was also a significant interaction between gender and self-efficacy such that the performance of low- versus high-efficacy males differed, whereas there was no such difference for females.

The relationship of self-efficacy to performance has also been investigated in tennis competition (Barling & Abel, 1983). This study required active tennis players to assess their perceived self-efficacy and participate in tennis competition. Results indicated that there was a significant and positive relationship between self-efficacy and tennis performance. Generally, self-efficacy moderates competition-induced stress in such a way as to result in superior performance by those with high self-efficacy.

Weinberg, Gould, Yukelson, and Jackson (1981) investigated the effect of preexisting and manipulated self-efficacy on a competitive muscular endurance task. The authors manipulated self-efficacy by altering explanations regarding the individual against whom subjects were to compete. Results indicated that the preexisting self-efficacy expectations the individual had prior to performance influenced their performance on the first trial, but the self-efficacy manipulation had more impact on subsequent trials. The authors suggested that this might imply possibilities for altering self-efficacy in relation to motor performance.

Summary. A summary of the research related to moderators of group-level stressors is provided in Table 5.3. As the table shows, little research has been

TABLE 5.3
Group-Level Stressors

Stressor	*Moderator*	*Effect*	*Authors*
Crowding	Arousal level	Arousal moderated the effects of stress in response to crowding.	Evans, 1979
	Perceived control	Perceived control ameliorates frustration aftereffects due to crowding stress.	Sherrod, 1974
Competition stress	Self-efficacy	Preexisting self-efficacy expectations were found to influence performance on the first trial, but manipulated self-efficacy had a greater influence on the second.	Weinberg, Gould, Yukelson, & Jackson, 1981
	Self-efficacy	There was a positive relationship between self-efficacy and performance.	Barling & Abel, 1983
	Self-efficacy	Individuals with high self-efficacy performed better than those with low self-efficacy.	Weinberg, Yukelson, & Jackson, 1980

performed regarding the effects of group level stressors and variables that might moderate the performance effects of such stressors. However, based on the available literature some tentative conclusions can be formed. In a manner similar to that observed with regard to moderators at the individual stress level, moderators of group-level stressors also appear to be related to the coping or control perception of the members involved. Generally, it appears that any variable that acts to increase the level of perceived control or coping capacity of individuals will act to moderate the deleterious effects of stress. Although crowding and competition are group phenomena by definition, past investigations have focused on individual performance. There is a clear need for investigations of the effects of group-level stressors on group and team performance as well. Another potentially fruitful line of research involves determining the degree to which the coping abilities of the team as a whole are determined by the coping skills of the individuals that compose the team. In addition, the extent to which coordination improves or diminishes performance deserves further study. Because groups of individuals are playing an increasingly important role in the workplace, research involving the factors that influence team performance is expected to increase, perhaps leading to answers to some of these questions.

Organizational-Level Stressors

Main Effects. Stressors at this level derive from the characteristics of the organization in which one is employed or that are inherent within the occupation of choice. For example, it is commonly accepted that medical personnel are exposed to high levels of stress because of the nature of their occupation. Other characteristics that have been found to be associated with stress in occupational settings are continuous operations and shift work. Continuous operations and shift work are typically investigated in relation to the individual's decreased ability to perform effectively. Occupations associated with continuous work and shift work are also considered to be quite stressful. Generally, findings from studies of shift work have indicated that initially better complex performance (requiring high memory load) is associated with night-shift work, whereas better performance of simple tasks (requiring low memory load) is associated with day-shift work (Monk & Folkard, 1983). Given the inherently stressful nature of shift work and continuous work, it is fortunate that researchers have attempted to discover factors which might alter the deleterious performance effects of these stressors.

Individual-Level Moderators. Mossholder, Bedeian, and Armenakis (1982) investigated the moderating impact of self-esteem on the relationship between group process and work outcomes in a sample of nurses. The results indicated that peer group interactions had a stronger effect on individuals with low self-esteem than on individuals with high self-esteem. In particular, high self-esteem individuals performed better under conditions of lower peer group interaction, and low self-esteem individuals performed best under conditions of high peer group interaction. Medical personnel have also been utilized to assess the moderating capabilities of understanding, prediction, and control on perceived stress, satisfaction, and psychological well-being (Tetrick & LaRocco, 1987). The results of this study of physicians, dentists, and nurses indicated that understanding and control moderated the negative relationship between perceived role stress and satisfaction. The authors hypothesized that understanding and control might also be effective in moderating the effects of organizational conditions on job-related attitudes and strains.

Another job-related characteristic that has been investigated in relation to performance is continuous work. Continuous operations have been defined as "uninterrupted schedules of nonstop activity" (Krueger, 1989, p. 129). One such study (Morgan, Coates, & Alluisi, 1975) investigated the relationship between performance on a synthetic job and the circadian rhythm under continuous-work conditions. The results indicated that performance in the continuous-work period was significantly affected by the circadian rhythm; that is, average performance decrements followed a pattern that approximated the circadian rhythm. Superior performance was demonstrated by crews starting continuous operations during daytime hours, and the worst performance was exhibited by crews starting during nighttime hours. Maximum decrements were exhibited by crews beginning continuous work at 2400 hours (midnight) and minimum decrements were exhibited by crews beginning continuous work at 1200 hours (noon).

Snook (1971) investigated the effects of age and physique on the physical performance of a manual handling task under continuous work circumstances—results indicated that continuous work capacity does not decrease with increased age. However, it was discovered that physique had a greater effect on continuous work capacity lifting heavier loads or performing slower paced tasks. In contrast, the physique of older subjects had a greater effect on the performance of tasks with lighter loads or faster paces. The buffering effect of physical fitness on job stress was confirmed by Tucker, Cole, and Friedman (1986).

The relationship of shift work to circadian rhythms has been studied by Ostberg (1973). Individuals working discontinuous alternating shifts were designated as morning, middle, or evening groups based on their work schedules. Significant differences were detected between the groups and the shifts, and a significant interaction of group by shift was also found. The authors concluded that the morning group had a much more difficult time adjusting to the shift schedule and the evening group experienced the least difficulty. The study indicates that circadian rhythms might be an important moderating factor in occupations requiring shift work.

Summary. As shown in Table 5.4, the literature regarding organizational stressors has been primarily related to moderators at the individual level. As was

TABLE 5.4
Organizational Level Stressors

Stressor	*Moderator*	*Effect*	*Authors*
Occupational stress	Self-esteem	Peer group interaction had a stronger effect on individuals with low self-esteem than on those with high self-esteem.	Mossholder, Bedeian, & Armenakis, 1982
	Control	Understanding and control moderated the negative effects of occupational stress.	Tetrick & LaRocco, 1987
	Physical fitness	Stress level is related to physical fitness in middle-aged men. Some physical exertion on the job was associated with less stress.	Tucker, Cole, & Friedman, 1986
Continuous work	Age, physique	Continuous-work capacity does not decrease with increased age. Physique has a greater effect on continuous-work capacity during heavier, slower tasks for younger subjects and lighter, faster tasks for older subjects.	Snook, 1971
Shift work	Circadian type	Circadian type significantly effected the subjects ability to adjust to shift work.	Ostberg, 1973

true with other stressors, variables that contribute to an individual's coping capability appear to be effective at moderating the detrimental effects of stress. For example, it is not surprising that physical fitness would raise the threshold at which job-related stress would impact performance of a manual handling task. This is not meant to imply, however, that all of the relationships discussed are completely intuitive. For instance, it is intuitive that increased age would have a negative impact on continuous work capacity. However, the research cited earlier did not support this relationship. It might be hypothesized that older persons learn cognitive strategies over their life span that enable them to adapt to even physically challenging tasks. That is, as coping is altered at one level (physical fitness, here), another individual capacity intercedes to maintain the baseline level of coping to which the individual is accustomed. However, it might be that there are limits to the degree that acquired strengths in one area can compensate for weaknesses in others.

It would be interesting to establish the degree to which other classes of moderators function to alter the impact of organizational stressors. It would seem that a number of other variables (e.g., social support) should be effective at moderating these effects.

SUMMARY AND CONCLUSIONS

Although the research in this area is relatively sparse, it is evident that moderators can have a substantial influence on the performance effects of various types of stressors. The categorization scheme adopted in this review might be helpful in identifying moderators that are effective in reducing the negative effects of these stressors. This method of organizing the moderator literature also clearly indicates areas in which additional research is needed. Table 5.5 summarizes the moderator literature in relation to the classes of stressors defined within the chapter. Within the table, a (+) denotes an effect that is beneficial and a (−) denotes an effect that is detrimental. It is evident that individual-level stressors and moderators have been the most thoroughly investigated.

The implications of achieving an adequate understanding of factors that moderate the relationship between stress and performance are relevant for researchers and practitioners alike. There are several benefits to researchers becoming well informed regarding moderators of stress. First, it is apparent that by accounting for these variables, researchers will be better able to obtain an unobscured view of the effects of stressors; that is, failure to consider the presence of factors that moderate the effects of stress leads to the presence of error variance in research designs. An increased understanding of the variables which serve to moderate stress will enable researchers to account for these factors through design and statistical control. Attempts to obtain a clear picture of particular stress effects without accounting for these variables might well result in equivocal findings. Another benefit to researchers becoming aware of moderating effects is increased specificity and clarity in discussions of stress and its effects. That is, an awareness of factors that alter the effects of stress will allow better communication and understanding among and between stress researchers. It is well documented by stress researchers that definitional problems plague the stress field. By accounting for moderator variables it will be possible to arrive at a more accurate representation of the stress phenomenon.

One important area in which researchers might be interested in investigating moderators of stress is that involving stress interventions. The research regarding factors that moderate the effects of stress is ripe with implications for the development of stress interventions. Therefore, researchers and practitioners might make a great contribution by interfacing at this point. By further investigating the effects of particular moderators and achieving an accurate understanding of their effects, researchers would be uniquely qualified to inform practitioners regarding factors that might ameliorate the performance effects of stress. For example, self-

TABLE 5.5
Stressors and Their Moderators

Types of stressors	Types of moderators		
	Individual	*Group*	*Environmental/Org.*
Physical environment	Noise (+) Knowledge/results (−) Alcohol intake (+) Perceived control (+) Experience (+) Incentive (+) Self-control (+)		
Individual level	Type A personality (−) Anxiety (−) Reactivity (−) Internal locus of control (+) Self-control (+) Perceived control (+) Participation in decision making (+) Experience (+)	Supervisor support (+) Social support (+) Group cohesion (+) Group cohesion × comman style (+)	Organizational level (+) Organizational structure (flat) (+) Role ambiguity × low leader consideration (−)
Group level	High arousal level (+) Self-efficacy (+)		
Organizational level	Self-esteem × low peer group interaction (+) Control (+) Age × physique × task Physical fitness (+) Circadian type		

Note. (+) denotes beneficial effects, (−) denotes detrimental effects.

efficacy appears to be especially promising as an enhancement to maintaining performance under stress. Furthermore, a recent line of research has demonstrated the trainability of self-efficacy (Gist, Schwoerer, & Rosen, 1989). Likewise, the understanding that stressor experience enables persons to function more effectively under stressful conditions has yielded an interest in stress-inoculation training (Meichenbaum, 1985). These are excellent examples of moderating factors that can be used to develop effective interventions to enhance performance under stress. It is recommended that self-efficacy training be combined with stress-inoculation training and tested as a practical intervention for moderating the performance effects of stressors.

It is also critical for practitioners to be well informed concerning moderator variables. It appears that practitioners can benefit particularly in the areas of training and selection. For example, when selecting for occupations associated with particular stressors, practitioners might select for worker characteristics known to be associated with coping in particular circumstances. That is, a worker with low trait anxiety might perform better under ambiguous occupational circumstances than one high in trait anxiety. The implications are perhaps even greater for training situations. As previously noted, researchers can greatly benefit the stress field by indicating those factors that might serve as effective targets of interventions. However, it is up to the practitioner to implement the suggestions by deriving interventions and testing them in applied environments. Thus, the practitioner can act as a source of feedback to researchers while improving the state of the art in stress intervention.

Although it is apparent from the earlier discussion that moderators might have potential for increasing the breadth and depth of understanding of stress and related areas, our understanding of moderators and their effects is still far from complete. In order to move toward the aforementioned possibilities, researchers must begin to delineate and investigate variables that moderate stress. That is, in order to gain the potential benefits of stress moderators, researchers must first become aware of the existence of moderators, control for their presence, and then systematically investigate their effects in controlled studies. It has been noted several times previously that factors which moderate stress appear to be related to the individual's perception of control. Although this is more evident for some variables than others, it is hypothesized that the ultimate determination of stress is related to the degree to which the individual feels in control of the situation.

Although the idea that stress results when persons feel unable to cope is not new (Lazarus, 1966), a review of moderator effects appears to provide support for this position. Moderator variables often appear to represent conditions that allow individuals to cope in particular situations. In other words, many of the moderators described here act by contributing to (or reducing) the resources that the individual can bring to bear in coping with stressors. In fact, it might be the case that there is substantial shared variance among the moderators we have considered. In order to test this hypothesis it is necessary to first establish the construct validity of these variables (e.g., self-efficacy, locus of control, trait anxiety) and test their interrelationships. By investigating the relationships of these factors it will be possible to determine the degree to which the constructs overlap.

In summary, this review has indicated the existence and importance of moderators of the performance effects of stress. Furthermore, the importance of the implications of understanding these factors has been stressed. However, although it is apparent that there are many benefits to be gained by developing a thorough understanding of moderators of stress, it is also obvious that to do so will take concerted effort by stress researchers. Therefore, it is recommended that researchers in this area investigate potential moderators more aggressively. Furthermore, it is recommended that research in the area of stress and human performance include a continuing discussion of moderator effects so that a full understanding of moderators can be developed.

REFERENCES

Abdel-Halim, A. A. (1981). Personality and task moderators of subordinate responses to perceived leader behavior. *Human Relations, 34,* 73–88.

Archer, R. P. (1979). Relationships between locus of control, trait anxiety, and state anxiety: An interactionist perspective. *Journal of Personality, 47,* 305–316.

Azer, N. Z., McNall, P. E., & Leung, H. C. (1972). Effects of heat stress on performance. *Ergonomics, 15,* 681–691.

Bandura, A. (1977). Self-efficacy: Toward a unifying theory of behavioral change. *Psychological Review, 84,* 191–215.

Bandura, A. (1988). Self-efficacy conception of anxiety. *Anxiety Research, 1,* 77–98.

Barling, J., & Abel, M. (1983). Self-efficacy beliefs and tennis performance. *Cognitive Therapy and Research, 7,* 265–272.

Beehr, T. A. (1976). Perceived situational moderators of the relationship between subjective role ambiguity and role strain. *Journal of Applied Psychology, 61,* 35–40.

Brand, J. J., Colquhoun, W. P., Gould, A. A., & Perry, W. L. M. (1967). 1-hyosine and cyclizine as motion sickness remedies. *British Journal of Pharmacological Chemotherapy, 30,* 463–469.

Brictson, C. A., McHugh, W., & Naitoh, P. (1974). Prediction of pilot performance: Biochemical and sleep–mood correlates under high workload conditions. *AGARD Conference Proceedings No. 146.*

Caplan, G., & Killilea, M. (1976). *Support systems and mutual help.* New York: Grune & Stratton.

Caplan, R. D., & Jones, K. W. (1975). Effects of workload, role ambiguity, and Type A personality on anxiety, depression, and heart rate. *Journal of Applied Psychology, 60,* 713–719.

Clark, R. E., & Cohen, A. I. (1960). Manual performance as a function of rate of change in hand-skin temperature. *Journal of Applied Physiology, 15,* 496.

Driskell, J. E., Mullen, B., Johnson, C., Hughes, S., & Batchelor, C. (1991). *Development of quantitative specifications for simulating the stress environment* (Tech. Rep. No. AL–7R–1991–01–09). Wright–Patterson AFB, OH: Armstrong Laboratory.

Evans, G. W. (1979). Behavioral and physiological consequences of crowding in humans. *Journal of Applied Social Psychology, 9,* 27–46.

Evans, P. D., & Moran, P. (1987). The Framingham Type A Scale, vigilant coping, and heart-rate reactivity. *Journal of Behavioral Medicine, 10,* 311–321.

Eysenck, H. J. (1983). Stress, disease, and personality: The inoculation effect. In C. L. Cooper (Ed.), *Stress Research* (pp. 121–146). New York: Wiley.

Folkman, S. (1984). Personal control and stress and coping processes: A theoretical analysis. *Journal of Personality and Social Psychology, 46,* 839–852.

Freedman, J., Klevansky, S., & Ehrlich, P. (1971). The effect of crowding on human task performance. *Journal of Applied Social Psychology, 1,* 7–25.

French, J. R. P., & Kahn, R. L. (1962). A programmatic approach to studying the industrial environment and mental health. *Journal of Social Issues, 18,* 1–47.

Gal-or, Y., Tenenbaum, G., Furst, D., & Shertzer, M. (1985). Effect of self-control and anxiety on training performance in young and novice parachuters. *Perceptual Motor Skills, 60,* 743–746.

Gawron, V. (1982). Performance effects of noise intensity, psychological set, and task type complexity. *Human Factors, 24,* 225–243.

Gaydos, H. F., & Dusek, E. R. (1958) Effects of localized hand cooling versus total body cooling on manual performance. *Journal of Applied Physiology, 12,* 377–380.

Geen, R. (1989). Alternative conceptions of social facilitation. In P. B. Paulus (Ed.), *Psychology of Group Influence* (pp. 15–51). Hillsdale, NJ: Lawrence Erlbaum Associates.

Gist, M. E., Schwoerer, C., & Rosen, B. (1989). Effects of alternative training methods on self-efficacy and performance in computer software training. *Journal of Applied Psychology, 74,* 884–891.

Glass, D. C., & Singer, J. E. (1972). *Urban stress: Experiments on noise and social stressors.* New York: Academic Press.

Glass, D. C., Singer, J. E., & Friedman, L. N. (1969). Psychic costs of adaptations to an environmental stressor. *Journal of Personality and Social Psychology, 12,* 200–210.

Griffith, J. (1989). The army's new unit personnel replacement and its relationship to unit cohesion and social support. *Military Psychology, 1,* 17–34.

Gurin, G., Veroff, J., & Feld, S. C. (1960). *Americans view their mental health.* New York: BasicBooks.

Hockey, G. (1978). Effects of noise on human work efficiency. In D. May (Ed.), *Handbook of noise assessment* (pp. 335–372). New York: Van Nostrand Reinhold.

Holt, R. (1982). Occupational stress. In E. L. Goldberger & S. Breznitz (Eds.), *Handbook of stress: Theoretical and clinical aspects* (pp. 419–444). New York: Macmillan.

House, J. S. (1974). Occupational stress and coronary heart disease: A review and theoretical integration. *Journal of Health and Social Behavior, 15,* 12–27.

Houston, B. K. (1972). Control over stress, locus of control, and response to stress. *Journal of Personality and Social Psychology, 21,* 249–255.

Ivancevich, J. M., & Donnelly, J. H. (1975). Relation of organizational structure to job satisfaction, anxiety–stress, and performance. *Administrative Science Quarterly, 20,* 272–280.

Ivancevich, J. M., & Matteson, M. T. (1980). *Stress and work.*Glenview, IL: Scott, Foresman.

Jackson, S. E. (1983). Participation in decision making as a strategy for reducing job-related strain. *Journal of Applied Psychology, 68,* 3–19.

Jex, S. M., & Beehr, T. A. (1991). Emerging theoretical and methodological issues in the study of work-related stress. *Research in Personnel and Human Resources Management,* 311–365.

Johnston, W. A., & Briggs, G. E. (1968). Team performance as a function of team arrangement and work load. *Journal of Applied Psychology, 52,* 89–94.

Kanfer, F. H., & Seidner, M. L. (1973). Self-control: Factors enhancing tolerance of noxious stimulation. *Journal of Personality and Social Psychology, 25,* 381–389.

Katchmar, L. T., Ross, S., & Andrews, T. G. (1958). Effects of stress and anxiety on performance of a complex verbal-coding task. *Journal of Experimental Psychology, 55,* 559–563.

Kirmeyer, S. L., & Dougherty, T. W. (1988). Workload, tension and coping: Moderating effects of supervisor support. *Personnel Psychology, 41,* 125–139.

Klonowicz, T. (1989). Reactivity, experience, and response to noise. In C. D. Spielberger, I. G. Sarason, & J. Strelau (Eds.), *Stress and anxiety* (Vol. 12, pp. 123–139). New York: Hemisphere.

Koeske, G. F., & Koeske, R. D. (1989). Workload and burnout: Can social support and perceived accomplishment help? *Social Work, 34,* 243–248.

Krueger, G. P. (1989). Sustained work, fatigue, sleep loss and performance: A review of the issues. *Work & Stress, 3,* 129–141.

Langer, E. J., & Saegert, S. (1977). Crowding and cognitive control. *Journal of Personality and Social Psychology, 35,* 175–182.

Lazarus, R. S. (1966). *Psychological stress and the coping process.* New York: McGraw-Hill.

Lee, C. (1982). Self-efficacy as a predictor of performance in competitive gymnastics. *Journal of Sport Psychology, 4,* 405–409.

Mackworth, N. H. (1961). Research on the measurement of human performance. In H. W. Sinaiko (Ed.), *Selected papers on human factors in the design and use of control systems* (pp. 174–331). New York: Dover.

Meichenbaum, D. (1985). *Stress inoculation training*. New York: Pergamon.

Miller, K. I., Ellis, B. H., Zook, E. G., & Lyles, J. S. (1990). An integrated model of communication, stress, and burnout in the workplace. *Communication Research, 17*, 300–326.

Monk, T. H., & Folkard, S. (1983). Circadian rhythms and shiftwork. In Robert Hockey (Ed.), *Stress and fatigue in human performance* (pp. 97–121). New York: Wiley.

Mossholder, K. W., Bedeian, A. G., & Armenakis, A. A. (1982). Group process–work outcome relationships: A note on the moderating impact of self-esteem. *Academy of Management Journal, 25*, 575–585.

Morgan, B. B., Jr., Coates, G. D., & Alluisi, E. A. (1975). *Final report on the effects of continuous work and sleep loss on sustained performance and recovery during continuous operations* (Tech. Rep. No. PR–75–5). Norfolk, VA: Old Dominion University.

Norris, F. H., & Murell, S. A. (1988). Prior experience as a moderator of disaster impact on anxiety symptoms in older adults. *American Journal of Community Psychology, 16*, 665–683.

Ostberg, O. (1973). Interindividual differences in circadian fatigue patterns of shift workers. *British Journal of Industrial Medicine, 30*, 341–351.

Pepler, R. D. (1958). Warmth and performance: An investigation in the tropics. *Ergonomics, 2*, 63–88.

Perrewe, P. L. (1986). Locus of control and activity level as moderators in the quantitative job demands–satisfaction/psychological anxiety relationship: An experimental analysis. *Journal of Applied Social Psychology, 16*, 620–632.

Ramsey, J. D. (1983). Heat and cold. In G. R. J. Hockey (Ed.), *Stress and fatigue in human performance* (pp. 33–60). New York: Wiley.

Rosenbaum, M. (1980). Individual differences in self-control behaviors and tolerance of painful stimulation. *Journal of Abnormal Psychology, 89*, 581–590.

Rosenbaum, M., & Rolnick, A. (1983). Self-control behaviors and coping with seasickness. *Cognitive Therapy and Research, 7*, 93–98.

Sanders, M., & McCormick, E. (1987). *Human factors in engineering and design*. New York: McGraw-Hill.

Schuler, R. S. (1975). Role perceptions, satisfaction, and performance: A partial reconciliation. *Journal of Applied Psychology, 60*, 683–687.

Sherrod, D. (1974). Crowding, perceived control, and behavioral aftereffects. *Journal of Applied Social Psychology, 4*, 171–186.

Snook, S. H. (1971). The effects of age and physique on continuous-work capacity. *Human Factors, 13*, 467–479.

Solomon, S., Holmes, D. S., & McCaul, K. D. (1980). Behavioral control over aversive events: Does control that requires effort reduce anxiety and physiological arousal? *Journal of Personality and Social Psychology, 39*, 729–736.

Spielberger, C. D., (1972). Anxiety as an emotional state. In C. D. Spielberger (Ed.), *Anxiety: Current trends in theory and research* (Vol. 1, pp. 24–26). New York: Academic Press.

Steiner, M., & Neumann, M. (1978). Traumatic neurosis and social support in the Yom Kipper War returnees. *Military Medicine, 143*, 866–868.

Strelau, J., & Maciejczyk, J. (1977). Reactivity and decision making in stress situations in pilots. In C. D. Spielberger & I. G. Sarason (Eds.), *Stress and anxiety* (Vol. 4, pp. 29–42). Washington, DC: Hemisphere.

Tetrick, L. E., & LaRocco, J. M. (1987). Understanding, prediction, and control as moderators of the relationships between perceived stress, satisfaction, and psychological well-being. *Journal of Applied Psychology, 72*, 538–543.

Tole, J. R., Stephens, A. T., Harris, R. L., & Ephrath, A. R. (1982). Visual scanning behavior and mental workload in aircraft pilots. *Aviation, Space and Environmental Medicine, 53*(1), 54–61.

Tucker, L. A., Cole, G. E., & Friedman, G. M. (1986). Physical fitness: A buffer against stress. *Perceptual and Motor Skills, 63*, 955–961.

Tziner, A., & Vardi, Y. (1982). Effects of command style and group cohesiveness on the performance of effectiveness of self-selected tank crews. *Journal of Applied Psychology, 67*, 769–775.

Weaver, J. L., Morgan, B. B., Jr., Adkins-Holmes, C., & Hall, J. (1992). *A review of potential moderating factors in the stress-performance relationship* (Tech. Rep. No. 92–012). Orlando, FL: Naval Training Systems Center.

Weinberg, R. S., Gould, D., Yukelson, D., & Jackson, A. (1981). The effect of preexisting and manipulated self-efficacy on a competitive muscular endurance task. *Journal of Sports Psychology, 4*, 345–354.

Weinberg, R. S., & Ragan, J. (1978). Motor performance under three levels of trait anxiety and stress. *Journal of Motor Behavior, 10*, 169–176.

Weinberg, R. S., Yukelson, D., & Jackson, A. (1980) Effect of public and private efficacy expectations on competitive performance. *Journal of Sport Psychology, 2*, 340–349.

Weinstein, N. (1977). Noise and intellectual performance: A confirmation and extension. *Journal of Applied Psychology, 62*, 104–107.

Whiteside, T. C. D. (1965). Visual illusions of movement. *Brain, 88*, 193–210.

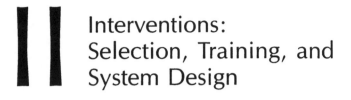

II Interventions: Selection, Training, and System Design

6 Selection of Personnel for Hazardous Performance

Joyce Hogan
Michael Lesser
University of Tulsa

Each year in the United States, thousands of men and women apply for jobs as police officers and firefighters. They are tested for cognitive and physical ability and given background checks. In some jurisdictions, candidates are interviewed by personnel officials, whereas others require clinical assessments of psychopathology. A medical examination insures current physical health and a residency investigation insures the correct address. Successful completion of these procedures results in a list of candidates eligible to perform hazardous work in service to the taxpayers. After this extensive evaluation, however, are the eligible candidates really able to handle the dangers, risks, and perils inherent in the emergencies to which they must respond? We do not know.

Personnel selection for hazardous jobs performance is problematic in two ways. The first concerns the meaning of "hazard" in hazardous work, and the second concerns defining the individual differences that predict effective performance in hazardous work. How should we define hazardous work, how can we evaluate the requirements of such work, and how can we translate these requirements into personnel assessment procedures? In trying to deal with the definitional problem, we should note that jobs are not hazardous because the people performing them experience stress. If this were true, then we might conclude such jobs as secretary, book editor, meter reader, and university professor are hazardous.

DEFINITIONS AND EXAMPLES

In a series of studies of stress among Navy divers, Biersner and his colleagues categorized diver performance as hazardous, because it involved substantial physical risks for the incumbent (Biersner, Dembert, & Browning, 1980; Biersner & Larocco, 1983, 1987). This definition, which focuses on the physical risks to which a person is exposed, would identify as hazardous such positions as test pilot, bomb disposal technician, firefighter, and infantry soldier. Butcher (1985) defined hazardous occupations as those that expose the incumbent to substantial occupational stress, personal risk, and personal responsibility. Such occupations might include air traffic controller, nuclear power plant operator, or astronaut.

More recently, Siegel (1988) proposed that hazard is not a property of work, but is a construct underlying performance. Conditions leading to the evaluation of work as hazardous include (a) a hostile job environment that potentially affects the safety of the incumbent, or (b) a job in which inadequate performance could result in a risk to public safety. This concept combines both themes of earlier definitions. Siegel's view is that hazardousness is a unique requirement and should be studied as a special attribute. Under this definition, such jobs as Arctic explorer, mercenary soldier, and safari guide are hazardous occupations.

Various government agencies concerned with labor issues have also provided definitions. For example, the U.S. Office of Personnel Management (1991) considers hazardous work as:

> duty performed under circumstances in which an accident could result in serious injury or death, such as duty performed on a high structure where protective facilities are not used or an open structure where adverse conditions such as darkness, lighting, steady rain, or high wind velocity exist. (p. 11735)

Under this definition, such jobs as construction/structural worker, refinery pipefitter, and oil platform roustabout are hazardous.

Physiological psychologists interested in industrial toxicology and medicine have also provided some useful definitions and distinctions. O'Connor (1990) defined hazard as an intrinsic property of a device, agent, or activity that is known to result in harm when contacted by others. A hazard creates the potential for injury or danger (Bentall, 1990). O'Connor's definition explains that, because the properties of a hazard are intrinsic to it, they cannot be altered. However, *risk* is a separate issue—it is not an intrinsic property of a device, agent, or activity and can be altered. Risk concerns the probability of being injured by a hazard, and can range from being nonexistent to being serious. Hazards can be avoided or risks associated with a hazard can by manipulated through engineering (e.g., protective equipment) or administrative (e.g., safety training) controls (Fraser, 1989).

We define a hazardous job as one in which inadequate performance will lead to physical harm to one's self, others, or both. We distinguish hazardous work from stressful work in that hazard is a function of potential for physical harm, whereas stress is a function of individual perception and concur with Lazarus' (1966) view that stress is in the eye of the beholder. It is argued here that stress and hazard are independent characteristics and people in hazardous jobs may or may not feel stressed by their work. In fact, we hypothesize that the most effective performers in hazardous jobs are those who experience the least stress.

Accidents are not, in and of themselves, indices of hazardous work—although they may be indices of job stress. Consistent with O'Connor's (1990) view of their intrinsic nature, hazards and hazardous work are defined by requirements of the job and not by characteristics of personnel who hold those jobs. Such work may also be influenced by the physical environment because the context in which the work is done may result in physical harm to self and others. Moreover, the degree to which a job has potential for hazard can depend on task demands and characteristics (e.g., overload, time sharing, response time, unpredictability) that interact with human operator capability. These task demands are often called *job stressors,* but we will not use this phrase in order to avoid confusion with the concept of stress as a property of the worker. In our view, the potential for harm can be exaggerated by the demands of the task environment, making some jobs inherently more hazardous than others.

ASSESSING REQUIREMENTS OF HAZARDOUS WORK

For industrial psychologists, the process of developing personnel selection procedures begins with a job analysis. Because the goal of personnel selection is to predict job performance, a thorough understanding of the job and its requirements are essential. Although the variations in approach, descriptions, ratings, and content inclusion make job analysis ambiguous, a carefully crafted job analysis is nonetheless useful for developing ideas about individual differences in performance. Traditional job analysis methods emphasize identifying actual behaviors required by the job; Dunnette (1966) provided an elegant presentation of this view. More recent conceptions of job analysis typically include statements regarding the tasks a worker does as well as the competencies a worker needs to accomplish the tasks. The distinction, although blurred as Sparks (1988) so insightfully pointed out, is labeled "job-oriented" versus "worker-oriented" job analysis approaches. Job-oriented information is contained in the tasks necessary for performing a job. Examples include such statements as, "Operates pressure or chemical fire extinguisher by carrying it to objective, aiming, and discharging at seat of the fire," and "Sets up portable screens, showers, and equipment, to provide for decontamination of personnel and equipment and to dispose of contaminated clothing and materials." Worker-oriented information is contained in statements about the knowledge, skill, and ability required to perform the tasks of the job. Examples include such statements as, "Knowledge of hazards and procedures associated with bottled and natural gas piped to fire building or adjoining buildings," "Ability to make skillful coordinated movements of the hand(s) and arms(s) to grasp, place, or assemble objects," and "Skill in driving motorized apparatus such as engines and trucks."

Both approaches are potentially useful for understanding the requirements of hazardous jobs. Job-oriented information concerns the tasks that involve hazards; worker-oriented information concerns the personal resources necessary to do hazardous work. Our goal is to link what the job requires with what it takes to fulfill the requirements; this means specifying the relation between job requirements and personal capabilities. Once this is done, it is still necessary to translate the linkages to individual difference assessments. Identifying the constructs common to the job and the performance domains requires insight and experience; the test of successful construct identification rests in the validity coefficients associated with construct measures.

Structured Job Analysis Methods

The well-recognized job analysis methods reveal, at best, the physical characteristics of job hazards (e.g., toxins, physical agents, chemical environment) and, at worst, no information at all. These analyses tend to treat job hazards as a minor issue and evaluate hazardous requirements in an ad hoc way. There are two exceptions: The Threshold Traits Analysis (Lopez, 1986) and The Critical Incident Technique (Flanagan, 1954). Levine, Ash, Hall, and Sistrunk (1983) concluded that Critical Incidents, Task Inventory/CODAP, and Functional Job Analysis (FJA) are best suited for analyzing job hazards. Unfortunately, CODAP and FJA do little more than identify the hazard and parameters of the task. These limitations also apply to the *Handbook for Analyzing Jobs* (U.S. Department of Labor, 1991), which identifies physical demands and environmental conditions that are inherently dangerous, and the Position Analysis Questionnaire, which includes subject matter experts' ratings of "possibility of occurrence" for some potential hazards.

The *Job Analysis Handbook for Business, Industry, and Government* (Gael, 1988) dedicates a chapter to identification of potential job hazards where hazards are defined as a hostile job environment or may be the result of improper or inadequate job performance (Siegel, 1988). Using observations and interviews to develop a job inventory, hazards are identified by incumbents who complete inventory task ratings using scales appropriate for the industry and hazard. Siegel suggested that combining task rating results from 7-point rating scales for the dimensions "Frequency of Performance" and "Consequences of Inadequate Task Performance" produce the frequently performed tasks possessing the highest public risk potential. When this procedure is applied to tasks performed by maintenance mechanics in nuclear power plants, the most hazardous tasks concerned removing and installing pumps, piping, seals, and valves.

From a job-oriented approach, if the goal is to identify hazards, it might be more useful to consider methods used by industrial hygienists; they evaluate the workplace for potential problems and identify their extent and rely on survey methods to collect data on existing and potential hazards. A survey method proposed by Fraser (1989) provides an example of the scope of this analysis. First, considering the end product and working backward, the hygienist identifies the byproducts, raw materials, processes used, and characteristics of exposed employees. Then, if there is a hazard or a potential for one, the evaluations include: (a) identifying its nature (e.g., chemical, biological, physical, ergonomic); (b) determining exposure and determining any foreseeable change in exposure; (c) determining engineering and administrative controls; and (d) assessing the adequacy and use of personal protective equipment. Findings from the survey are used for hazard control, which attempts to minimize the risks associated with the work environment.

From a worker-oriented approach, the Threshold Traits Analysis System (TTAS; Lopez, 1986) is a structured job analysis method used to address worker characteristics associated with hazardous performance. This method attempts to develop a job classification scheme that matches traits to job functions. Five categories of personal qualities, or traits, are included and these are subsequently divided into "can do" and "will do" factors. In TTAS, hazardous performance is a "will do" trait in the motivational category and it is needed to perform "dangerous" job functions. Supervisors are asked to evaluate the trait "Adaptability to Hazards" for such dimensions as importance, relevancy, level of trait needed for acceptable performance, level of trait needed for superior performance, and number of employees who possess the trait. Lopez (1988) reported that the results from TTAS are useful for designing personnel selection procedures; however, without additional job information, it is difficult to see the links to assessments. Evidence that people whose work is "dangerous" need to be adaptable to hazards is uninformative, at best, and incorrect, at worst. For example, absence of adaptability and strict adherence to procedures may be best for dealing with many workplace hazards (e.g., bombs to be defused).

Critical Incidents of Hazardous Performance

Although most job analysis methods describe what jobs require, few describe the behavior or characteristics necessary to perform effectively. This is a serious problem in the study of hazardous jobs, because it is not sufficient to know that the job involves risks and danger. What is needed is information about the characteristics and actions that lead to successful performance. Perhaps the best method for understanding successful hazardous job performance is Flanagan's (1954) critical incident method.

Flanagan defined as incident as:

> any observable human activity that is sufficiently complete in itself to permit inferences and predictions to be made about the person performing the act. To be critical, an incident must occur in a situation where the purpose or intent of the act seems fairly clear to the observer and where its consequences are sufficiently definite to leave little doubt concerning its effects. (p. 327)

Using the method, observers are asked to record critical incidents and to describe (a) what led up to the incident, (b) what exactly the person did that was so effective or, conversely, ineffective, and (c) the consequences of the behavior. Typically, a large number of incidents are collected and content analyzed into a reduced set of descriptive categories. The categories are then used to obtain information about errors (Flanagan, 1947) to develop task behavior checklists (Kirchner & Dunnette, 1957).

In hazardous jobs, there seems to be a ceiling on performance where good performance is characterized by an absence of errors. In a sense, there may be only one way to perform correctly, but many ways to fail. Much of this information can be captured in critical incidents. Dunnette, Bownas, and Bosshardt's (1981) evaluation of nuclear power plant operator jobs is a good example of how to use critical incidents to study performance in a hazardous job. Dunnette et al. wanted to identify job behavior associated with safety and security risks in nuclear plants. Supervisors and managers from U.S. operating companies provided 158 critical incidents of inappropriate, unreliable, or aberrant operator behavior. These incidents were content analyzed, 18 behavior categories were defined, and performance rating scales were developed for each category. Supervisors then used these scales to rate the performance of their operators. The ratings were correlated, the resulting matrix was factor analyzed, and six factors were defined by the authors.

The first factor, "Argumentative Hostility Toward Authority," involves antisocial conduct such as refusing to work as part of a team, challenging authority, arguing and fighting, and taking actions to get back at others. The second factor, "Irresponsibility and Impulsivity," involves refusing to comply with rules and procedures, engaging in horseplay, and playing pranks on others. The third factor, "Defensive Incompetence," is characterized by withdrawal from the work group, avoiding others, failing to pass along information, trying to cover up mistakes, and refusing to take responsibility in the work group. The fourth factor, "Psychopathology," is defined by moodiness, depression, panic in emergency situations, unprovoked emotionality, and frustration over rules and procedures. The fifth factor, "Compulsive Incompetence," involves refusing to delegate duties, checking everything excessively, and "cracking" from the inability to do all the work of the team. The final factor, "Substance Abuse," involves chemically related instability and erratic behavior.

Although this seems to be a unique study of ineffective performance in a hazardous job, we suspect that these dimensions will generalize to other situations involving personal risk. The categories are not industry specific and may characterize aberrant job behavior in general. The most surprising result of this study is that ineffective performance is only marginally related to psychopathological behavior. The constructs underlying four of the six behavioral dimensions are characterological and resemble personality dysfunctions. Moreover, the "Irresponsibility and Impulsivity" factor may be a function of poor person–job fit where bright, energetic, and curious young employees find themselves engaged in a boring, 10-hour-per-day control room vigilance task. Consequently, they engage in pranks and horseplay as a way of staying interested and alert.

In a more limited evaluation of successful and unsuccessful performance in hazardous work, Hogan and Hogan (1989) collected 72 critical incidents from Navy explosive ordinance disposal (EOD) technicians, all of whom had completed at least two tours of duty. EOD technicians dismantle or "render safe" live ordnance in both military and civilian settings. After successful completion of technical training, they are qualified to disarm ordnance ranging from Civil War munitions to present day conventional and nuclear devices to terrorist bombs.

The incidents were assigned to one of four categories according to the following content themes. First, a number of incidents concerned deficient technical training that appeared to be errors of judgment or action (e.g., an EOD technician pick up a piece of live ordnance and carried it over to another technician to find out what it was). Second, some incidents concerned being defensive about one's technical deficiencies (e.g., an "experienced" technician would not listen to safety procedures for moving a rocket offered by a newly trained technician). Third, a number of incidents involved irresponsibility—either on the part of the supervising officer or technician. In one example, a firing range supervisor failed to require the technicians to stamp expended ordnance as safe and merely segregated ordnance into piles. The contractor who had purchased the expended ordnance picked up the wrong pile and two civilians were seriously injured when handling the "safe" ordnance. Finally, some incidents clearly reflected impulsivity. Figure 6.1 shows an incident that describes such a problem. Failure to check the identifying color of a bomb before rigging and removing it is a deadly shortcut.

There are some overlapping themes in the studies of nuclear power plant operators and EOD technicians. Common to both is the requirement for thoughtful, thorough, and comprehensive action; ineffective performance appears, in part, as irresponsible and impulsive behavior. An additional area of concern is technical competence. This includes both sufficient technical knowledge to deal with problems and the ability to acquire technical knowledge from others without being defensive.

Summary

Campbell, McCloy, Oppler, and Sager (1993) defined job performance as task behavior—something people do or the actions people take. For any given job, a number of performance components exist that can be identified though job analysis. For hazardous jobs, we favor the latent factor model. This model focuses on the latent structure of job requirements by identifying constructs underlying performance (Campbell, 1990b). The advantage of this approach is that the

Directions:
Please think back over your years of experience working as an EOD technician in a detachment and try to remember noteworthy examples of things EOD technicians have done in their jobs that illustrate HAZARDOUS or POOR job performance.

1. What were the circumstances leading up to this example?

Fifteen EOD technicians were doing a range sweep. A Mark 83 was found covered up to the tailfin with earth. The bomb was pulled out of the ground with a frontend Loader. while the bomb was suspended from the ground the chain broke. The bomb came Loose and fell to the ground. The bomb turned out to be live with an armed fuse.

2. Describe exactly what was done that makes this a noteworthy example of POOR job performance.

No one actually checked the color of the bomb before pulling it out of the ground. All 15 of the technicians could have been killed.

3. What were the consequences of these actions?

In this instance none. The bomb did not explode

4. Please provide the following information:

a. Your current job assignment: instructor

b. Please assign a rating of performance level of the example you provided:

(1) 2 3 4 4 6 7
Poor — Superior

FIG. 6.1 EOD technician performance example description form.

criterion and predictor domains can be aligned according to constructs underlying both domains. Although this strategy seems obvious (e.g., see Pulakos, Borman, & Hough, 1988), it is seldom used because little attention is normally paid to the construct validity of job performance indicators. Different performance factors will be predicted by different construct measures; in this section, we have tried to identify constructs underlying performance of hazardous work.

Specifically, these constructs concern poor person–job match, technical incompetence, physical inability to perform safely, and personality dysfunctions.

Our next task is to use the constructs underlying job performance to specify measurement of the predictor domain. What are the most promising assessments that could be used to identify personnel who can work effectively in hazardous jobs?

ASSESSING PERSONNEL FOR HAZARDOUS PERFORMANCE

Our performance theory suggests that personnel who are effective in hazardous work are (a) interested in the nature of the work, (b) technically competent, (c) physically capable, and (d) psychologically well-suited. (Psychological suitability entails not feeling stress associated with the risks of the work.) Each of these characteristics is associated with well-developed assessment tools.

Occupational Interest

Measures of vocational and occupational interests assess a person's preference for activities and jobs. If a person is interested in an activity or job, then he or she is more likely to be attracted to and satisfied with that activity than a person who has limited or no interest. Moreover, attraction to an occupation is likely to result in tenure in the position or specialty. Interest should not be confused with talent because interest does not guarantee talent and vice versa. However, the most productive career choices arise when occupational interests and talent are aligned; this match may be problematic for careers in hazardous work.

The first standardized instruments to assess occupational interests were published in the 1920s by Clarence Yoakum at the Carnegie Institute for Technology. In 1927, Edward K. Strong published a practical device to assess occupational preferences—the Strong Vocational Interest Blank—which is still available today. In 1934, G. Frederick Kuder published the Kuder Preference Record, a simple and easily scored instrument, which is still used. However, noting that these assessments are atheoretical, Cronbach (1960) predicted, "The next development that may be forecast is the systematic interpretation of interests in terms of more fundamental constructs" (p. 436). Six years later John Holland published his *Psychology of Vocational Choice*, where he integrates vocational interests with personality and organizational environments in an explicit theory of careers. This theory is particularly appropriate for understanding interest patterns of those who will be suited for hazardous work.

Holland's (1966) typology organizes data about different work environments and explains job satisfaction and vocational achievement. He conceptualized interests in terms of six types, which are actually ideal personality types. The assessments, therefore, estimate a person's resemblance to each of the types and most of the preference variance is summarized using three of the six types. In Holland's (1979) developmental model, cultural and personal experiences result in preference for some activities over others; preferred activities become strong interests. In adulthood, "interests and competencies create a particular personal disposition that leads to thinking, perceiving, and acting in special ways" (p. 4). Holland's (1966, 1975, 1985) six occupational types, each with distinctive interests, preferences, and personality characteristics, are (a) Realistic (engineer), (b) Investigative (scientist), (c) Artistic (writer), (d) Social (counselor), (e) Enterprising (manager), and (f) Conventional (banker). Gottfredson and Holland (1989) provided the following summary:

- Occupations classified as Realistic (R) tend to involve concrete and practical activity involving machines, tools, or materials.
- Occupations classified as Investigative (I) tend to involve analytical or intellectual activity aimed at problem solving, trouble shooting, or the creation and use of knowledge.
- Occupations classified as Artistic (A) generally involve creative work in the arts: music, writing, performance, sculpture, or other relatively unstructured and intellectual endeavors.
- Occupations classified as Social (S) typically involve working with people in a helpful or facilitative way.
- Occupations classified as Enterprising (E) tend to involve working with people in a supervisory or persuasive way to achieve some organizational goal.
- Occupations classified as Conventional (C) typically involve working with things, numbers, or machines in an orderly way to meet the regular and predictable needs of an organization or to meet specified standards. (p. 6)

Figure 6.2 presents the hexagonal model for interpreting these types, where the distances between the types are inversely proportional to the size of the correlations between them. The spatial arrangement enhances interpretation of the adjacent versus distance types—adjacent types are more psychologically similar than distant types. Also seen in interpreting the model in Fig. 6.2., the categories are not distinct, indicating no pure types. The categories blend together at some point and the distinctions among lower educational level occupations are the most blurred (Gottfredson & Holland, 1989). For this reason, occupations (and people) will resemble more than one type and Holland (1985) reported occupational results using three types. His model is capable of classifying every job in the *Dictionary of Occupational Titles* (Gottfredson & Holland, 1989).

What is the interest profile of persons who prefer to work in hazardous jobs? One answer is to consider the classifications in which hazardous jobs appear and detect interest commonalities. A second strategy is to consider the types that rarely appear in classifications of hazardous jobs. A final strategy is to collapse

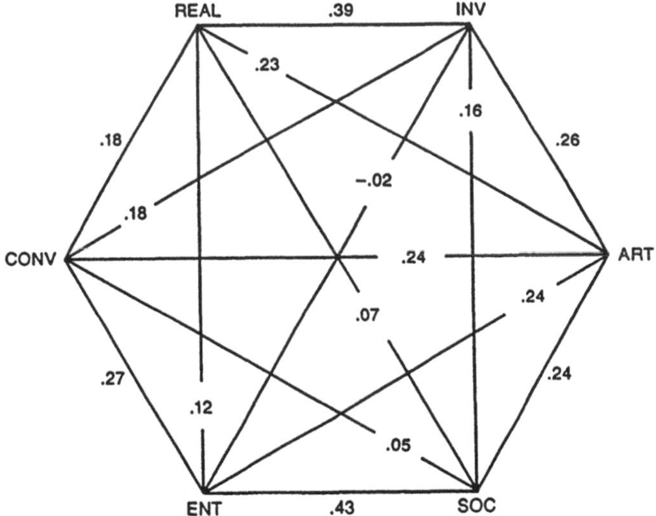

FIG. 6.2 A hexigonal model for interpreting inter- and intra-class relationships of occupational types. Correlations are between the SDS scales (1985 Revision) for 175 women age 26–65. From Holland (1985). Reprinted by permission.

these two initial strategies; we chose this final approach. Table 6.1 presents a sample listing of hazardous jobs and their respective Holland type codes. The modal type for all occupations listed is Realistic, with Social, Conventional, and Enterprising types also represented with lesser frequency. This suggests that hazardous jobs involve practical activity carried out in a systematic way in order

TABLE 6.1
Sample Hazardous Jobs and Their Corresponding Holland Types

Occupational Title	*Code*	*Occupational Title*	*Code*
Air Traffic Control Coordinator	SER	Explosives Operator	CRE
Air Traffic Control Specialist, Station	SCE	Firefighter	RES
Air Traffic Control Specialist, Tower	SER	Firefighter, Crash, Fire, and Rescue	RES
Airplane Pilot, Commercial	RIE	Infantry Weapons Crewmember	RES
Airplane Pilot	IRE	Nuclear Critical Safety Engineer	IRC
Atomic Fuel Assembler	RCE	Nuclear Engineer	IRE
Bomb Loader	RES	Oiler	RCS
Body Guard	ESC	Oil Well Services Operator Helper	RCS
Bouncer	ESR	Pilot, Ship	REI
Diver	RES	Plant Operator	RIE
Electrical Line Splicer	RCS	Police Officer	SER

to help people or to accomplish an organizational goal. Notably absent in Table 6.1 are any classifications involving Artistic types, which suggests that hazardous jobs do not involve creative work. Further, although interest measures are rarely used as predictors in validation research, this rational analysis is corroborated by findings that bomb disposal divers have their highest scores for Realistic, Investigative, and Social interests and the Self-Directed Search Realistic scores were significantly related to success in diving training ($r = .21, p < .05$), in bomb disposal training ($r = .24$, $p < .01$), and chief petty officers' ratings ($r = .24$, $p < .01$) (Hogan & Hogan, 1989). Artistic interests were negatively related to course success ($r = -.17$, $p < .05$).

There is a message in this analysis. Successful job performance will depend, in part, on the congruence between occupational requirements and personality characteristics. Persons with artistic interests will be unsuited for work that requires concrete, practical, and systematic activities. Artistic types are disorderly, emotional, impractical, impulsive, and nonconforming—characteristics undesirable for a nuclear power plant operator. Unfortunately, artistic persons are sometimes attracted to hazardous jobs because of the thrilling image the media creates about these occupations. In fact, creative people who enter Realistic and Conventional jobs are quickly dissatisfied with the routine, careful, and self-controlled activities required. In the best case, these people are unhappy and turnover in the job; in the worst case, they make mistakes, cause accidents, and become hazards themselves. Although there are a variety of types of people working successfully in hazardous jobs, some types will appear more frequently than others and we conclude that ones most suited for the work will have predominantly Realistic, Conventional, and Social interests.

Technical Competence

Hazardous work requires people who will profit from training and continue to learn techniques while on the job. The complexity and the increasing high-tech nature of hazardous work environments require comprehensive initial training and almost continuous retraining to maintain personnel efficiency (Nauta, Ward, & D'Ambrosia, 1983). Traditionally, selection and placement decisions have been based on tests of intelligence and general aptitude, and for good reason such measures predict performance in educational settings, as Binet originally intended. This is important for our purposes because most hazardous jobs are preceded by extensive training programs. Firefighters spend up to 4 months in classroom training and 1 year on probation before attaining permanent status. Nuclear power plant operators spend a full year in training, with extensive exposure to work simulations and to possible incidents that must be resolved. Cognitive tests predict performance in job training and the data to support this conclusion are substantial (Hunter, 1983).

To what degree are cognitive ability assessments useful for selecting and placing individuals in hazardous jobs? Schmidt and Hunter (1977, 1981) provided data showing that cognitive ability tests are valid predictors of performance in all jobs. Hunter and Hunter (1984) asserted that there are no alternatives to cognitive ability tests for predicting job performance. We disagree with this generalization and argue instead that the correct analysis will result in a theory of job performance, much like the one we are considering here.

Project A: The U.S. Army Selection and Classification Project (Campbell, 1990a) contains the most extensive study of the usefulness of cognitive ability tests for predicting performance in hazardous jobs. This 7-year study was designed to develop an organization-wide selection and classification system focused on 19 prototype occupational specialties. Eight of the nine jobs targeted for extensive field validation were combat specialties. These included infantryman, cannon crewman, tank crewman, medical care specialist, and military police.

Soldiers whose ASVAB scores were available completed a battery of cognitive ability and temperament/personality measures. They were evaluated on five components of job performance including core technical proficiency, general soldiering proficiency, effort and leadership, personal discipline, and physical fitness and military bearing. These criteria resulted from LISREL VI tests of performance models based on 200 indicators from 9,000 soldiers (Campbell, McHenry, & Wise, 1990). Correlational analyses indicated that the ASVAB and the experimental cognitive ability tests were the best predictors of the proficiency criteria. ASVAB composites correlated with Core Technical Proficiency (mean $R = .63$) and General Soldiering Proficiency (mean $R = .65$), but they did not predict the other performance factors very well (McHenry, Hough, Toquam, Hanson, & Ashworth, 1990). This supports the job performance model where cognitive ability is the latent structure underlying both the predictor and criterion domains.

It is interesting to note that all the predictors used in this study were logically related to job performance components. Cognitive tests predicted proficiency criteria. Temperament/personality measures predicted the noncognitive criteria. The only performance factor not well predicted was Physical Fitness and Military Bearing, which could be expected because Project A contain no physical ability tests. Campbell et al. (1990) demonstrated that the job performance model might generalize to the population of hazardous jobs; Campbell et al. (1990) suggested that technical proficiency, effort and leadership, and personal discipline are common to all jobs. For law enforcement occupations, cognitive ability tests predict training (validities ranged from .41 to .75) and job proficiency (validities ranged from .17 to .31) criteria, and this corroborates the military job performance model (Hirsh, Northrop, & Schmidt, 1986).

Physical Capability

Since the late 1970s, people have studied individual differences in physical abilities almost exclusively in the context of personnel selection. This is likely to continue over the next 10 years in both military and civilian settings because the military is concerned with maximizing human capability in an all-volunteer force, and civilian employers are concerned with the Americans with Disabilities Act of 1990 (ADA). The ADA will have a greater impact on employment in physically demanding work than Title VII of the Civil Rights Act, although issues of sex discrimination are still of enormous concern particularly in such hazardous positions as firefighter, police officer, and correctional officer.

The fundamental problem in studying physical performance in hazardous work is a lack of an adequate taxonomy of (a) work requirements and (b) individual differences associated with those requirements. The source of the problem is at least twofold. First, relative to the study of cognitive abilities, there has been only a small amount of research on physical abilities in the workplace. Historically, considerations other than performance tests were used to make hiring decisions. Such considerations, which included gender, physical characteristics, and manual labor restrictions, were prohibited under Title VII of the Civil Rights Act of 1964. The use of physical ability tests to select people for physically demanding jobs is relatively recent. Second, research in physical performance is interdisciplinary and draws on physiology, biomechanics, and ergonomics. Although an interdisciplinary approach will produce new insights, it is also divisive in that it is difficult to generate a consensus about a framework for taxonomic representation.

The characteristics necessary for successful performance depend on what a job requires and, for hazardous work, the evaluation should include a description of the tasks and abilities that involve strength, endurance, and movement quality (Hogan, 1985, 1991a, 1991b). Researchers are primarily concerned with developing predictors that are maximally related to important criteria, which requires focusing on the physical ability constructs that underlie both predictor and criteria domains. The most recent example using this strategy is a study of physical ability tests and their relation to police officer performance (Arvey, Landon, Nutting, & Maxwell, 1992); this is a model for the future because of its research strategy. Although the state of the art in physical ability research is not well advanced, three points are relevant for personnel selection in hazardous, physically demanding jobs.

First, physical ability tests are valid predictors of job performance when the critical tasks are physically demanding. At least three reviews support this assertion. Lewis (1989) computed a meta-analysis of 24 physical ability selection studies including police officers, firefighters, refinery operators, and miners, where he classified the predictors as muscular strength tests, anthropometric measures, muscular power tests, and muscular endurance tests, and the criterion measures as work sample measures, supervisor ratings, and training performance. Table 6.2 presents his results and, as can be seen, the highest mean validity ($\rho = .82$) is between muscular strength tests and work sample measures. Also, note that, in general, the work sample criterion yields higher validities than rating or training criteria. The overall average validity across muscular tests and criteria is .39.

TABLE 6.2
Summary Validities of Physical Ability Tests

Lewis (1989) Meta-Analysis of Physical Tests

	Criterion								
	Work sample			*Supervisor rating*			*Training criteria*		
Predictor	ρ	*N*	*K*	ρ	*N*	*K*	ρ	*N*	*K*
Muscular strength	.816	2064	10	—	—	—	.23	396	3
Anthropometric	.49	2176	12	.255	1426	6	.23	396	3
Muscular endurance	.367	1740	6	.23	2022	9	.30	750	5
Muscular power	—	—	—	.26	1699	5	—	—	—

Hogan (1991a) Mean Validity Coefficients of Physical Tests

	Criterion		
Predictor	*Objective*	*Subjective*	*Work sample*
Muscular strength	.36	.30	.79
Endurance	.21	.13	.49
Movement quality	.38	.11	.44

Blakely, Quinones, & Jago (1992) Strength Test Mean Validity Coefficients

	Criterion	
Job	*Supervisor ratings*	*Work sample*
Police officer	.36	.33
GAS service	.33	.75
Construction	.29	.72
Pipefitter	.24	.69
Maintenance	.29	.50
Utility	.17	.48
Average Validity	.34	.68

Note. ρ = corrected mean validity; *N* = number of subjects; *K* = number of samples.

Hogan (1991a) reviewed 14 empirical physical ability test validation studies that were conducted between 1979 and 1988. Samples studied included refinery and chemical plant operators, police officers, miners, divers, and infantry soldiers. The predictors were classified according to the physical performance dimensions listed in Fig. 6.3 and the criteria were labeled as objective (e.g., training time, injuries sustained), subjective (i.e., supervisor ratings), or work sample scores. The strength tests yielded the highest average validities regardless of criterion measure used and those correlations ranged from .30 to .79. Lower

MUSCULAR STRENGTH

MUSCULAR TENSION: requires exerting muscular force against objects. It is used to push, pull, lift, lower, and carry objects or materials.

MUSCULAR POWER: requires exerting muscular force quickly.

MUSCULAR ENDURANCE: requires exerting muscular force continuously over time while resisting fatigue.

CARDIOVASCULAR ENDURANCE

CARDIOVASCULAR ENDURANCE: requires sustaining physical activity that results in increased heart rate.

MOVEMENT QUALITY

FLEXIBILITY: requires flexing or extending the body limbs to work in awkward or contorted positions.

BALANCE: requires maintaining the body in a stable position, including resisting forces that cause loss of stability.

COORDINATION: requires sequencing movements of the arms, legs, and/or body to result in skilled action.

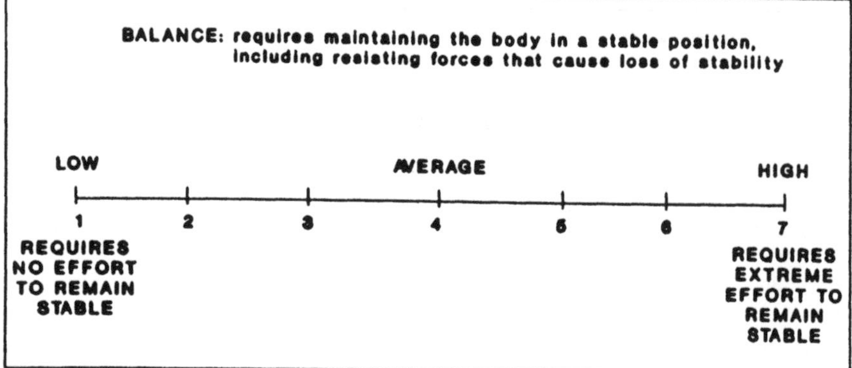

FIG. 6.3. Physical performance construct definitions and sample rating scale. From Hogan, J. (1991b). Adapted by permission.

validities also were obtained for the endurance measures; the average correlations were .21 and .49 for the objective and work sample criteria, respectively. The average validities for the movement quality tests, as a group, were .38 and .44 with the objective and work sample criteria, respectively. Mean validities were lowest for tests correlated with supervisor ratings and, consistent with Lewis' analysis, were highest with the work sample measures. These results also appear in Table 6.2.

Blakley, Quinones, and Jago (1992) summarized studies of the jobs of police officer, customer gas service technician, construction worker, pipefitter, pipeline construction and maintenance worker, and utility worker. The predictors were four isometric muscular strength tests and the criterion measures were supervisors' ratings of physical performance and work sample scores. Across the jobs in the total sample (N = 1276), the average validities for the summed strength test scores with supervisors' ratings and work sample simulations were .34 and .68, respectively (see Table 6.2).

A second point regarding physical ability assessment is that we now understand the structure of predictor batteries derived from job analysis data. Hogan (1991b) analyzed the structure of seven physical performance test batteries used across a wide range of jobs and concluded that test performance can be described in terms of three dimensions—muscular strength, endurance, and movement quality.

Perhaps the most comprehensive study available is Denning's (1984) analysis of the physical requirements of 112 production jobs and 17 maintenance jobs in chemical, plastics, synthetics, and paint processing manufacturing. Many of these jobs would be considered hazardous using the definitions provided earlier. Based on a thorough job analysis, she identified nine physical tests for the experimental battery in a concurrent validation strategy. She tested 853 male and 203 female employees across 10 plant locations. The test battery consisted of body fat skinfold (a proxy for cardiovascular endurance), static strength (push, pull, upper extremity cable pull), medicine ball throw (muscular power), sit-and-reach (flexibility), arm-cranking (muscular endurance), balance, and eye–hand coordination tests. The results of varimax rotated principal components analysis of the test correlation matrix appear in Table 6.3. The first factor was defined by the muscular strength tests, with the muscular tension and power tests loading highest. Factor two was defined by the coordination, flexibility, and balance tests and factor three was defined exclusively by the skinfold measure. Body fat, which is what skinfold measures indicate, is inversely related to cardiovascular fitness (McArdle, Katch, & Katch, 1991). The remaining studies analyzed by Hogan (1991b) also reveal these dimensions. The structure of test performance appears to be uninfluenced by test type, as long as the predictor battery is sufficiently complex. Also, results appear not to be a function of statistical method; nearly identical results were obtained after both orthogonal and oblique rotations. In addition, we see no persuasive argument to assume that the structure of physical performance requirements for hazardous jobs differs from that for other jobs with physical demands.

A third point is that, because muscular strength, endurance, and movement skill tests are valid predictors of job performance, and because these measures are statistically independent, we would expect incremental validity as each of these measures is added to a regression equation. This proposition is supported when complex experimental test batteries are evaluated. Denning (1984) reported

TABLE 6.3
Principal Components Analysis of Physical Performance Tests: Chemical, Plastics, Synthetics, and Paint Processing Maintenance and Production Technicians[a]

Test	*Factor 1*	*Factor 2*	*Factor 3*
Skinfold	−.20	.08	−.87
Balance	.10	.42	.38
Static push	.88	−.15	.06
Static pull	.89	.00	−.17
Cable pull	.79	.08	.25
Coordination	−.08	−.77	.26
Sit & reach	−.05	.68	.11
Medicine ball throw	.82	.07	.29
Cranking (arm)	.63	.24	.26
Percentage of variance	40	19	12

Note. Based on data compiled by Denning (1984). From Hogan (1991b). Adapted by permission.

a multiple *R* of .45 with the medicine ball (muscular strength), skinfold (endurance), and coordination (movement quality as independent variables and overall supervisors' ratings as the dependent variable. She does not report the variance accounted for by each variable; however, the zero-order correlations for the aforementioned variables were .26, −.17, and −.13, respectively. Similarly, Reilly, Zedeck, and Tenopyr (1979), who studied pole climbing jobs, combined experimental predictors using stepwise regression to predict time-to-complete training in pole climbing school. They report a multiple *R* of .45 with measures of body density (endurance), balance (movement quality), and static strength defining the equation. These studies suggest that tests from the three performance domains made significant independent contributions to the prediction of the criterion space. In terms of the relation between physical performance tests and other domains, we speculate that physical ability is unrelated to cognitive ability and modestly related to some factors in the psychomotor and personality domains.

Pscyhological Suitability

Constructs underlying psychological suitability can be identified from critical incidents of job performance discussed earlier. Merging this information with the domain of personality characteristics, performance in hazardous jobs can be described with a five-factor taxonomy used by a number of researchers (McCrea & Costa, 1987; Peabody & Goldberg, 1989; Wiggins, 1973). This taxonomy, referred to as the "Big-Five" personality factors emerged from Allport and Odbert's (1936) surveying of English language trait terms. The factor analytic work of Cattell (1943) followed by Fiske (1949), Tupes and Christal (1961), Norman (1963), and Digman and Takemoto-Chock (1981) reveals five large personality factors: (I) Surgency, (II) Agreeableness, (III) Conscientiousness, (IV) Emotional Stability, and (V) Intellect/Openness to Experience.

Surgency. The first factor, surgency, may be irrelevant for performance in hazardous jobs. John (1990) contended that this factor has at least five distinguishable components, including activity level, dominance, sociability, expressive under control, and positive emotionality. With the exception of the adventurous theme that is part of the expressive under control component, it appears that surgency is not an individual difference characteristic associated with performance where performance entails risk.

Agreeableness. The second factor, agreeableness, also may be inconsequential for performance in hazardous work. This dimension includes such components as warmth, sympathy, giving and helping, and trust. Although agreeable personnel will be more likable coworkers, this may have little to do with actual work effectiveness. In fact, if hazardous duty involves administering or enforcing any security procedures, persons who are seen as disagreeable (e.g., stern, hard-hearted, and unfriendly) may be effective. Montgomery, Butler, and McPhail (1987) found that the Hogan Personality Inventory (HPI; Hogan, 1986; Hogan & Hogan, 1992) Likeability scale, a measure of the Big-Five agreeableness factor, was significantly and negatively related to performance effectiveness as a health security technician in nuclear power plants.

Conscientiousness. The third factor, conscientiousness, is extremely important for performance in hazardous jobs; however, the reason this factor is important is not as obvious as it might seem. Conscientiousness includes at least two major components, one of which is characterized by virtuousness and conformity, and the other of which is characterized by impulse control and attitudes toward authority. Some researchers argue that measures of conscientiousness will be valid predictors of performance in all jobs (Barrick & Mount, 1991; Goldberg, 1992; Tett, Jackson, & Rothstein, 1991). However, we believe that this is only partially true and performance in hazardous jobs illustrates the exception. Pilots, police officers, firefighters, munitions specialists, and special forces personnel like risks and challenging assignments involving physical danger. This preference for and display of excitement- and experience-seeking behavior (e.g., low conscientiousness) is the conceptual opposite of the routine, ordered, and deliberate behavior required for many hazardous jobs. On the other hand, conscientious personnel will have positive attitudes toward authority and will be willing to follow rules.

There is some empirical validity for this curious and improbable syndrome. In a study of highly rated police officers, R. Hogan (1971) administered the California Psychological Inventory (CPI; Gough, 1957) and collected evaluations of performance in the field after one year. One of the best predictors of performance ($r = .52$, $p < .01$) was the Intellectual Efficiency scale that Gough (1968) described as:

> (a) self-confidence and self-assurance, freedom from unsubstantiated fears and apprehensions; (b) effective social techniques and adjustment, sense of social acceptability without dependence on others; not suspicious, touchy, or overly sensitive; (c) good physiological functioning, absence of minor debilitating symptoms and complaints; and (d) liking and respect for intellectual pursuits; wide range of interests. (p. 71)

However, the specific scale pattern of correlations for Self-Control (r) $= .53$, $p < .01$), Responsibility ($r = .30$, $p < .05$), and Achievement via Independence ($r = .32$, $p < .05$) shares the themes of the syndrome we described earlier. Self-Control and Responsibility reflect lack of impulsivity and conformity, whereas Achievement via Independence reflects openness to experience, wide-interests, and curiosity.

In a study of the personality characteristics associated with effective performance in bomb disposal, Navy students completed the HPI. HPI scores were then correlated with explosive ordnance disposal training course completion and performance in the fleet (Hogan & Hogan, 1989). As a group, these personnel were self-confident, introverted, and aggressive; the most effective trainees and fleet divers were nonconforming but not reckless. Consistent correlations ($r = .22–.28$, $p < .01$) across training and field performance criteria for a cluster of conscientiousness items called "*Not* Thrill Seeking" suggest that these divers do not take risks for the thrill of taking risks. Correlations for "Impulse Control" ($r = -.30$, $p < .01$) and "Not Spontaneous" ($r = -.23$, $p < .05$) also showed a degree of impulsivity among the successful Navy trainees. This may be a function of immaturity because these correlations were nonsignificant in the bomb disposal diver sample. Nevertheless, this curious combination of nonconformity and self-control characterizes those who dismantle bombs.

Finally, in a study of the personality characteristics associated with ineffective performance as municipal law enforcement officers, Hargrave and Hiatt (1989) compared CPI profiles of 45 incumbent officers who had experienced serious job problems to those of 45 matched controls who had experienced no disciplinary action. Serious job problems included providing drugs to inmates, conducting illicit relationships with inmates, conviction for the use of illegal drugs, using unnecessary force, fighting with fellow officers, and violating department procedures where such violations resulted in escape of inmates. The CPI scales of Self-Control ($F = 8.53$, $p = .004$) and Socialization ($F = 7.02$, $p = .01$) were most powerful in differentiating these two groups. The themes of impulsivity, irresponsibility, and, in this case, hostility are central in discriminating conscientious performers from the organizational delinquents and misfits.

Emotional Stability. The fourth Big-Five factor, emotional stability, is also an important element of effective performance in hazardous work. Emotional stability is a major issue in jobs where improper performance poses potential hazards to the public. Positions in nuclear power plants, law enforcement, aircraft cockpits, air traffic control, and commercial navigation are examples of jobs in which the incumbent may subject the public to considerable risk. These personnel are usually screened for psychopathology (Butcher, 1985; Dunnette et al., 1981). The emotional stability factor contains three major components for which there are appropriate evaluation instruments.

The first component concerns emotional stability as positive adjustment and stress resistance. This is reflected in self-confidence, self-esteem, and an absence of anxiety and these can be assessed using any well-validated measure of adjustment from a measure of normal personality. Well-validated adjustment scales are related statistically to critical indices of job performance.

For example, Hogan (1971) found that highly rated police officers were distinguished from others by their scores on the CPI "Well-Being" scale ($r = .38, p < .05$). Pugh (1985) used the CPI in a 4.5 year follow-up of police officer field performance and found that "Well-Being" significantly differentiated between high, average, and low performers. Hogan and Hogan (1989) found that HPI measures of "Self-Confident" and "Not Depressed" significantly (*r*s ranged from .17 to .22) predicted completing second-class diving and bomb-disposal schools. Cortina, Doherty, Schmitt, Kaufman, and Smith (1992) used the Inwald Personality Inventory, designed especially for use in the selection of corrections officers in a predictive validation study of state police recruits. They composed a Big-Five neuroticism factor (emotional stability) based on the inventory scales for Hyperactivity, Illness Concerns, Treatment Programs, Anxiety, Phobic Personality, Obsessive Personality, and Depression and found that this composite correlated with probationary job performance ratings ($r = -.15$, $p < .05$), peer evaluations ($r = -.22, p < .05$), academy training ratings ($r = -.23, p < .05$), and job turnover ($r = .19$, $p < .05$).

The second component of emotional stability is an absence of psychopathology. This differs from positive adjustment in that positive adjustment concerns characteristics that are present, to some degree, in all individuals. Assessments of normal-range personality characteristics are useful because all candidates will obtain scores that are meaningful and interpretable. However, some personality characteristics are only present in some personnel and such characteristics are identifiable by high scores on psychopathology measures. The most widely used standardized measures of psychopathology include the Minnesota Multiphasic Personality Inventory (MMPI; Hathaway & McKinley, 1951) and the Clinical Analysis Questionnaire, where scales scores on these inventories are usually interpreted as being either "maladjusted" or "normal range." Butcher (1985) concluded that tests such as the MMPI are appropriate for personnel selection in occupations where extreme personality characteristics may make a person vulnerable to stress, may result in abuses of public trust, or may impair judgment or action in positions of public responsibility.

For personnel selection, attributes required for job performance are evaluated using measures of characteristics likely to operate against effective performance. In practice, a clinician reviews test results, usually the MMPI, and based on evaluation of scores that fall into the "maladaptive" range, an applicant is excluded from further consideration. These decisions are based on expert judgments and, in some cases, expert consensus decision rules. However, clinical assessments are rarely validated empirically using traditional criterion-related strategies. Butcher conceded that little research exists showing that the MMPI identifies maladaptive characteristics for "sensitive or stress-vulnerable positions." However, Cortina et al. (1992) composed an MMPI neuroticism index from Hypochondriasis, Depression, and Psychasthenia scales and, in a sample of police officers, this index correlated with probationary job performance ratings ($r = -.19$, $p < .05$), academy training ratings ($r = -.24$, $p < .05$), and job turnover ($r = .27$, $p < .05$).

The third component of emotional stability concerns the absence of personality disorders. The Diagnostic and Statistical Manual of Mental Disorders (DSM–III–R) of the American Psychiatric Association (1987) includes personality traits that when "inflexible and maladaptive and cause either significant impairment in social or occupational functioning or subjective distress . . . constitute Personality Disorders" (p. 305). Many personality researchers consider the personality disorders to be maladaptive exaggeration of normal personality characteristics. Of the 11 personality disorders, 4 appear to be associated mathematically with emotional stability (Wiggins & Pincus, 1989). These are Borderline, Passive–Aggressive, Dependent, and Avoidant and it is conceivable that these, rather than gross psychopathology, may be responsible for job-related emotional instability. For example, consider the dimensions of emotional instability identified by Dunnette et al. (1981) in nuclear power plant operators. Under stress, "Hot-Temperedness," defined as fits of temper and arguing and fighting with others when problems arise resembles the borderline personality. "Hostility toward Authority," defined as resentment of supervision and taking actions to get back at the company, resembles the passive–aggressive personality. The dependent personality is reluctant to act without direct orders to offer advice or expertise to others. The avoidant personality is seen in the dimensions of "Interpersonal Isolation," where the operator refuses to work as part of a team and "Reaction to Crises," where the operator disappears when faced with a sudden crisis.

Intellect/Openness to Experience. Intellect and openness to experience is also important for effective performance in hazardous jobs. John (1990) explained that this factor "describes the depth, complexity, and quality of a person's mental and experiential life" (p. 71). He considered this construct to be different from IQ and originality, creativity, and cognitive complexity. Our research suggests that measures of intellect/openness to experience are moderately related to cognitive ability, but also include range of interests, experience seeking, and curiosity (Hogan & Hogan, 1992). One theme common for employment in hazardous jobs is intensive, and usually challenging, technical training. Effective personnel should be curious about the subject matter and enjoy the intellectual challenge. In addition, the nature of the jobs are such that those who do well will be interested in continued learning as new ideas, innovations, and technical developments emerge.

Two studies illustrate the usefulness of measures of intellect/openness to experience in predicting training performance. First, in their study of bomb disposal technician training, Hogan and Hogan (1989) tested Navy students (n = 97) with the HPI prior to training and then collected performance data as students progressed through training. Successful completion of second class diver training was related to high emotional stability scores and low conscientiousness scores. However, one facet of the HPI Intellectance scale, Curiosity, correlated modestly (r = .14) with diver training success. Successful completion of the explosive ordnance disposal (EOD) training course was also associated with high emotional stability scores and low scores on most facets of conscientiousness. In addition, the HPI Intellectance scale, Science Ability, which reflects interest in science, correlated (r = .14) with completion of the 42 week EOD course. Second, Gregory (1992) used the HPI to predict the performance of Army enlisted personnel in training to maintain the TOW and DRAGON missile systems. These students had been selected for training using scores on the ASVAB. Immediately prior to the course, they (n = 182) completed a battery of noncognitive measures including the HPI. These students were followed through training and grades for academic performance were obtained at the end of the course. Gregory split the group into a primary sample (n = 57) and a hold-out sample (n = 59) for cross-validation. For the primary sample, the best predictor of academic training was School Success (r = .55), whereas School Success correlated .34 in the hold-out sample. Facets that compose School Success consist of small scales for Education, Math Ability, Good Memory, and Reading. These reflect interests, values, and personality styles rather than aptitude.

CONCLUDING REMARKS

Returning to our opening example of police officer and firefighter selection, what have we learned about how to determine whether job applicants can handle the dangers, risks, and perils associated with hazardous work? First, risk and stress are independent of hazardous duty. Risk concerns the probability of being harmed by a hazard and risk can be controlled. Stress results from not being suited for hazardous work, where the dangers of the work erode one's confidence in one's ability to perform. A theme common to definitions of hazardous jobs is that inadequate performance will affect the safety of the incumbent, other workers, the public, or any combination thereof.

Second, structured job analysis methods are not very helpful for studying hazardous performance. An analyst could recover the same level and detail of information from casual observation or an informal interview with a worker. Our view is that structured job analysis procedures, which are predominantly inventory methods, provide little useful information for understanding effective performance of hazardous work.

So, third, this forces us to rely on our clinical judgment, about which we may feel less than confident. The critical incident method is well-suited for studying hazardous work because it (a) elicits examples of effective versus ineffective performance and (b) reveals low baserate behavior. The disadvantages of the method include the large number of incidents that must be collected, the problem of how to categorize the incidents, and (c) how to interpret the category dimensions. The Dunnette et al. (1981) study of aberrant and unreliable behavior in nuclear power plant operators illustrates the power of critical incidents. Interpretation of critical incidents is the most valuable source of information about constructs underlying effective performance in hazardous jobs—where "good" performance is defined as an absence of errors. Poor performance may be due to any number of problems, but underlying many of them is impulsivity when performing technical and procedural tasks.

Fourth, the most effective way to select personnel for hazardous work is to construct a model of job performance and assess all of its components. From the critical incident results, this model, minimally, should consider evaluations of job suitability, technical competence, physical capability, and psychological suitability. Researchers should avoid picking and choosing among various components and use all that are relevant. Physical capability assessments will be inappropriate for some jobs. Application of this model will result in personnel who like the work, can be trained, will continue to learn about technical developments, and will not exhibit psychologically dysfunctional behavior due to perceived job stress. We know a substantial amount about assessments and their validity for predicting occupational suitability, cognitive performance in training and on the job, and physical performance. We know less about how to use measures of personality, personality disorders, and psychopathology; nevertheless, we suspect that gains in predicting effective performance will come from advances in understanding personality and personality disorders associated with persons who are attracted to hazardous work, but who are characterologically unsuited.

Fifth, some research suggests a role for the Big-Five personality factors—specifically, conscientiousness, emotional stability, and intellect/openness to experience. The later is important because the training and job content in many hazardous jobs requires personnel who are open, curious, analytic, and interested in new ideas and experiences. Conscientiousness reflects a subtle, but essential syndrome. Effective personnel are risk takers who like excitement; nevertheless, they are cautious, careful, attentive to detail, and not impulsive. Emotional stability is important not because it entails psychopathology at the extremes, but because, under stress, maladaptive or inflexible behavior (personality disorders) can emerge and can result in disastrous job performance. Many of the critical incidents cited by Dunnette et al. (1981) concern employees with personality disorders.

So much for the academic view of personnel selection for hazardous work. What is the advice from the field? Consider the following *New York Times* report from Kuwait City, October 15, 1992 on the "sappers" responsible for deactivating land mines and munitions left behind by retreating Iraqi troops:

> About a third of the 100,000 or so tons of munitions dropped by the allies over Kuwait never exploded, either because they were duds or were swallowed by the sand. When sappers try to deactivate some of these devices, like the V-69 antipersonnel mine, a spike cone that shoots out trip wires and jumps to waist level before exploding, they have to get down on their stomachs and delicately maneuver a pin into the works. The men, who earn about $90,000 a year, say the best attributes for the work are cool nerves and a nimble step. (Hedges, 1992, p. A4)

ACKNOWLEDGEMENT

We wish to thank Mary Ellen O'Connor and Darryl Stark for their helpful comments on various parts of the manuscript.

REFERENCES

Allport, G. W., & Odbert, H. S. (1936). Trait-names: A psycho-lexical study. *Psychological Monographs, 47,* (1, Whole No. 211).

American Psychiatric Association. (1987). *Diagnostic and statistical manual of mental disorders* (3rd ed., rev.). Washington, DC: Author.

Americans with Disabilities Act of 1990, Pub. L. No. 101–336, 104 Stat. 327 (1990).

Arvey, R. D., Landon, T. E., Nutting, S. M., & Maxwell, S. E. (1992). The development of physical ability tests for police officers: A construct validation approach. *Journal of Applied Psychology, 77,* 996–1009.

Barrick, M. R., & Mount, M. K. (1991). The "Big Five" personality dimensions and job performance: A meta-analysis. *Personnel Psychology, 44,* 1–26.

Bentall, R. H. C. (1990). Electromagnetic energy: A historical therapeutic perspective; its future. In M. E. O'Connor, R. H. C. Bentall, & J. C. Monahan (Eds.), *Emerging electromagnetic medicine* (pp. 1–17). New York: Springer-Verlag.

Biersner, R. J., Dembert, M. L., & Browning, M. D. (1980). Comparisons of performance effectiveness among divers. *Aviation, Space, and Environmental Medicine, 51,* 1193–1196.

Biersner, R. J., & LaRocco, J. M. (1983). Personality characteristics of U.S. Navy divers. *Journal of Occupational Psychology, 56,* 329–334.

Biersner, R. J., & LaRocco, J. M. (1987). Personality and demographic variables related to individual responsiveness to diving stress. *Undersea Biomedical Research, 14,* 67–73.

Blakely, B. R., Quinones, M. A., & Jago, I. A. (1992, May). *The validity of isometric strength tests: Results of five validity studies.* Paper presented at the 7th annual Conference of the Society for Industrial and Organizational Psychology, Inc., Montreal, Canada.

Butcher, J. N. (1985). Personality assessment in industry: Theoretical issues and illustrations. In J. J. Bernardin & D. A. Bownas (Eds.), *Personality assessment in personnel selection* (pp. 277–309). New York: Praeger.

Campbell, J. P. (1990a). An overview of the Army selection and classification project (Project A). *Personnel Psychology, 43,* 313–333.

Campbell, J. P. (1990b). Modeling the performance prediction problem in industrial and organizational psychology. In M. D. Dunnette & L. M. Hough (Eds.), *Handbook of industrial and organizational psychology* (Vol. I, pp. 39–74). Palo Alto, CA: Consulting Psychologists Press.

Campbell, J. P., McHenry, J. J., & Wise, L. L. (1990). Modeling job performance in a population of jobs. *Personnel Psychology, 43,* 313–333.

Campbell, J. P., McCloy, R. A., Oppler, S. H., & Sager, C. E. (1993). A theory of performance. In N. Schmitt & W. Borman (Eds.), *Personnel selection in organizations* (pp. 35–70). San Francisco: Jossey-Bass.

Cattell, R. B. (1947). Confirmation and clarification of primary personality factors. *Psychometrika, 12,* 197–220.

Cortina, J. M., Doherty, M. L., Schmitt, N., Kaufman, G., & Smith, R. G. (1992). The "Big Five" personality factors in the IPI and MMPI: Predictors of police performance. *Personnel Psychology, 45,* 119–140.

Cronbach, L. J. (1960). *Essentials of psychological testing* (2nd ed.). New York: Harper & Row.

Denning, D. L. (1984, August). *Applying the Hogan model of physical performance of occupational tasks.* Paper presented at the 92nd Annual Convention of the American Psychological Association, Toronto, Canada.

Digman, J. M., & Takemoto-Chock, N. K. (1981). Factors in the natural language of personality. *Multivariate Behavioral Research, 16,* 149–170.

Dunnette, M. D. (1966). *Personnel selection and placement.* Belmont, CA: Brooks/Cole.

Dunnette, M. D., Bownas, D. A., & Bosshardt, M. J. (1981). *Prediction of inappropriate, unreliable, or aberrant job behavior in nuclear power plant settings.* Minneapolis, MN: Personnel Decisions Research Institute.

Fiske, D. W. (1949). Consistency of the factorial structures of personality ratings from different sources. *Journal of Abnormal and Social Psychology, 44,* 329–344.

Flanagan, J. C. (1947). *The aviation psychology program in the Army Air Forces (AAF Aviation Psychology Program Research Report,* No. 1). Washington, DC: U. S. Government Printing Office.

Flanagan, J. C. (1954). The critical incident technique. *Psychological Bulletin, 51,* 327–358.

Fraser, T. M. (1989). *The worker at work.* London: Taylor & Francis.

Gael, S. (Ed.). (1988). *Job analysis handbook for business, industry, and government.* New York: Wiley.

Goldberg, L. R. (1992, May). *Basic research on personality structure: Implications of the emerging consensus for applications to selection and classification.* Paper presented at the Army Research Institute Conference on Selection and Classification for the U.S. Army, Alexandria, VA.

Gottfredson, G., & Holland, J. L. (1989). *Dictionary of Holland occupational codes* (2nd ed.) Odessa, FL: Psychological Assessment Resources.

Gough, H. G. (1957). *Manual for the California Psychological Inventory* (Rev. Ed., 1964). Palo Alto, CA: Consulting Psychologists Press.

Gough, H. G. (1968). An interpreter's syllabus for the California Psychological Inventory. In P. McReynolds (Ed.), *Advances in psychological assessment* (Vol. 1, pp. 55–79). Palo Alto, CA: Science and Behavior Books.

Gregory, S. (1992, May). *Noncognitive measures for Army technical training placement.* Paper presented at the Seventh Annual Conference of the Society for Industrial and Organizational Psychology, Inc., Montreal, Canada.

Hargrave, G. E., & Hiatt, D. (1989). Use of the California Psychological Inventory in law enforcement officer selection. *Journal of Personality Assessment, 53,* 267–277.

Hathaway, S. R., & McKinley, J. C. (1951). *Manual for the Minnesota Multiphasic Personality Inventory* (rev.). New York: Psychological Corporation.

Hedges, C. (1992, October 15). Bang! Bang! Bang! Cleanup crews tackling gulf war's deadly debris. *The New York Times,* p. A4.

Hirsh, H. R., Northrop, L. C., & Schmidt, F. L. (1986). Validity generalization results for law enforcement occupations. *Personnel Psychology, 39,* 399–420.

Hogan, J. (1985). Tests for success in diver training. *Journal of Applied Psychology, 70,* 219–224.

Hogan, J. (1991a). Physical abilities. In M. D. Dunnette & L. M. Hough (Eds.), *Handbook of industrial & organizational psychology* (Vol. 2, pp. 751–831). Palo Alto, CA: Consulting Psychologists Press.

Hogan, J. (1991b). Structure of physical performance in occupational tasks. *Journal of Applied Psychology, 76,* 495–507.

Hogan, J., & Hogan, R. (1989). Noncognitive predictors of performance during explosive ordnance disposal training. *Journal of Military Psychology, 1,* 117–133.

Hogan, R. (1971). Personality characteristics of highly rated policemen. *Personnel Psychology, 24,* 679–686.

Hogan, R. (1986). *Manual for the Hogan Personality Inventory.* Minneapolis: National Computer Systems.

Hogan, R., & Hogan, J. (1992) *Hogan Personality Inventory manual (2nd ed.).* Tulsa, OK: Hogan Assessment Systems.

Holland, J. L. (1966). *Psychology of vocational choice.* Lexington, MA: Ginn.

Holland, J. L. (1975). *Manual for the vocational perference inventory.* Palo Alto, CA: Consulting Psychologists Press.

Holland, J. L. (1979). *The self-directed search professional manual.* Palo Alto, CA: Consulting Psychologists Press.

Holland, J. L. (1985). *Making vocational choices: A theory of careers.* Englewood Cliffs, NJ: Prentice-Hall.

Hunter, J. E. (1983). *Test validation for 12,000 jobs: An application of job classification and validity generalization analysis to the General Aptitude Test Battery* (USES Test Research Rep. No. 43). Washington, DC: U.S. Employment Service, U.S. Department of Labor.

Hunter, J. E., & Hunter, R. F. (1984). Validity and utility of alternative predictors of job performance. *Psychological Bulletin, 96,* 72–98.

John, O. P. (1990). The "Big Five" factor taxonomy: Dimensions of personality in the natural language and in questionnaires. In L. A. Pervin (Ed.), *Handbook of personality: Theory and research* (pp. 66–100). New York: Guilford.

Kirchner, W. K., & Dunnette, M. D. (1957). Identifying the critical factors in successful salesmanship. *Personnel, 34,* 54–59.

Lazarus, R. S. (1966). Psychological stress and the coping process. New York: McGraw-Hill.

Levine, E., Ash, R., Hall, R., & Sistrunk, F. (1983). Evaluation of job analysis methods by experienced job analysts. *Academy of Management Journal, 26*(2), 339–348.

Lewis, R. E. (1989). *Physical ability tests as predictors of job-related criteria: A meta-analysis.* Unpublished manuscript.

Lopez, F. M. (1986). *The threshold traits analysis technical manual.* Port Washington, NY: Lopez & Associates.

Lopez, F. M. (1988). Threshold traits analysis system. In S. Gael (Ed.), *Job analysis handbook for business, industry, and government* (Vol. 2, pp. 880–901). New York: Wiley.

McArdle, W. D., Katch, F. I., & Katch, V. L. (1991). *Exercise physiology.* Philadelphia: Lea & Febiger.

McCrea, R. R., & Costa, P. T. (1987). Validation of the five-factor model of personality across instruments and observers. *Journal of Personality and Social Psychology, 52,* 81–90.

McHenry, J. J., Hough, L. M., Toquam, J. L., Hanson, M. A., & Ashworth, S. (1990). Project A validity results: The relationship between predictor and criterion domains. *Personnel Psychology, 43,* 335–354.

Montgomery, J., Butler, S., & McPhail, S. M. (1987). *Development and validation of a selection test battery for radiation protection technicians.* Houston, TX: Jeanneret & Associates.

Nauta, F., Ward, K. H., & D'Ambrosia, W. B. (1983). *Maintenance training technology research; manpower, personnel, and training related factors influencing maintenance performance.* Washington, DC: Logistics Management Institute.

Norman, W. T. (1963). Toward an adequate taxonomy of personality attributes: Replicated factor structure in peer nomination personality ratings. *Journal of Abnormal and Social Psychology, 66,* 574–583.

O'Connor, M. E. (1990). Safety issues in electromagnetic medicine. In M. E. O'Connor, R. H. C. Bentall, & J. C. Monahan (Eds.), *Emerging electromagnetic medicine* (pp. 291–298). New York: Springer-Verlag.

Peabody, D., & Goldberg, L. R. (1989). Some determinants of factor structures from personality-trait descriptors. *Journal of Personality and Social Psychology, 57,* 552–567.

Pugh, G. (1985). The California psychological inventory and police selection. *Journal of Police Science and Administration, 13(2),* 172–177.

Pulakos, E. D., Borman, W. C., & Hough, L. M. (1988). Test validation for scientific understanding: Two demonstrations of an approach to studying predictor–criterion linkages. *Personnel Psychology, 41,* 703–716.

Reilly, R. R., Zedeck, S., & Tenopyr, M. L. (1979). Validity and fairness of physical ability tests for predicting craft jobs. *Journal of Applied Psychology, 64,* 262–274.

Schmidt, F. L., & Hunter, J. E. (1977). Development of a general solution to the problem of validity generalization. *Journal of Applied Psychology, 62,* 529–540.

Schmidt, F. L., & Hunter, J. E. (1981). Employment testing: Old theories and new research findings. *American Psychologist, 36,* 1128–1137.

Siegel, A. I. (1988). Identification of potential hazards and job stress. In S. Gael (Ed.), *Job analysis handbook for business, industry, and government* (Vol. I, pp. 796–803). New York: Wiley.

Sparks, C. P. (1988). Legal basis for job analysis. In S. Gael (Ed.), *Job analysis handbook for business, industry, and government* (Vol. I, pp. 37–47). New York: Wiley.

Tett, R. P., Jackson, D. N., & Rothstein, M. (1991). Personality measures as predictors of job performance: A meta-analytic review. *Personnel Psychology, 44,* 703–742.

Tupes, E. C., & Christal, R. E. (1961). *Recurrent personality factors based on trait ratings* (Rep. No. ASD–TR–61–97). San Antonio, TX: Personnel Laboratory USAF, Lakeland Air Force Base.

U.S. Department of Labor, Employment and Training Administration. (1991). *Handbook for analyzing jobs* (rev.). Washington, DC: U.S. Government Printing Office.

U.S. Office of Personnel Management. (1991). Title 5, Code of Federal Regulations: White collar general schedule employees. Federal Register, 56, No. 86.

Wiggins, J. S. (1973). *Personality and prediction.* Reading, MA: Addison-Wesley.

Wiggins, J. S., & Pincus, A. L. (1989). Conceptions of personality disorders and dimensions of personality. *Psychological Assessment: A Journal of Consulting and Clinical Psychology, 1,* 305–316.

7 Training for Stress Exposure

Joan Hall Johnston
Janis A. Cannon-Bowers
Naval Air Warfare Center
Training Systems Division

Since the 1980s, the implementation of stress-coping training programs to enhance employee attitudes and performance has accelerated (Ivancevich, Matteson, Freedman, & Phillips, 1990). For example, Gebhardt and Crump (1990) reported that "in 1987 there were 50,000 companies (with 100 or more employees) providing some type of employer-sponsored health promotion program" (p. 263). In order to justify this enormous capital investment in employee welfare, such programs should show evidence of training effectiveness such as improved productivity. However, empirical confirmation for worksite stress coping training success is lacking. In a recent review, Ivancevich et al. found that most studies consisted of subjective and anecdotal evaluations that were based on an "atheoretical" foundation. They concluded that a disproportionate focus on individual attitudes had slowed progress in the development of a model of stress coping training effectiveness and recommended that future research should be based on a sound theoretical framework which includes evaluations of relevant organizational stressors and performance variables. In particular, they recommended that the effects of situational or "naturalistic" stressors in the work environment should be studied in a systematic fashion in order to identify the causes of performance problems. Consequently, training programs should be designed to address performance outcomes associated with specific stressors.

Indeed, such disasters as Three Mile Island, Chernobyl, and the USS Vincennes have underscored the importance of developing training interventions to offset the effects of real world stressors on complex cognitive tasks. Although little research in this area exists to date, there are current efforts to addressing this issue (Cannon-Bowers, Salas, & Grossman, 1991). For example, the USS Vincennes incident instigated the creation of a research program to develop training for enhancing the tactical decision making performance of Navy combat information center (CIC) operators. The technological complexity in the CIC environment plus the rapidly unfolding events of warfare scenarios create such stressors as high workload, information ambiguity, severe time pressure, and sustained operations. Therefore, the CIC operator's tactical decision making must be quickly responsive and accurate in order to achieve military success.

The aforementioned example highlights the need to design training that incorporates naturalistic stressors so that performance is enhanced. We include as naturalistic stressors both high- and low-demand task situations. For example, sustained operations, ill-structured problems, complex tasks, uncertainty, time pressure, high stakes, multiple players, interpersonal conflicts, and physical danger imply high-demand task situations. However, underutilization of skills and highly routine, boring tasks can be stressors as well.

Research has shown that training interventions that include task specific stressors have been successful in improving performance (Larsson, 1987; Meichenbaum, 1985; Novaco, 1988; Siegel et al., 1981). For example, promising "cognitive/behavioral" stress-coping training programs such as Meichenbaum's (1985) Stress Inoculation Training, Smith's (1980) Stress Management Training, and Suinn's (1990) Anxiety Management Training gradually expose trainees to stressors while they practice stress-coping skills (Meichenbaum, 1985).

Since its inception in the early 1970s, cognitive/behavioral stress-coping training has generally remained in the clinical domain. Most of the research has reported application of this training to alleviate physical pain, anxiety, depression, and anger. Some studies have reported success in application of the training to a number of occupational groups (e.g., nurses, police officers, oil-rig trainees), and to enhance athletic performance. In addition, results of a meta-analysis by Saunders, Driskell, Johnston, and Salas (in press) has demonstrated that this type of training is an "effective means for reducing state anxiety, reducing skill-specific anxiety, and enhancing performance under stress".

However, it is not sufficient to suggest that training developers adopt the strategies currently used in the programs described earlier. The training methodology tends to be limited in its application because the research has had a narrow focus (e.g., sports and clinical psychology). Second, similar to the work-site stress-management literature, little work has been done to integrate research findings into a conceptual framework that can be used to guide the development of (a) ways to incorporate relevant stressors into training and (b) coping skills training for complex cognitive tasks.

Based on these issues, it is clear that more work is needed to determine how training for stress exposure can be designed to enhance performance in various task environments. Therefore, we use "Stress Exposure Training" (SET) as a term that extends the focus of "cognitive/behavioral" stress coping training beyond its original clinical domain. We propose that SET has three main objectives: (a) to build skills that promote effective performance under stress, (b) to build performance confidence, and (c) to enhance familiarity with the stress environment (Driskell, Hughes, Hall, & Salas, 1992). In addition, a solid conceptual foundation for SET methodology should be developed in order to apply it to realistic complex and stressful environments.

To this end, the remainder of the chapter examines the contribution of SET research toward understanding training and performance under stress. Figure 7.1 summarizes chapter organization. First, a detailed account of the origins of SET is provided. Second, ideas from models of training effectiveness (Tannenbaum, Cannon-Bowers, Salas, & Mathieu, 1992; Tannenbaum, Mathieu, Salas, & Cannon-Bowers, 1991) were used to establish a conceptual model of SET. Third, a set of research questions were developed from the discussion of SET model components. A search for an answer to these questions resulted in a review of the empirical literature. SET guidelines were then developed for current training applications in light of review results. Last, suggestions for future research and theory development are provided.

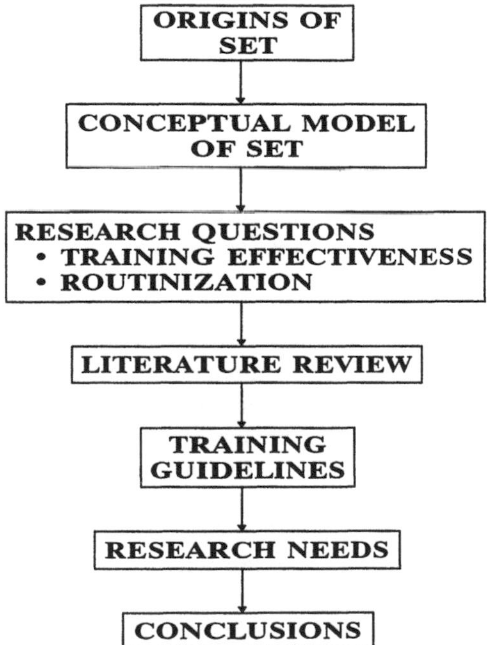

FIG. 7.1. Chapter organization.

ORIGINS OF STRESS EXPOSURE TRAINING

Originally, the "stress-exposure" aspect of SET came from Wolpe's (1959) work on systematic desensitization to help alleviate clinical patients of conditioned fears (Deffenbacher & Suinn, 1988). The systematic desensitization approach exposes an individual to anxiety-arousing situations while practicing relaxation strategies. However, by the 1970s, many clinical researchers had become dissatisfied with the behavioral modification programs of that time period, which did not attempt to address the cognitive mechanisms of anxiety problems (Meichenbaum, 1985). Therefore, new clinical methods were developed that included "systematic rational restructuring" training aimed at reducing the irrational thinking that accompanies anxiety attacks. Consequently, most current SET programs, such as Meichenbaum's stress-inoculation training, have incorporated both relaxation techniques and thought restructuring strategies.

Although much has been written regarding SET methodology, attention to developing a complete theoretical model has lagged behind. In general, however, most SET researchers subscribe to a transactional model of stress to explain the mechanism by which SET has worked (Meichenbaum, 1985). For example, Meichenbaum (1993) has often referred to the Lazarus and Folkman (1984) transactional model: "stress occurs whenever the perceived demands of a situation tax or exceed the perceived resources of the system (individual, group, community) to meet those demands" (p. 8). Based on this theory, Meichenbaum (1993) has proposed that SET provides the skills that should reduce the imbalance between the adaptive demands of stressful situations and the individual's coping resources.

Although this proposal serves to justify using SET methods, it lacks the detail required in a theoretical model to support research hypotheses, or to advance training applications. In order to describe the most salient aspects of SET for developing a conceptual model, we next provide a generic, but detailed, description of SET.

SET is composed of two components: stress-coping training and instructional design (Meichenbaum, 1985). The stress-coping training component features developing skills that reduce potential cognitive and psychomotor performance deficiencies resulting from specific stressors (Driskell & Salas, 1991; Hall, Driskell, Salas, & Cannon-Bowers, 1992). The instructional design component features a process of gradual exposure to realistic stressors that enhances learning of coping skills, and is based on basic principles of training design for skills acquisition (Meichenbaum, 1985; Smith, 1980; Suinn, 1990).

Integrating the two components has resulted in a three-phase process. Table 7.1 describes the three-phase SET, training objectives, and expected outcomes (Meichenbaum, 1985). Each training phase is listed across the top of the table. Training objectives and outcomes are described below each phase.

Typically, the first phase involves a discussion of common reactions people

TABLE 7.1
Stress Exposure Training Design, Objectives, and Expected Outcomes

	Stress Exposure Training Design		
	Phase 1 *Presentation of requisite knowledge*	*Phase 2* *Skill practice with feedback*	*Phase 3* *Skill practice with stressors*
Objectives	• Knowledge of typical reactions to stressors	• Develop metacognitive skills — Positive coping thoughts and behaviors — Use relaxation techniques to calm physiological reactions • Develop cognitive skills — Use problem solving skills	• Use phase 2 skills while exposed to stressors
Outcomes	• Increased perceived efficacy in dealing with stressors	• Reduced negative attitudes toward self and stressors • Increased use of positive thoughts and behaviors • Reduced blood pressure, heart rate, and increased psychomotor steadiness • Successful coping skill performance	• Reduced anxiety • Increased efficacy • Successful application of skills while exposed to stressors • Improved cognitive and psychomotor performance under stress

have to specific stressors they encounter. For example, a typical focus of SET researchers has been on alleviating test anxiety and improving test performance (Meichenbaum, 1993). The objective is to help the individual understand that they will be able to stop the negative thoughts and behaviors that contribute to test-anxiety stress. It is expected that as the individual's anxiety is reduced, and test performance is improved, their efficacy in dealing with problem stressors is improved.

The second phase focuses on learning stress coping skills through practice and feedback. The objective is to train the individual to maintain an awareness of stress reactions in order to invoke appropriate skills to reduce stress. The act of maintaining an awareness of thoughts and actions is referred to as "metacognition" (Glaser & Bassok, 1989). Therefore, metacognitive skills support the execution of competent performance.

Typical metacognitive skills are thought restructuring, problem solving, and physiological control. Thought restructuring involves replacing negative thoughts and reactions, that are triggered by a stressor, with positive coping thoughts and reactions. Problem solving is used to reduce task performance errors. Physiological control involves using deep breathing and muscle relaxation methods to calm physiological reactions to stressful encounters (Meichenbaum, 1985).

The third phase involves practicing the coping skills in a setting that simulates or reproduces the problem stressors. For example, the individual takes a test while practicing thought restructuring and deep breathing, to counter the negative thoughts and rapid heartbeat that escalate fear of exam failure. The emphasis in the last phase is on developing skills to transfer what is learned to the real world without relapse into counterproductive thoughts and behaviors. Training simulations may employ memory recall, behavioral rehearsal, role playing, and modeling (Meichenbaum, 1985). Expected outcomes from this phase are reduced anxiety, increased perceived efficacy toward performance, and improved cognitive and psychomotor performance.

The framework in Table 7.1 provides a basis of how SET is currently designed, and its expected effects. However, it is necessary to proceed beyond a prescriptive matrix, and develop a conceptual model of SET that can be used to identify ways to establish training effectiveness. Therefore, prior to examining the research on SET, we first describe a conceptual model of how SET can influence attitudes and performance. Discussion of model components focuses on the development of research questions for the literature review.

A CONCEPTUAL MODEL OF STRESS EXPOSURE TRAINING

As we mentioned earlier, many researchers have used the transactional model of stress to explain SET effects. To date, however, evaluations of the SET literature have tended to focus on the overall effectiveness of the intervention, rather than extending the basic transactional model (Meichenbaum, 1985, 1993). Likewise, the broader stress-coping training literature only provides prescriptive frameworks that identify organizational (e.g., structure and work conditions) and individual (e.g., time management and depression) factors for which stress-coping training may be targeted (Ivancevich & Matteson, 1987; Ivancevich et al., 1990; Reynolds & Shapiro, 1991).

However, a number of researchers have presented theoretical models of training effectiveness that are useful in extending a model of SET (Salas, Dickinson, Converse, & Tannenbaum, 1992; Tannenbaum et al., 1992; Tannenbaum et al., 1991). Generally, these models have a systems-theory orientation because they specify inputs, throughput, outputs, and feedback loops (Berrien, 1976; Reynolds & Shapiro, 1991). Relationships of organizational and individual factors are specified, and the effects of such factors on development of training interventions and performance outcomes are identified. Feedback loops show relationships of performance outcomes and training interventions with organizational and individual factors (Tannenbaum et al., 1992).

Figure 7.2 presents an integrative model of SET effectiveness based on previous SET research and theory, and the Tannenbaum et al. (1992) model of training effectiveness. Incorporated into the model are variables relevant to our discussion of SET. The major categories of variables are organizational and task stressors, individual characteristics, SET interventions, and individual changes in attitudes and performance. Specific variables are noted as bullets in each box. The variables are meant to be representative of the broader categories, and are not an exhaustive list.

Illustrated in the model are the relationships among the variables. Organizational and task stressors, and individual factors are expected to have a direct impact on individual performance and attitudes. SET researchers recommend that an effective SET program should result from an analysis of these three factors. Effective SET should result in improved task performance, reduced anxiety, and increased efficacy, which then has an impact on such organizational outcomes as organizational efficiency and effectiveness (e.g., goal accomplishment). Conversely, individuals receive feedback regarding the positive or negative contribution of their effort and performance on organizational objectives. Feedback from these outcomes should have a direct impact on SET training design. Finally, SET, individual outcomes, and organizational outcomes can also

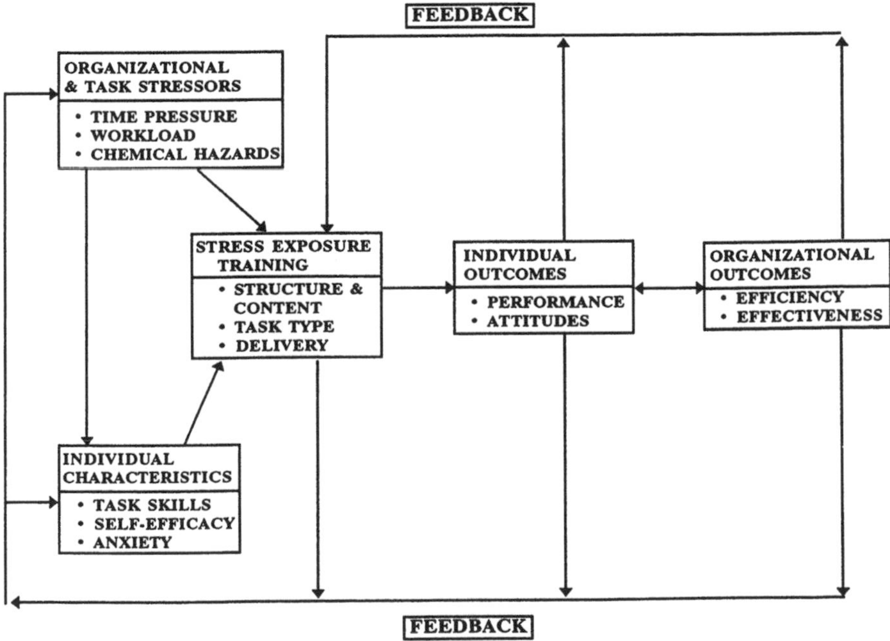

FIG. 7.2 Model of Stress Exposure Training effectiveness.

have an impact on organizational and task stressors (either by reducing or exacerbating them), and on individual characteristics (by raising or lowering self-efficacy and anxiety, and by improving task skill).

Next is a discussion of major model elements. Because the purpose of this chapter is to demonstrate the contribution of SET research toward understanding training and performance under stress, the discussion focuses on identifying relevant research questions for the literature review.

Organizational and Task Stressors

Considerable evidence has shown that stressors (e.g., time pressure, chemical hazards, and organizational structure) cause decrements in attitudes and performance (Driskell et al., 1992; Ivancevich & Matteson, 1980; Jex & Beehr, 1991; Tannenbaum et al., 1992; Weaver, Morgan, Adkins-Holmes, & Hall, 1992). However, less is known about specific effects of stressors on different types of performance (Riddle, Hall, Cannon-Bowers, & Salas, 1993). Consequently, to insure that SET is effective, researchers have recommended that it incorporate specific stressors encountered by the individual (Meichenbaum, 1985). Therefore, a number of questions that need to be addressed are (a) how should stressors be identified, (b) how should stressors be introduced into training, (c) when should stressors be introduced during training, (d) and can incorporation of realistic stressors into training design enhance performance and attitudes?

Individual Characteristics

Considerable research has shown that such individual characteristics as task skill, anxiety, efficacy, and perceived control influence task performance (Driskell et al., 1992; Ivancevich & Matteson, 1980; Jex & Beehr, 1991; Tannenbaum et al., 1992; Weaver et al., 1992). Therefore, SET researchers have emphasized that screening individuals for specific problems with stressors (e.g., text anxiety or anger and hostility) will enhance SET effectiveness (Meichenbaum, 1985). On the other hand, in order to apply SET to a wide variety of populations, it is just as important to know whether subjects with "normal" ranges of individual characteristics will benefit from SET. Therefore, the main question here is: Can SET be applied to populations within "normal" ranges of individual characteristics?

Stress Exposure Training Interventions

Research Design

In light of the study design problems Ivancevich et al. (1990) encountered in the worksite stress-coping training studies, at least two questions that should be addressed are: (a) How well can SET research methods support valid inferences regarding training effects and (b) how long do SET effects last?

Training Structure and Content

Researchers have suggested that SET is most effective when it combines a three-step training strategy (gradual exposure to stressors) with the development of metacognitive, cognitive, psychomotor, and physiological control skills (Meichenbaum, 1985). Meichenbaum (1993) has suggested that the three-step aspect of the training may improve performance because it raises beliefs in performance success, and reduces negative attitudes toward the stressor. Therefore, a primary question is: Does the three-step training structure (education, skills training, practice and feedback with stress exposure) enhance performance and attitudes? Second, is a combination of skills training (e.g., relaxation, cognitive restructuring, problem solving) more effective in enhancing performance than providing each skill training strategy by itself?

Task Type

Another issue of concern is that the clinical focus of the SET literature has limited the types of performance addressed by this intervention. The typical stressor in most SET studies has been the requirement to perform a difficult but single task (e.g., tests, rock descending, smoke diving, volleyball service reception, and giving a speech). Although the literature indicates that a variety of tasks have been incorporated into SET, little work has been done to identify SET effects on complex cognitive and psychomotor tasks. Therefore, a question that arises is, what effect does SET have on tasks that have complex cognitive and psychomotor requirements? Second, how should SET be designed so that it can enhance complex cognitive and psychomotor performance?

Training Delivery

Another issue of concern has been the clinical focus of SET. Most SET studies have employed instructors with advanced degrees in counseling, and have taken many sessions to deliver the training (Saunders et al., in press). A question raised by this issue is: Can training delivery variables be changed (i.e., fewer sessions, self-instruction) in order to make SET more applicable to varied training needs (e.g., distance training and limited budgets)?

Individual Outcomes

The primary focus of SET interventions has been to improve performance, reduce anxiety, and improve performance confidence (i.e., self-efficacy). A number of questions raised are: (a) To what degree does SET improve attitudes and performance, (b) can SET improve performance and attitudes better than other interventions, (c) what aspects of SET specifically enhance performance, reduce anxiety, and improve performance confidence, and (d) how does performance enhancement influence confidence and anxiety (and vice versa) under SET?

Organizational Outcomes

An obvious concern in evaluating SET effectiveness is whether SET interventions actually lead to improved organizational accomplishments. Of particular concern here is: Will SET principles transfer from the training to the operational environment, and which factors facilitate or inhibit this transfer?

Feedback

Feedback of training results has been demonstrated to enhance trainee performance (Goldstein, 1993). In addition, many researchers have recommended that

TABLE 7.2
Summary of SET Research Questions on Training Effectiveness and Routinization

Training Effectiveness

- How well do SET research methods support valid inferences regarding training effects?
- How long do SET effects last?
- How should stressors be identified?
- How should stressors be introduced into training?
- When should stressors be introduced during training?
- Can incorporation of realistic stressors into training design enhance performance and attitudes?
- Can the three-step training structure enhance performance and attitudes?
- Is a combination of skills training more effective than providing each skill training strategy by itself?
- To what degree does SET improve attitudes and performance?
- What aspects of SET specifically enhance performance, reduce anxiety, and improve performance confidence?
- How does performance enhancement influence confidence and anxiety (and vice versa) under SET?
- What effect does SET have on tasks that have complex cognitive and psychomotor requirements?
- How should SET be designed so that it can enhance complex cognitive and psychomotor performance?
- How can feedback of performance results improve SET effectiveness?
- How can transfer of SET skills to the operational environment be optimized?

Routinization

- Can SET be applied to populations with "normal" ranges of individual characteristics?
- Can training delivery variables be changed in order to make SET more applicable to varied training environments?

training results be used to enhance training design, to determine organizational effectiveness, and to alter the organizational environment, if needed (Goldstein, 1993). Therefore, one question is: How can feedback of performance results improve SET effectiveness?

Research Questions

Research questions raised in each of the model components cluster into two major categories: (a) demonstrating the effectiveness of SET and (b) "routinization"; that is, extending application of SET beyond the clinical scope of the literature (Larsson, 1987). Table 7.2 is a summary of these questions according to these two categories.

Although not all these questions can be answered based on the current research base, findings from the SET literature are systematically evaluated according to these two issues in order to develop training guidelines, and implications for future research and theory.

LITERATURE REVIEW

A literature search was conducted on SET articles published since 1970. Studies were included if they tested the effectiveness the three-phase training strategy on attitudes and performance. For comparison with SET, information is provided regarding alternative interventions on the same attitude and performance variables.

The search process yielded 37 articles that met our criteria. Five studies were reported in the 1970s, 27 studies were reported in the 1980s, and 5 studies have been reported since 1990. In addition, the results of a meta-analysis of SET studies by Saunders et al. (in press) is included in the discussions that follow.

Table 7.3 presents a summary of the empirical articles reviewed. The Table 7.3 Key is the last section of the table, and it gives details about the table heading abbreviations and coding of results. Following is a discussion of the results of analyses.

Training Effectiveness

Research Design

Overall, in contrast to the worksite stress management research, most of the SET studies used relatively sound methodological design. The majority (84%) of the studies had randomly assigned subjects to treatment conditions, and 73% of them were able to determine the extent of training effects by having measured attitudes and performance both pre- and postintervention. Fourteen percent of the studies conducted follow-up evaluations of attitudes and performance. Although

TABLE 7.3
Summary of Empirical Results From SET Research

Reference	*Set Type*	*AT*	*CG*	*Training Time*	*Design*	*Trainer Level*	*Subjects/ Screened*	*Task*	*Significant Findings*
Altmaier et al. (1982)	C3	0	1	3–90 min sessions	E PP	P	65 CS/A	Speech	SET with cognitive restructuring only, SET with relaxation only, and SET with both training approaches produced fewer anxious behaviors compared with the no-treatment control group, but there was no effect on self-reported anxiety. SET with relaxation and cognitive restructuring improved self-efficacy.
Backer (1989)	C1	1	1	3–60 min sessions	E PP/POST	PP	84 M MR	Recruit training	No treatment effects on any performance or anxiety measures.
Blackmore (1983)	C1	1	1	NR	Q PP	PP	17 M CS/A 17 F CS/A	College exams	SET and test skills training groups had an increase in semester grade point average, but no effect on anxiety.
Bloom & Hautaluoma (1990)	C3	1	0	1–120 min session	E PP/POST	NR	26 M CS/A 54 F CS/A	Computer spreadsheet	Compared to the alternative group, SET with relaxation skills only, SET with cognitive coping skills only, and SET with both approaches reduced error rates and task times. SET with relaxation training only reduced general anxiety.

Cecil & Forman (1990)	C1	1	1	6–90 min sessions	E PP/POST	NR	52 FT 2 MT	Teacher classroom performance	No effect of treatments on motoric manifestations of anxiety, but SET reduced anxiety.
Crocker (1989)	C1	0	0	Not applicable	E PP	NR	7 M V 7 F V	Volleyball service reception	Six months following treatment, the SET group had better service reception performance, see Crocker et al. (1988). Female volleyball players exposed to SET had reduced anxiety and increased self-efficacy 6 months later. Male volleyball players exposed to SET had increases in self-efficacy 6 months later.
Crocker et al. (1988)	C1	0	1	8–60 min sessions	Q PP	PP	16 M V 15 M V	Volleyball service reception	The SET group had better service reception performance than the control group, but there was no effect of treatment on anxiety.
Deffenbacher & Hahnloser (1981)	C3	0	1	4–50 min sessions	E PP	P	31 F CS/A 16 M CS/A	Cognitive tests and classroom exams	None of the treatments were found to affect two cognitive tests or classroom exam performance. Compared to the control group, SET with relaxation skills only, SET with cognitive coping skills only, and SET with both approaches reduced anxiety immediately following treatment and 5 weeks later.

(Continued)

TABLE 7.3
(Continued)

Reference	*Set Type*	*AT*	*CG*	*Training Time*	*Design*	*Trainer Level*	*Subjects/ Screened*	*Task*	*Significant Findings*
Deikis (1982)	C2	0	1	8–15 min sessions	Q PP/POST	PP	59 M CS 12 F CS	Scuba diving test (Ditch & Don)	No difference in scuba diving test performance or anxiety among the groups. But self-efficacy increased in the SET group compared to the control group.
Finger & Galassi (1977)	C3	0	1	8–45 min sessions	E PP	PP	48 CS/A	Exam	No effects of treatment on exam performance, but SET with relaxation skills only, SET with cognitive restructuring skills only, and SET with both approaches reduced anxiety compared with the control group.
Foley (1987)	C3	0	1	1–180 min session	E PP	PP	64 F CS/A	Exam	None of the 3 SET treatments had an effect on nursing test performance. However, SET with relaxation skills only, SET with cognitive restructuring skills only, and SET with both approaches reduced anxiety and negative thoughts, and increased self-efficacy compared to the control group.
Fremouw & Zitter (1978)	C1	2	1	5–60 min sessions	E PP	P	19 M CS/A 27 F CS/A	Speech	Skills training, SET with cognitive restructuring and relaxation, and the discussion placebo groups had fewer inappropriate speech behaviors than the

									waiting-list control group. The skills-training group reported less anxiety compared to the control group.
Gist (1989)	C1	1	0	1–210 min session	E PP	NR	43 M MGR 16 F MGR	Idea generation	SET with cognitive modeling generated more idea generation and idea divergence than the lecture and practice method. The cognitive modeling method increased self-efficacy compared with the lecture and practice method.
Hayslip (1989)	C1	1	1	5–60 min sessions	E PP	NR	358 Senior Citizens	Cognitive tests	Both SET and the math training groups obtained higher test scores than the control group. No self-report measures were administered.
Hussian & Lawrence (1978)	C2	1	1	3–50 min sessions	E PP	PP	39 F CS/A 9 M CS/A	Exam	There was no effect of any of the treatments on exam performance, however, both the test-specific SET and generalized SET groups reported less anxiety compared to the control and discussion groups immediately after, 3 weeks following, and 8 months following treatment.
Hytten et al. (1990)									
Exp.I	C1	0	1	60 min	E PP/POST	P	77 M OOW 10 F OOW	Smoke diving exercise	The SET group received less help than the control group, but reported higher anxiety during diving.

(Continued)

TABLE 7.3
(Continued)

Reference	*Set Type*	*AT*	*CG*	*Training Time*	*Design*	*Trainer Level*	*Subjects/ Screened*	*Task*	*Significant Findings*
Exp. II	C1	0	1	60 min	E PP/POST	P	53 M OOW 9 F OOW	Freefall lifeboat exercise	No effect of SET on reasoning test or on anxiety.
Jaremko (1980)	C1	0	1	2–60 min sessions	Q PP	P	62 CS/A	Speech	No effect of treatment on anxiety behaviors, but the SET group reported less anxiety and increased self-efficacy compared with the control group immediately following treatment, and at the end of the semester.
Jaremko et al. (1980)	S3	0	1	2–60 min sessions	Q PP	P	31 CS/A	Speech	No effect of treatment on anxiety behaviors, but both treatment groups (SET with education phase only, and SET with both skill and education phases) reported higher self-efficacy than the control group.
Kooken & Hayslip (1984)	C1	1	1	Not reported	E PP	NR	35 senior citizens	Cognitive tests	SET, attention placebo, and wait-list control groups had lower vocabulary scores and lower anxiety.
Mace & Carroll (1985)	S2	1	1	7–45 min sessions	E PP/POST	NR	19 M CS/H 21 F CS/H	Rock descending	SET, SET without the practice phase, and practical training produced fewer anxiety behaviors. Self and observer ratings of perceived stress prior to rock descent were correlated. The SET group reported less anxiety on the second rock-

									descent task than the control group and the practical training group.
Mason (1988)	C1	0	1	8–90 min sessions	E PP	P	36 F CS 18 M CS	Classroom exams	SET did not have an effect on academic test performance, however, the SET group reported a decrease in test anxiety compared with the control group.
Meichenbaum (1972)	C1	1	1	8–60 min sessions	E PP	PP	15 M CS/A 6 F CS/A	Cognitive tests	Both the systematic desensitization and SET groups had improved digit symbol test performance and grade point average compared with the control group. Both treatment groups showed a decrease in debilitating anxiety. SET had an effect on increasing facilitating anxiety.
Moon & Eisler (1983)	C1	3	0	5 sessions	E PP	PP	40 M CS/AN	Role play	The SET, social skills, and problem solving groups were rated lower on aggressive behaviors compared with the minimal attention group. The social skills group was rated lower on aggressive behaviors than the SET and problem-solving groups. Social skills and problem-solving groups were rated higher on assertiveness behaviors than the minimal attention and SET groups. All treatment groups reported fewer anger provoking inci-

(*Continued*)

TABLE 7.3
(*Continued*)

Reference	*Set Type*	*AT*	*CG*	*Training Time*	*Design*	*Trainer Level*	*Subjects/ Screened*	*Task*	*Significant Findings*
									dents. The SET group reported less anger compared to the other treatment groups.
Novaco (1980)	C1	0	1	7–90 min sessions	Q PP/POST	P	6 M PR 3 F PR	Case history exam and interview task	On the case history test, the SET group was consistently rated higher than the control group in terms of anger assessment and treatment plan. On the intervew test, SET subjects were rated higher on poise, effectiveness in understanding, and rapport, and in understanding of anger problems, than the control group. The SET group reported higher self-efficacy than the control group.
O'Neill et al. (1982)	C2	2	0	104–90 min sessions	E POST	NR	86 PO	Police work performance	No effects of training on police performance. No self-report measures were taken.
Pruitt (1986)	C1	2	0	6–50 min sessions	E PP	PP	35 F ES/A 16 M ES/A	Cognitive tests	SET, systematic desensitization, and the placebo control produced a gain in math test peformance, with the placebo control group achieving the greatest gain. All groups reported a decline in anxiety over time.
Register et al. (1991)	C2	0	2	4–week period	E PP	PP	38 M CS/A 83 F CS/A	Classroom exams	SET with bibliotherapy did not affect classroom exam or se-

									mester grade point average performance, however, SET with bibliotherapy did reduce self-reported anxiety.
Salovey & Haar (1990)	C1	1	1	8–90 min sessions	E PP	PP	43 CS/A	Writing task	SET and the no-treatment control group showed an improvement in performance on essay composition quality. The control group and the writing process group had improved performance on essay completion. SET and the writing process group showed a significant decrease in reported anxiety.
Sarason et al. (1979)	C1	0	1	6–120 min sessions	E PP/POST	PP	10 M PT 8 F PT	Five simulated police scenarios	The SET group was rated by observers as performing better than the control group during two of the scenarios. The SET group reported greater test anxiety and hostility than the control group.
Schneider (1989)	C1	1	1	6–60 min sessions	E PP	PP	30 F CS/A 15 M CS/A	Math tests	The SET group obtained higher course grades than the systematic desensitization treatment group, and the SET group had a reduction in anxiety compared with the systematic desensitization and control groups.
Schuler et al. (1982)	S2	0	0	3–75 min sessions	E PP	NR	11 F CS/A 11 F CS/A	Speech	Both treatments reduced speech anxiety behaviors, but the SET group that had the education phase reported less anxiety and higher speech self-efficacy compared to the SET group without the education phase.

(*Continued*)

TABLE 7.3
(*Continued*)

Reference	*Set Type*	*AT*	*CG*	*Training Time*	*Design*	*Trainer Level*	*Subjects/ Screened*	*Task*	*Significant Findings*
Sharp & Forman (1985)	C1	1	1	8–120 min sessions	E PP	P	49 F T/A 11 M T/A	Classroom performance	Compared to a control group, the classroom management training and SET groups were rated lower on anxiety behaviors, and were rated higher on classroom performance. The SET group reported less anxiety than the control group.
Smith (1989)	C1	0	1	5–60 min sessions	E PP	P	28 F CS/A 14 M CS/A	Exam	The SET group had higher levels of academic performance than the control group. A reduction in test anxiety was correlated with an increase in academic performance in the SET group, but not in the control group. The SET group reported less anxiety and higher self-efficacy than the control group. A reduction in anxiety was correlated with an increase in self-efficacy in the SET group, but not in the control group.
Smith & Nye (1989)	C2	0	1	5–60 min sessions	E PP	P	42 F CS/A 21 M CS/A	Exam	SET with the induced affect approach resulted in an improvement of academic performance over time compared to the control group. Improvement in state anxiety scores was asso-

									ciated with improvement in exam scores in the induced affect condition, whereas improvement in state anxiety was related to poorer exam scores in the covert rehearsal condition. Both treatment groups reported less anxiety and increased self-efficacy. Anxiety reduction was correlated with an increase in self-efficacy in both treatment groups.
Sweeney & Horan (1982)	C3	1	1	6–60 min sessions	E PP	P	25 M CS/A 24 F CS/A	Piano recital	Cue-controlled relaxation (CCR) improved musical performing competence (MPC) behaviors. Cognitive restructuring (CR) improved the behavioral index of anxiety ratings (BIA). A combination of CCR and CR improved MPC and BIA. CCR and the combined CCR/CR groups reported less anxiety than the control group.
Wernick (1984)	C1	0	1	9–60 min sessions	Q POST	PP	128 F CS 2 M CS	Attrition	Fewer students in the SET group left school. No self-report measures were taken.
Zeidner et al. (1988)	C1	0	1	5–60 min sessions	E PP	PP	249 M ES 248 F ES	Exam	SET increased performance on digit symbol coding, vocabulary, and math tests, but had no effect on anxiety.

(*Continued*)

TABLE 7.3
(Continued)

Reference	*Set Type*	*AT*	*CG*	*Training Time*	*Design*	*Trainer Level*	*Subjects/ Screened*	*Task*	*Significant Findings*
	Cn = number of SET interventions tests (Content) Sn = number of instructional design components tested (Structure)	Number of alternative treatments tested	Number of no treatment control groups in study	Amount of time devoted to training interventions				Types of tasks evaluated in each study	Reported effects of study interventions on attitudes and performance

Study Design/Measurement	*Trainer Experience Level*	*Subjects/Screened*
E = Experiment Q = Quasi-experiment (nonrandom assignment of subjects to conditions) PP = performance measures taken both pre- and postintervention POST = performance measures taken only postintervention PP/POST = attitude measures taken pre- and postintervention/performance measures taken only postintervention	P = Professional PP = Paraprofessional (conducted counseling without advanced degree)	Gender: F = Female M = Male Type: CS = College Student ES = Elementary School Student MGR = Managers MR = Marine Recruits OOW = Offshore Oil Worker PO = Police Officer PR = Probation Officer PT = Police Officer Trainee T = Teachers V = Volleyball Players Screened: /A = Measured Anxiety Level /AN = Measured Anger Level /H = Experience with Heights

stronger research designs are needed, these findings improve the validity and reliability of interpreting SET results.

Outcomes

Performance. Two-thirds of the studies (67%) demonstrated that SET significantly improved performance, and a wide variety of performance effects were found. Twenty-seven percent of the studies showed significant improvements in task performance (e.g., number of volleyball receptions, computer spreadsheet errors, piano recital errors, amount of help during smoke diving, errors in role-played police scenarios). For example, Crocker, Alderman, and Smith (1988) found that SET improved competitive volleyball player service reception performance. In addition, 24% of the studies reported significant improvements in paper and pencil test performance.

Support for SET enhancing performance also comes from the results of a meta-analysis of selected SET studies connected by Saunders et al. (in press). They found a small, but significant effect ($r = .225$, $p < .01$) of SET on task performance.

Nineteen percent of the studies showed significant reductions in anxiety-related behaviors (e.g., speech, rock descending, teaching). For example, Sharp and Forman (1985) found that SET reduced motoric manifestations of anxiety in classroom teachers. In support of this finding, Saunders et al., (in press) found a significant effect ($r = .362$, $p < .001$) of SET on skill-specific anxiety.

Attitudes. SET was found to significantly improve self-efficacy, or other self-confidence variables, in all of the 10 studies that evaluated the relationship. SET significantly reduced self-reported anxiety in 72% of the studies that had evaluated anxiety effects. Of the 29 studies that had measured both self-reported anxiety and performance, 45% showed a significant reduction in anxiety, and a significant improvement in performance as a result of SET. Just three studies evaluated the relationship between attitudes and performance, and all of them resulted in a significant correlation between reduction in anxiety and performance improvement. The meta-analysis conducted by Saunders et al. (in press) supported a significant effect ($r = .296$, $p < .001$) of SET on reducing self-reported anxiety.

Of the 14 studies that had used a longitudinal research design, 36% showed that anxiety had continued to decrease over time, and 14% demonstrated that self-efficacy had increased over time. For example, Crocker (1989) conducted evaluations of attitudes and performance of volleyball players 6 months following treatment with SET. They found that both female and male volleyball players experienced increased self-efficacy. In addition, the female volleyball players show reduced anxiety.

SET Interventions and Alternative Treatments

Seventy-eight percent of the studies that included alternative training interventions reported a significant effect of SET on performance. Seventy percent of the studies indicated that SET improved performance and attitudes about as well as the other treatments. For example, Sharp and Forman (1985) found that, compared to a control group, both classroom management training and SET groups were rated lower on anxiety behaviors, and rated higher on classroom performance.

Routinization

The positive findings regarding effects of SET on performance and attitudes are encouraging for training development. Of equal importance, however, is the applicability of SET to other training arenas. Following are the results gathered regarding the routinization of SET. Findings are discussed in terms of the relationship of each variable with significant performance effects. Because comparisons were not based on statistical tests, caution should be taken in drawing inferences from them.

Length of Treatment

Eighty-nine percent of the studies reported the exact amount of training time for SET. The average training time for studies demonstrating a significant effect of SET on performance was approximately 6 hours and 30 minutes. The mode was about 4 hours and 30 minutes. Average training time for studies not finding a SET effect on performance was 4 hours and 30 minutes, and the mode was about 2 hours and 30 minutes.

Trainer Experience Level

Sixty-two percent of the studies with a professional trainer reported significant effects of SET on performance. Similarly, 60% of the studies with a nonprofessional trainer reported a significant effect of SET on performance. For example, Bloom and Hautaluoma (1990) had a computer spreadsheet instructor provide training in relaxation and cognitive coping skills, and the training successfully reduced computer errors and task response time. In addition, trainees rated the instructor as more helpful when coping-skills training was incorporated into the technical-skills training.

Study Participants

Most of the SET studies (65%) were based on student subject populations (college undergraduates and two groups of elementary school children). Sixty-three percent of these studies demonstrated a significant improvement in performance. Of the 11 studies involving occupational, military, and sports groups (e.g., oil-rig workers, marine recruits, office managers, volleyball players, probation counselors, correctional officers, police officers and trainees, and teachers), 64% demonstrated a significant improvement in performance from SET. The two senior-citizen groups also benefited by improved cognitive test performance as a function of SET.

Another issue of concern was the applicability of SET to "normal" groups. Over half (57%) of all studies reported that results were based on subjects that had been screened prior to treatment. In most of these studies (90%), subjects had been screened for high anxiety levels. Therefore, significant findings for studies that screened for an individual difference variable (e.g., anxiety, aggression, experience) were compared with those that did not screen. Sixty-seven percent of the screening studies found significant effects of SET on performance. A similar proportion of studies that did not screen subjects (63%) were also found to have a significant effect of SET on performance.

Summary

In general, the SET research has been adequate in helping us determine the efficacy of SET effects, and for supporting the model of SET effectiveness. The majority of the research shows that SET positively influences a variety of performance behaviors, improves perceived self-efficacy, and reduces perceived anxiety. The Saunders et al. (in press) meta-analysis indicates there has been enough consistency in study design to determine the size of training effects on anxiety and performance. An encouraging aspect of the SET research was that most of the studies attempted to use multiple measures of attitudes and performance to evaluate training effects.

In addition, the findings are encouraging for routinizing SET. Treatment effects can occur in as short a time as 2 hours of training, SET can be taught by paraprofessionals, and it influences a variety of subject populations regardless of age, occupation, or anxiety level. These findings indicate that SET research should be more widespread, because treatment should not be limited by level of instructor training, or subject characteristics (e.g., anxiety level).

STRESS EXPOSURE TRAINING GUIDELINES

The current SET model and the research findings provide a framework from which training guidelines can begin to be developed. Therefore, viable guidelines were extracted from the studies based on the details of effective SET procedures. We used the method developed by Swezey and Salas (1992) to format the guidelines.

The guidelines are categorized according to four major training design features: (a) fidelity, (b) sequencing and training content, (c) delivery, and (d) evaluation. These features highlight the most important aspects of SET that will enable the building of skills to promote effective performance under stress, and the enhancement of familiarity with stress (Driskell et al., 1992). Comments are provided to clarify each set of guidelines.

Fidelity

Fidelity refers to the degree to which characteristics of the training environment are similar to those of the actual setting. The general training literature suggests that the training environment should incorporate aspects of the task environment to enhance skill acquisition (Swezey & Salas, 1992). Likewise, the SET research indicates that incorporating relevant stressors into the training environment leads to improved performance and attitudes. Therefore, in order to enhance training fidelity, the following guidelines outline a systematic effort to identify stressors, performance deficiencies, and effective coping skills prior to SET development.

- Conduct a step-by-step stressor analysis, similar to a task analysis, to identify typical stressors encountered.
- Identify performance deficiencies due to stressors in terms of psychomotor and cognitive processes.
- Identify knowledge, skills, and abilities (KSAs) required to promote technical performance while exposed to the stressors.
- Identify coping skills training that develop identified KSAs.
- Identify specific cues in the environment that trigger use of effective stress coping skills.

Comments: Fidelity

Using these guidelines prior to SET development should improve training design and fidelity. For example, test anxiety has been studied extensively in the SET literature. The atmosphere created by examinations can result in serious performance decrements on the exam, which, consequently, leads to lowered efficacy, and heightened emotional anxiety in future exam situations. To enhance training fidelity, most studies expose trainees to the examination situation while having trainees practice coping skills (Meichenbaum, 1985). For example, Smith (1989) identified cognitive-coping-skills training and relaxation training as specific coping skills to address both the situational and behavioral aspects of test anxiety. In both cases, trainees learn to identify potential stressors and stress reactions as cues to invoke both types of coping skills. For example, Smith & Nye (1989) trained participants to use, in response to a stress-provoking situation, stress-reducing self-statements during the inhalation phase of the breathing cycle. During exhalation, cue-controlled relaxation with the covert command "relax" was then applied to control physiological arousal from anxiety.

Sequencing and Training Content

The three-step training sequence of SET was designed to ensure that exposure to relevant situational stressors did not interfere with coping-skills acquisition. SET should be structured so that exposure to stressors is gradual, thus coping-skills acquisition is maximized prior to full exposure to real stressors. The following guidelines outline recommendations for the content of training at each step of SET:

Step 1

- Lecture, discussion, and example should be used to explain how operational stressors can be addressed with stress-coping skills. At this stage, SET program goals and procedures should be discussed with trainees in order to foster a mental model of performance expectations.

Step 2

- During this phase, coping skills should be taught using modeling, practice, and feedback.
- SET should be designed to provide coping skills training that addresses the situational, cognitive, physiological, and behavioral components of stressors typically encountered by the trainee.
- SET should train participants to use relaxation and cognitive restructuring skills in combination to create an integrated coping response to control arousal, to prevent affect-eliciting statements, and to facilitate task-relevant responses.
- Relaxation skill training should include practice in the acquisition and use of deep muscle relaxation, deep breathing methods, and practice in the use of cue words and images to trigger relaxation methods.

Step 3

- During step three, coping skills should be practiced under gradual increases in simulated and/or real stressful conditions.
- Role-playing typical stressful situations is an effective means for trainees to practice coping skills.
- During role play sessions, SET trainers should coach trainees: (a) to identify critical points during exposure to task stressors that should trigger positive task focused thoughts and use of relaxation skills, (b) in how to use appropriate coping skills at those critical events, and (c) to engage in self-reward for using appropriate coping skills.

- Mental imagery of stressful scenarios can be used by the trainee to rehearse ways to deal with stressors and to practice coping responses to be used in the actual stressful situation.
- Trainees should keep a daily log in order to monitor negative and coping self-statements.
- Trainees should have homework assignments in order to monitor their reading of SET materials and to practice skill acquisition. Homework should be reviewed at the next training session to ensure follow through of skill learning.
- Trainers should encourage trainees to practice cognitive coping skills and relaxation at least once per day outside of the training environment and preferably during typical stressful situations.

Sequencing and Training Content: Comments

Careful evaluation of training fidelity should be followed by the development of SET that incorporates the three-step process: education, skills training, and stress exposure. For example, Bloom and Hautaluoma (1990) conducted computer anxiety training using the three-step process. The first step began with a discussion of anxiety-provoking factors based on an assessment of stressors typically encountered by novice computer users. Next, technical learning and practice sessions on the computer were conducted that were followed by stress-coping skills training (relaxation and positive self-statements) session. Another set of technical learning and practice sessions ensued, and participants were encouraged to use their coping-skills training during these technical sessions. The result was reduced spreadsheet error rates and time to complete the task.

Delivery

The SET research indicates that certain training delivery methods (e.g., audiotape, videotape) can be effective in modeling appropriate responses to stressful situations. The following guidelines list recommended training delivery methods:

- SET should include videotapes of people modeling appropriate coping skills behaviors, preferably people in roles similar to trainees, and modeling behaviors in situations similar to those the trainees encounter.
- Role playing of scenarios that represent typical stressors is effective in enhancing trainee acquisition of coping skills.
- Videotaped and/or live demonstrations of how performance deficiencies can develop from losing control of attention due to stressors can be followed by demonstrations of how operator performance can be enhanced by using coping skills.

- Trainees should practice modeled skills in conjunction with feedback from the instructor and other trainees.
- Audiotapes of relaxation methods can enhance practice and transfer of relaxation skills.
- SET training time should last at least 1 hour.
- SET should be administered by an instructor that has had adequate training in the delivery of SET.

Delivery: Comments

The design of training delivery for SET is especially important because it addresses both the cognitive (e.g., cognitive restructuring) and emotional aspects (e.g., physiological control) of learning. Modeling of appropriate behaviors *and* thought processes are essential to the individual's understanding of how both thoughts and actions influence stress reduction.

Evaluation

The importance of taking multiple measures of attitudes and performance in order to confirm training effects cannot be understated. Training evaluation is useful in assessing trainee reactions and performance, to measure achievement of training objectives, and to provide immediate feedback to trainees.

- If trained in a group or team setting, trainees should provide immediate coping skills performance feedback to each other.
- Trainers should hold group discussions of trainee experiences in identifying, practicing, and applying the various stress-coping skills procedures.
- Measures of performance should be taken during the training so that appropriate and immediate feedback can be provided to the trainees in order to enhance their coping skills strategies while they are actually exposed to the stressors.
- Measures of trainee attitudes and performance should be taken in order to determine SET effects; multiple measures of performance and attitudes over time should be attempted.
- Measures of performance and attitudes should be taken prior to and after SET in order to determine changes in attitude and performance.
- SET instructors should be provided with standard outlines of SET and observe and/or participate in demonstrations of SET procedures in order to provide accurate and timely feedback to trainees.
- Measures of organizational outcomes (e.g., improved goal accomplishment) should be taken to ensure that the program is addressing crucial performance needs.

Evaluation: Comments

Technical and coping skills measures should be developed to provide feedback to trainees on how their coping-skills performance enhances effective technical performance. Because stressors typically induce negative thought patterns and actions, trainees should have as many opportunities as possible to demonstrate successful performance under stressful conditions. Therefore, instructor training is critical to trainee understanding and acceptance of SET procedures. Instructors should be skilled in providing immediate feedback to trainees in order to develop positive performance expectations.

Summary

These guidelines are an initial effort to develop training for stress exposure that is task-specific and operationally focused. Overall, this information can provide the people involved in the design and development of training programs with practical advice for creating effective training for stressful work conditions.

RESEARCH NEEDS

In this section, the research findings are summarized in terms of the questions raised regarding training effectiveness and routinization. Second, opportunities for future research and theory development are discussed.

Effectiveness of SET

Research Design

Only a few of the studies conducted longitudinal evaluations, and although some positive transfer occurred for attitudes, none found that performance improvements lasted beyond immediate treatment. However, most of the longitudinal findings were based on small, self-selected subject populations, from which conclusions are limited. Therefore, future research should be designed to capture longitudinal evaluation of performance over time in which larger numbers of subjects participate. In addition, more work should be done to apply SET to other operational settings to determine the efficacy of it for other training arenas.

Training Design

Only a few of the questions raised about SET in previous sections were addressed by the literature. More work is needed regarding the development of a systematic approach to the design of SET. For example, research should determine a procedure to identify task stressors, performance deficiencies, effective knowledge, skills, and abilities that help adapt to stress, and ways to introduce stressors during SET. A systematic approach to developing SET will make conclusions drawn from future research findings more consistent. Another research issue is determining which aspects of SET specifically enhance performance, reduce anxiety, and improve performance confidence, how performance enhancement influences confidence and anxiety (and vice versa) under SET, and how SET should be designed so that it can enhance complex cognitive and psychomotor performance.

A related issue is determining the relative benefits of skills training in SET (e.g., cognitive restructuring), compared to other skills training (e.g., problem solving, physiological control). Also needed is a way to determine the combination of skills training (e.g., cognitive restructuring and relaxation) that is effective.

Another issue that needs to be addressed is the relative effectiveness of the three levels of phased training. The length of time and effort given to each phase should be determined, because certain operational stressors and tasks may require a longer period of lecture, skills, or transfer than others.

Evaluation, Feedback, and Performance Maintenance

Future research should develop more sensitive measures, and multiple measures of attitudes and performance in order to increase the power to detect differences in treatment effects. In addition, little is known about the lasting effects of treatment, or whether providing follow-up treatment encourages performance enhancement and attitude change. Finally, work needs to be done regarding identification of an effective means for providing performance feedback to enhance training effects.

Routinization of SET

Future research should continue to broaden the application of SET to other subject populations (e.g., military personnel, nuclear power plant operators). In addition, SET should be designed for team-level applications (e.g., surgical teams, military teams).

CONCLUSIONS

The purpose of this chapter was to evaluate the SET research in order to determine its contribution to understanding stress and performance. The result was the development of a conceptual model of SET effectiveness, an evaluation of SET literature to identify support for the SET model and future research issues, and, last, to begin development of SET guidelines for current training applications.

REFERENCES

Altmaier, E. M., Ross, S. L., Leary, M. R., & Thornbrough, M. (1982). Matching stress inoculation's treatment components to client's anxiety mode. *Journal of Counseling Psychology, 29,* 331–334.

Backer, R. A. (1989). *A stress inoculation training intervention for marine corps recruits.* Unpublished doctoral dissertation, California School of Professional Psychology, San Diego, CA.

Berrien, F. K. (1976). A general systems approach to organizations. In M. D. Dunnette (Ed.), *Handbook of industrial and organizational psychology* (pp. 41–62). Chicago: Rand McNally.

Blackmore, S. H. (1983). *A comparison of training in stress inoculation, effective test taking techniques and study skills as treatments for test anxiety.* Unpublished doctoral dissertation, Texas A&M University, Houston, TX.

Bloom, A. J., & Hautaluoma, J. E. (1990). Anxiety management training as a strategy for enhancing computer user performance. *Computers in Human Behavior, 6,* 337–349.

Cannon-Bowers, J. A., Salas, E., & Grossman, J. D. (1991, June). *Improving tactical decision making under stress: Research directions and applied implications.* Paper presented at the International Applied Military Psychology Symposium, Stockholm, Sweden.

Cecil, M. A., & Forman, S. G. (1990). Effects of stress inoculation training and coworker support groups on teachers' stress. *Journal of School Psychology, 28,* 105–118.

Crocker, P. R. E. (1989). A follow-up of cognitive–affective stress management training. *Journal of Sport and Exercise Psychology, 11,* 236–242.

Crocker, P. R. E., Alderman, R. B., & Smith, F. M. R. (1988). Cognitive affective stress management training with high performance youth volleyball players: Effects on affect, cognition, and performance. *Journal of Sport and Exercise Psychology, 10,* 448–460.

Deffenbacher, J. L., & Hahnloser, R. M. (1981). Cognitive and relaxation coping skills in stress inoculation. *Cognitive Therapy and Research, 5,* 211–215.

Deffenbacher, J. L., & Suinn, R. M. (1988, January). Systematic desensitization and the reduction of anxiety. *The Counseling Psychologist, 16,* 9–30.

Deikis, J. G. (1982). *Stress inoculation training: Effects on anxiety, self-efficacy, and performance in divers.* Unpublished doctoral dissertation, Temple University.

Driskell, J. E., Hughes, S. C., Hall, J. K., & Salas, E. (1992, May). *Guidelines for implementing stress training* (NAWCTSD Tech. Rep. in progress). Orlando, FL: Naval Air Warfare Center Training Systems Division.

Driskell, J. E., & Salas, E. (1991). Overcoming the effects of stress on military performance: Human factors, training, and selection strategies. In R. Gal & A. D. Mangelsdorff (Eds.), *Handbook of military psychology* (pp. 183–193). Chichester, UK: Wiley.

Finger, R., & Galassi, J. P. (1977). Effects of modifying cognitive versus emotionality responses in the treatment of test anxiety. *Journal of Counseling and Clinical Psychology, 45,* 280–287.

Foley, J. M. (1987). *Matching stress inoculation treatment to student nurses' primary mode of anxiety expression.* Unpublished doctoral dissertation, University of Iowa, Iowa City, IA.

Fremouw, W. J., & Zitter, R. E. (1978). A comparison of skills training and cognitive restructuring–relaxation for the treatment of speech anxiety. *Behavior Therapy, 9,* 248–259.

Gebhardt, D. L., & Crump, C. E. (1990). Employee fitness and wellness programs in the workplace. *American Psychologist, 45,* 262–272.

Gist, M. E. (1989). The influence of training method on self-efficacy and idea generation among managers. *Personnel Psychology, 42,* 787–805.

Glaser, R., & Bassok, M. (1989). Learning theory and the study of instruction. *Annual Review of Psychology, 40,* 631–636.

Goldstein, I. L. (1993). *Training in organizations* (3rd ed.). Monterey, CA: Brooks/Cole.

Hall, J. K., Driskell, J. E., Salas, E., & Cannon-Bowers, J. A. (1992). Development of instructional design guidelines for stress exposure training. In *Proceedings of the 14th Annual Interser-*

vice/Industry Training Systems Conference (pp. 357–363). Washington, DC: National Security Industrial Association.

Hayslip, B., Jr. (1989). Alternative mechanisms for improvements in fluid ability performance among older adults. *Psychology and Aging, 4,* 122–124.

Hussian, R. A., & Lawrence, P. S. (1978). The reduction of test, state, and trait anxiety by test-specific and generalized stress inoculation training. *Cognitive Therapy and Research, 2,* 25–37.

Hytten, K., Jensen, A., & Skauli, G. (1990). Stress inoculation training for smoke divers and free fall lifeboat passengers. *Aviation, Space, and Environmental Medicine,* 983–988.

Ivancevich, J. M., & Matteson, M. T. (1980). *Stress and work: A managerial perspective.* Glenview, IL: Scott, Foresman.

Ivancevich, J. M., & Matteson, M. T. (1987). Organizational level stress management interventions: A review and recommendations. *Journal of Organizational Behavioral Management, 81,* 229–248.

Ivancevich, J. M., Matteson, M. T., Freedman, S. M., & Phillips, J. S. (1990). Worksite stress management interventions. *American Psychologist, 45,* 252–261.

Jaremko, M. E. (1980). The use of stress inoculation training in the reduction of public speaking anxiety. *Journal of Clinical Psychology, 36,* 735–738.

Jaremko, M. E., Hadfield, R., & Walker, W. E. (1980). Contribution of an educational phase to stress inoculation of speech anxiety. *Perceptual and Motor Skills, 50,* 495–501.

Jex, S. M., & Beehr, T. A. (1991). Emerging theoretical and methodological issues in the study of work-related stress. In G. R. Ferns & K. M. Rowland (Eds.), *Research in personnel and human resources management* (Vol. 9, pp. 311–365). Greenwich, CT: JAI.

Kooken, R. A., & Hayslip, B., Jr. (1984). The use of stress inoculation in the treatment of test anxiety in older students. *Educational Gerontology, 10,* 39–58.

Larsson, G. (1987). Routinization of mental training in organizations: Effects on performance and well-being. *Journal of Applied Psychology, 72,* 88–96.

Lazarus, R. S., & Folkman, S. (1984). *Stress, appraisal, and coping.* New York: Springer.

Mace, R. D., & Carroll, D. (1985). The control of anxiety in sport: Stress inoculation training prior to abseiling. *International Journal of Sport Psychology, 16,* 165–175.

Mason, J. L. (1988). *The treatment of test anxiety using stress inoculation training.* Unpublished doctoral dissertation, Indiana State University, Terre Haute, IN.

Meichenbaum, D. H. (1972). Cognitive modification of test anxious college students. *Journal of Consulting and Clinical Psychology, 39,* 370–380.

Meichenbaum, D. (1985). *Stress inoculation training.* New York: Pergamon.

Meichenbaum, D. (1993). Stress inoculation training: A twenty year update. In R. L. Woolfolk & P. M. Lehrer (Eds.), *Principles and practice of stress management* (2nd ed., pp. 373–406). New York: Guilford.

Moon, J. R., & Eisler, R. M. (1983). Anger control: An experimental comparison of three behavioral treatments. *Behavior Therapy, 14,* 493–505.

Novaco, R. W. (1980). Training of probation counselors for anger problems. *Journal of Counseling Psychology, 27,* 385–390.

Novaco, R. W. (1988). *Stress reduction programs.* Irvine, CA: National Research Council Publications on Demand Program, National Academy Press.

O'Neill, M. W., Hanewicz, W. B., Fransway, L. M., & Cassidy-Riske, C. (1982). Stress inoculation training and job performance. *Journal of Public Science and Administration, 10,* 388–397.

Pruitt, P. L. (1986). *The treatment of test anxiety in middle school students.* Unpublished doctoral dissertation, University of South Carolina, Columbia, SC.

Register, A. C., Beckham, J. C., May, J. G., & Gustafson, D. J. (1991). Stress inoculation bibliotherapy in the treatment of test anxiety. *Journal of Counseling Psychology, 38,* 115–119.

Reynolds, S., & Shapiro, D. A. (1991). Stress reduction in transition: Conceptual problems in the design, implementation, and evaluation of worksite stress management interventions. *Human Relations, 44,* 717–733.

Riddle, D., Hall, J., Cannon-Bowers, J. A., & Salas, E. (1993). *Understanding the effects of ambient and task-related stressors on performance: A proposed framework.* Unpublished manuscript.

Salas, E., Dickinson, T. L., Converse, S. A., & Tannenbaum, S. I. (1992). Toward an understanding of team performance and training. In R. W. Swezey & E. Salas (Eds.), *Teams: Their training and performance* (pp. 3–29). Norwood, NJ: Ablex.

Salovey, P., & Haar, M. D. (1990). The efficacy of cognitive-behavior therapy and writing process training for alleviating writing anxiety. *Cognitive Therapy and Research, 14,* 513–527.

Sarason, I. G., Johnson, J. H., Berberich, J. P., & Siegel, J. M. (1979). Helping police officers to cope with stress: A cognitive–behavioral approach. *American Journal of Community Psychology, 7,* 593–603.

Saunders, T., Driskell, J. E., Johnston J. H., & Salas, E. (in press). The effect of stress inoculation training on anxiety and performance. *Journal of Occupational Health Psychology.*

Schneider, W. J. (1989). *Stress inoculation training in the treatment of math anxiety.* Unpublished doctoral dissertation, St. John's University, Jamaica, NY.

Schuler, K., Gilner, F., Austrin, H., & Davenport, D. G. (1982). Contribution of the education phase to stress inoculation training. *Psychological Reports, 51,* 611–617.

Sharp, J. J., & Forman, S. G. (1985). A comparison of two approaches to anxiety management for teachers. *Behavior Therapy, 16,* 370–383.

Siegel, A. I., Kopstein, F. F., Federman, P. J., Ozkaptan, H., Slifer, W. E., Hegge, F. W., & Marlowe, D. H. (1981). *Management of stress in Army operations* (A–19). Alexandria, VA: Army Research Institute for the Behavioral and Social Sciences.

Smith, R. E. (1980). A cognitive/affective approach to stress management training for athletes. In C. H. Nadeau, W. R. Halliwell, K. M. Newell, and G. C. Roberts (Eds.), *Psychology of motor behavior and sport—1979* (pp. 54–73). Champaign, IL: Human Kinetics.

Smith, R. E. (1989). Effects of coping skills training on generalized self-efficacy and locus of control. *Journal of Personality and Social Psychology, 56,* 228–233.

Smith, R. E., & Nye, S. L. (1989). Comparison of induced affect and covert rehearsal in the acquisition of stress management coping skills. *Journal of Counseling Psychology, 36,* 17–23.

Suinn, R. M. (1990). *Anxiety management training.* New York: Plenum.

Sweeney, G. A., & Horan, J. J. (1982). Separate and combined effects of cue-controlled relaxation and cognitive restructuring in the treatment of musical performance anxiety. *Journal of Counseling Psychology, 29,* 486–497.

Swezey, R. W., & Salas, E. (1992). Guidelines for use in team training development. In R. W. Swezey & E. Salas (Eds.), *Teams: Their training and performance* (pp. 219–245). Norwood, NJ: Ablex.

Tannenbaum, S. I., Cannon-Bowers, J. A., Salas, E., & Mathieu, J. E. (1992). Deriving theoretically based principles of training effectiveness to optimize training system design. In *Proceedings of the 14th Annual Interservice/Industry Training Systems Conference* (pp. 619–631). San Antonio, TX: National Security Industrial Association.

Tannenbaum, S. I., Mathieu, J. E., Salas, E., & Cannon-Bowers, J. A. (1991). Meeting trainees' expectations: The influence of training fulfillment on the development of commitment, self-efficacy, and motivation. *Journal of Applied Psychology, 76,* 759–769.

Weaver, J. L., Morgan, B. B., Jr., Adkins-Holmes, C., & Hall, J. K. (1992). *A review of potential moderating factors in the stress–performance relationship* (Rep. No. 92–012). Orlando, FL: Naval Training Systems Center.

Wernick, R. L. (1984). Stress management with practical nursing students: Effects on attrition. *Cognitive Therapy and Research, 8,* 543–550.

Wolpe, J. (1959). *Psychotherapy by reciprocal inhibition.* Stanford, CA: Stanford University Press.

Zeidner, M., Klingman, A., & Papko, O. (1988). Enhancing students' test coping skills: Report of a psychological health education process. *Journal of Educational Psychology, 80,* 1988.

8 Training Effective Performance Under Stress: Queries, Dilemmas, and Possible Solutions

Giora Keinan
Nehemia Friedland
Tel Aviv University

The training of soldiers, police officers, firemen, deep-sea divers, and athletes comprises just a few examples of the numerous instances where persons are trained to perform stressful tasks. In all these instances, the effectiveness of training is judged according to standards set in extremely stressful criterion situations (e.g., combat). These present problems and requirements that seldom exist when persons are trained to perform tasks that are not inherently stressful. Most notably, adequate training requires not only the acquisition of basic skills such as marksmanship, map reading, or pure oxygen breathing, but also the ability to cope with and withstand intense stressors.

The double requirement alluded to here justifies a distinction between *skills training* and *stress training*. Skills training should be, and normally is, conducted under conditions that are designed to promote skills acquisition and retention; that is, conditions that maximize learning. These include presentations in quiet and comfortable classrooms, the use of teaching aids, uniform presentations, and opportunities to practice skills and tasks repeatedly, under standards and predictable conditions, in environments insulated from task-irrelevant stimuli. However, skills acquired under such conditions might not be sufficiently transferable to the "noisy," dangerous, and unpredictable stressful environments in which they would have to be performed eventually. Therefore, skills training has to be combined or supplemented with stress training that aims at assuring the retention of effective performance under stress. To this end, trainees have to be exposed, in the course of training, to simulated aspects and features of the stressful environment.

Although skills training and stress training are essential elements of the overall training process, the requirements that have to be satisfied in order to maximize the effectiveness of each type are patently incompatible. The trainer who seeks to promote effective performance of stressful, dangerous tasks is thus confronted with critical queries and dilemmas. The gravity of these is exacerbated, of course, by the possibility that faulty training might extract a heavy psychological or physical cost, during training or in the criterion situation. These dilemmas, along with possible solutions that were drawn from research or from practical experience, are addressed in the present chapter.

THE INFORMATION DILEMMA: SHOULD TRAINEES BE FULLY INFORMED ABOUT ALL THAT MIGHT OCCUR IN THE CRITERION SITUATION?

Should persons who train to perform in highly stressful criterion situations be informed fully, in advance, of the stressors that they may expect to encounter in the criterion situation? Ideas presented by a number of authors suggest that the answer should be affirmative. Janis (1958) maintained that precise and specific information about an upcoming stressful event elicits anticipatory fears. These instigate, in turn, a "work of worrying" process that attenuates anger and distress when the anticipated aversive event materializes. Additionally, precise information may contribute to the alignment of expectations with actual occurrences and thus prevent the emotional distress that often results from discrepancies between the two (cf. Shipley, Butt, Horwitz, & Farbry, 1978). Information may also reduce stress by helping persons discriminate between dangerous and safe periods (see the "safety signal theory" formulated by Seligman, Maier, & Solomon, 1971); by enabling the preparation of adequate instrumental responses (Perkins, 1955, 1958); by reducing uncertainty and conflict (Berlyne, 1960); and by heightening a sense of control (Glass & Singer, 1972). The greater predictability resulting from advance information alleviates the need to monitor, constantly, new, threatening, or distracting stimuli, thus allowing greater attention to and investment in task performance (cf. Cohen, 1978).

Although the aforementioned arguments are quite compelling, the relevant empirical evidence is equivocal. Some studies showed that advance, preparatory information reduced tension during aversive events or immediately after their occurrence, and improved coping (Auerbach, Martelli, & Mercuri, 1983; Leventhal, Brown, Shachman, & Engquist, 1979). In one of the few studies conducted to test the effect of preparatory information on performance in an applied task environment, Inzana, Driskell, Salas, and Hall (1994) showed that the provision of such information resulted in reduced subjective stress, enhanced task confidence, and fewer errors. Other studies showed, on the other hand, that detailed early information did not lower arousal during exposure to a stressor (Egbert, Battit, Welch, & Bartlett, 1964; Vernon, 1971; Vernon & Bigelow, 1974). In some instances it even increased tension and impeded recovery from the stressful experience (Langer, Janis, & Wolfer, 1975; Miller & Mangan, 1983). These contradicting results suggest that information per se is not a panacea for stress reduction, and that additional conditions have to be satisfied to make information useful. What these conditions are we learned, partly, from laboratory and field studies, and partly from our own personal experience.

Information and Control

Detailed information about an upcoming stressful event is beneficial to the extent that it is instrumental; that is, to the extent that it facilitates the exercise of some choice or control over the environment (Averill & Rosen, 1972; Miller, 1981; Miller & Mangan, 1983). Thus, information given to trainees should include not only a description of the stressors that they are likely to experience in the criterion situation but also prescriptive ways wherefore these stressors can be coped with. These combined descriptive and prescriptive elements are likely to elicit realistic expectations about future stressors along with a sense of control, and thereby attenuate stress and improve performance. On the other hand, detailed information about future stressors that does not prescribe ways of coping would likely attune persons to dangers that cannot be averted, resulting in a sense of helplessness, heightened stress, and diminished performance.

The preceding argument was supported by a number of field studies. Datel and Lifrak (1969) produced a movie that aimed to create realistic expectations about army basic training. "This is How It is" depicted the hardships and stressors that recruits could expect to encounter in the course of basic training. It did not address, however, the ways in which to cope with such stressors. The results revealed no appreciable effect of the movie on the recruits' distress during basic training. In contrast, studies that combined the provision of information with instructions on how to cope with stressors showed positive results. Novaco, Cook, and Sarason (1983) showed army recruits a movie that portrayed the basic training environment, the stressors that the trainee could expect to encounter, and ways of coping with them. The movie raised efficacy expectations regarding certain parts of training, and strengthened the trainees' belief in their ability to control training stressors. Horner, Meglino, and Mobley (1979) used a similar procedure and reported a significant decrease in the attrition rate at the end of the enlistment period.

Trainers who seek to maximize the potentially beneficial effects of preparatory information should be aware of a useful distinction among different types of information and design their messages accordingly. Several authors (e.g., Inzana et al., 1994; Ludwick-Rosenthal & Neufeld, 1988; Taylor & Clark, 1986) distinguished between *sensory information* and *procedural information*. The former describes the feelings and sensations that persons are likely to experience under stress. Such information alleviates stress by aiding individuals in labeling their physiological reactions and in attributing them to identifiable causes (Worchel &

Yohai, 1979). Procedural information makes individuals aware of events that are likely to occur in the stressful environment and of performance errors that are typically committed under stress. Such information might facilitate avoidance of typical performance deficiencies (cf. Inzana et al., 1994). To these two types of information, a third type may be added. *Instrumental information* prescribes ways and means to cope with the stressors and their effects directly. This last type of information is particularly important as it contributes directly to one's actual and perceived control over the stressful environment.

Information and Reassurance

As already noted, the receipt of advance information about threats to one's physical integrity or psychological well-being might heighten arousal and tension. Accordingly, some authors recommend adding a reassuring message to the factual information offered to people. Leventhal et al. (1979) concluded their review of pertinent research with the assertion that, "Distress reduction is readily achieved by combinations of monitoring strategies (sensation information) and reassurance that the procedure will help" (p. 710).

Reassuring messages given to trainees can take a variety of forms. For example, trainees may be informed that their experience of stress is normal, that others have performed the tasks required from them successfully, despite difficulties, that people often have more coping resources than they themselves believe to possess, that certain tasks tend to be seen as more frightening from afar than when they are actually performed, and that the trainers are highly skilled and competent to provide the trainees with effective means of coping.

The use of reassuring messages is demonstrated convincingly in the training program developed by Novaco et al. (1983) for U.S. Army basic training:

> The messages related to the regulation of emotion begin with validation of the recruits' experience during the initial days. It is conveyed that fear, anger, disappointment, and worry are perfectly normal and quite common reactions among recruits. They are presented with the circumstances that have induced this distress and are told that despite their worry and confusion, thousands of recruits have felt the same way yet have ultimately succeeded in training. (p. 409)

Information About the Onset and Termination of the Stressor

Information about the onset and ending of an aversive event heightens predictability and perceived control, and may thus alleviate stress (Breznitz, 1989; Cohen, Evans, Stokols, & Krantz, 1986; Friedland & Keinan, 1992; Glass & Singer, 1972). Janis (1949) presented a compelling demonstration of this effect:

Information conveyed to bomber crews about the number of missions that they would have to fly over Germany reduced their stress symptoms dramatically. Similar evidence was presented by Breznitz (1989), who compared two groups of Israeli soldiers that took part in a 20-km forced march. One group was told, in advance, about the distance it would have to march. In addition, the group was informed, at the end of every 5 km of the distance remaining to the finish line. The proportion of soldiers who could not sustain the pace and dropped from the march was significantly lower in the informed than in the uninformed group (6.2% vs. 21%, respectively). Moreover, blood samples taken from the soldiers when they reached kilometers 14 showed significantly higher cortisol and prolactin levels in the uninformed than in the informed group. Breznitz concluded, "Knowing exactly when the ordeal will be over helped the soldiers both physically and mentally" (p. 257). It appears, therefore, that a training process which informs the trainee about stressors that may be expected in the criterion situation should also include, whenever possible, information about their onset and ending.

The Credibility of Information

Information about the stress that one might experience, given in advance, breeds expectations. If such information were violated—found to be erroneous or unrealistic—stress would likely intensify. As an illustration consider the comment made by a respondent to a survey conducted in Israel during the Gulf War concerning the missile attacks to which the country was subjected (Keinan, Carmil, Zomer, in preparation):

> Almost everything said about the Skuds was proven wrong eventually. We were told that the 6 hours it takes to prepare them for launching give ample time to detect and destroy them; that they are old and, even if they managed to take off, they could not reach Israel; that even if they reached Israel, the Airforce would intercept them. "Be calm," they told us, "everything is under control." Then, when [the missiles] started falling on our heads, and took all of us by surprise, stress was much greater. It would have been much better had they not told us anything.

The deleterious effect on coping with stress that unreliable information might have was demonstrated also by Breznitz' (1989) study, cited earlier. Aside from the two groups of soldiers described earlier, the study included two additional groups. One group was told that they would march a distance of 15 km. When the group reached kilometer 14 it was told that 5 km were added to the march. The other group was told that it would march 25 km, but when they reached kilometer 14 the distance was shortened to 20 km. Interestingly, the dropout rate was higher in both groups than in the group that received veridical early information. That is, violated expectations appear to be detrimental even when people find out that their lot is actually better than they had expected.

Summary

Research conducted to date indicates that advance, preparatory information can heighten the effectiveness of stress training. However, trainers should be cognizant of the adverse effect that improperly chosen information or information that is presented inadequately might have on individual's coping with stress.

In order to heighten the beneficial effect of preparatory information, and to avert pitfalls, trainers should avoid limiting themselves to mere descriptions of the stressor. Information should be given on feelings and sensations that are likely to be experienced under stress, on common effects of stress on performance, on signals that mark the onset and the ending of the stressing event, and on ways to cope with the stressor and with its effects. In general, such information improves stress training by heightening actual or perceived control. This effect can thus be strengthened by adding reassuring messages that emphasize the normalcy of the persons' reactions to stress and the effectiveness of the coping strategies that they were taught. Most important, preparatory information must be credible.

THE INTENSITY DILEMMA: SHOULD TRAINEES BE EXPOSED TO CRITERION-INTENSITY STRESSORS DURING TRAINING?

The second dilemma concerns the advisability of simulating, during training, intense stressors that are typical of the criterion situation. On the one hand, the work of several authors gives grounds to the assumption that exposure to high-intensity stressors during training is likely to improve performance in a stressful criterion situation. Presumably, such exposure enhances the trainee's familiarity with conditions that are likely to prevail in the criterion situation and thereby reduces uncertainty and fear of the unknown (Coleman, 1976; West, 1958). Exposure to intense stressors during training might erect psychological defenses against anxiety (Fenz & Epstein, 1968) and augment physiological toughness (Dienstbier, 1989; Ellersten, Johnsen & Ursin, 1978). In addition, stressors that simulate criterion conditions may be assumed to facilitate stimulus generalization and the transfer of skills acquired during training (Cascio, 1989; Terris & Rahhal, 1969; Wexley & Latham, 1981; Willis, 1967). On the other hand, propositions put forth by other authors cast doubt on the effectiveness of exposure to intense stressors. The experience of intense stress during training might interfere with task acquisition (Lazarus, 1966; Vroom, 1964), instill fears and enhance emotional sensitivity (Haggard, 1949; Janis, 1971), and promote a despair attitude (Kern, 1966). In addition, physiological–biochemical models of stress (Selye, 1956) suggest that repeated exposure to stress causes exhaustion and could thus hinder performance. These assumptions imply, then, that training under low-intensity stressors or under no stress at all might bear better outcomes.

Existing empirical evidence does not provide clear grounds for the clarification of the ambiguity portrayed earlier. Some studies suggest that task performance under stress improves subsequent performance of tasks under similar stress conditions (Glass & Singer, 1972; Vossel & Laux, 1978) and that training under relaxed conditions does not lead to improved performance of the tasks trained for when they are tested under stress (Zakay & Wooler, 1984). On the other hand, other results show the contrary; that is, training under no stress produces better outcomes than training under intense stress (e.g., Keinan & Friedland, 1984). It was also shown that training under intense stress resulted in a decrease in performance confidence (Driskell, 1984).

Considering the arguments for and against the exposure of trainees to high-intensity stressors, stress training procedures should be designed to satisfy three requirements:

1. Trainees should be given the opportunity to familiarize themselves with stressors characteristic of the criterion situation. Familiarity is needed in order to reduce uncertainty and to improve the transfer of learning.
2. Exposure to stressors, during training, should be carried out in a manner that does not hinder the trainees' confidence in their ability to cope with them.
3. Exposure to stressors, during training, should be done in ways that minimize interference with the acquisition of skills that training is designed to teach.

Two training procedures that satisfy these requirements are described in the following (see also Friedland & Keinan, 1992).

Graduated Training

This approach to training draws from the notion of stress inoculation (Epstein, 1983; Meichenbaum, 1977; Meichenbaum & Novaco, 1978; Orne, 1965), which suggests that the graduated exposure of trainees to stress, starting with mild levels at the beginning of training and reaching criterion-level intensities at its conclusion, might improve performance in the criterion situation. The graduated approach appears consistent with the three requirements set above: Exposure to low-intensity stress at the early stages of training reduces interference with the acquisition of the task for which one has trained. Exposure to gradually intensifying stress, with each stage being only slightly more intense than the preceding one, might enhance the trainee's sense of control and protect his or her self-confidence. Finally, the need to cope with criterion-intensity stressors at last stages of training is likely to establish the trainee's familiarity with such stressors.

Research results and practical experience suggest, however, two safeguards that should be implemented when graduated intensity training is applied. First, training should not progress from one level of intensity to a higher one unless the trainee achieved an acceptable degree of task proficiency. If at some stage a decrement in performance occurs, or if the trainee reports a sense of failure, training should revert to an earlier stage or remain at its current stage until better results are achieved. Epstein (1983) commented on this matter, stating:

> Individuals should not be exposed to increased levels of stress until they have demonstrated the ability to perform well under reduced levels of pressure. Thus, challenges should be increased in increments corresponding to the individual's gain in proficiency. Excessive challenge can be diagnosed by a regression of performance as a result of heightened arousal. A cost of exposing an individual to high levels of stress is that fixations may occur that are difficult to unlearn. (p. 54)

Second, trainees should be informed, in advance, of the highest level of intensity that stressors might reach in the course of training. In the absence of such information, trainees might develop exaggerated expectations about the severity of future stressors.

The importance of the second safeguard was demonstrated in an experiment conducted by Friedland and Keinan (1982, Study 1). Subjects in this experiment performed 10 training trials on a visual search task. In one condition, the subjects received a 1.5 mA electric shock, on trials, 1, 3, 7, and 10. In two other conditions, a graduated-intensity training procedure was used. Subjects in these conditions were exposed to shocks that intensified gradually from 0.2 mA to 1.5 mA, on trials 1, 3, 7, and 10. In one of the two conditions, referred to as *graduated intensity with ceiling,* the subjects were informed after Trial 10 that the intensity of subsequent shocks would not exceed that of the last shock that they had experienced. In the other condition, *graduated intensity without ceiling,* no such information was provided. The fourth condition served as a control. In it, no shocks were administered. After the last training trial (Trial 10), and a short break that followed it, the subjects performed two additional trials. They were not told that only two additional trials would be performed, or that these trials were to be considered criterion trials. These criterion trials were carried out under identical conditions that did not involve the administration of electric shocks.

Not surprisingly, the highest level of performance was reached by subjects in the control condition. These, it should be recalled, were not exposed to any shocks. Subjects in the *graduated intensity without ceiling* condition and subjects who were exposed to constant-intensity shocks of 1.5 mA performed at a significantly lower level than subjects in the control condition. However, the performance of subjects in the *graduated intensity with ceiling* condition was almost as proficient as that of subjects in the control condition. The comparison between the performance of subjects in the two graduated-intensity conditions suggests that this procedure is effective only if it contains a safeguard against the buildup of exaggerated expectations about the severity of future stressors.

Phased Training

An alternative solution to the intensity dilemma was proposed in our studies (Friedland & Keinan, 1986; Keinan, Friedland, & Sarig-Naor, 1990). We hypothesized that training effectiveness can be maximized by partitioning the training process into distinct phases. In one, the trainees acquire the task and its basic component skills without being exposed to any stressors. The second phase is designed to familiarize the trainees to criterion-like stressors without their being required to perform any task. The third phase combines the first two, such that the trainees are required to carry out the task under stressors characteristic of the criterion situation. We reasoned that this form of training satisfies optimally the requirement presented earlier. First, it allows the trainees an uninterfered with acquisition of the task. Second, it familiarizes the trainees with stressors that are likely to be encountered in the criterion situation and facilitates transfer of learning. Third, it prevents the formation of exaggerated or unrealistic expectations about stressors not yet experienced, as might be the case in graduated intensity training.

In a recent study (Keinan et al., 1990), subjects were trained to perform a visual search task, where training consisted of 10 trials. The effectiveness of the training was evaluated in one criterion trial. Stress was here induced with a cold pressor: The subject was required to immerse a hand in a bucket filled with ice water for the duration of any trial designated stressful.

The training procedures evaluated in the experiment consisted of different combinations of three elementary phases:

1. *Task Acquisition (TA):* The subject performed repeated trials of the visual search task without being exposed to stress.
2. *Stress Exposure (SE):* The cold pressor was applied without any task being performed. Specifically, the subject immersed his or her nondominant hand in ice water for a duration of 90 seconds.
3. *Practice Under Stress (PUS):* The subject performed repeated trials of the visual search task. The cold pressor was applied in some trials.

Different combinations of the three elementary phases resulted in three different training procedures:

1. *TA/SE:* The subjects performed 10 TA trials. These were followed by the SE procedure.
2. *TA/SE/PUS:* The subjects performed 5 TA trials, followed by the SE

procedure. In the third, PUS, phase, stress was induced on the first, third, and fifth of five trials.

3. *TA/PUS:* This procedure combined the first and third phases of the TA/SE/PUS procedure. This two-phased procedure is relevant under circumstances in which the experience of stress cannot be dissociated from the performance of the task.

The three phased-training procedures were compared to two single-phase procedures:

4. *TA:* The subjects performed 10 TA trials.
5. *PUS:* The subjects performed 10 training trials. Stress was induced on trials, 1, 3, 5, 7, and 9.

In all five conditions, the subjects were given a 5-minute break at the conclusion of training. After the break they performed a single criterion trial, which was conducted under stress.

The results showed that the criterion performance in the TA/PUS condition did not differ significantly from that in the TA/SE/PUS condition. Criterion performance in these two conditions was superior to performance observed in any of the remaining conditions.

The results led to a number of conclusions. First, it is possible to enhance the effectiveness of training by dividing it into phases that emphasize different elements. Second, the training process must include two elements: (a) a minimally interfered with process of task acquisition (TA), and (b) practice of newly acquired skills under stress (PUS). Third, stress exposure per se (SE) does not promote training effectiveness. This conclusion derives clearly from the finding that the TA/SE/PUS was not superior to the TA/PUS procedure.

Summary

The intensity dilemma is, perhaps, the most difficult of the dilemmas analyzed in this chapter. Arguments favoring the exposure of trainees to stressors that match the intensity of stressors characteristic of the criterion situation are as compelling as arguments against such exposure. In essence, the dilemma stems from the incompatibility of requirements for effective *skills training* and requirements that have to be met to promote effective *stress training*.

Two training methods were proposed to circumvent the dilemma. Graduated intensity training emphasizes skills training in the early stages of the training process and stress training toward its end. This procedure requires, however, the exposure of trainees to a schedule of gradually intensifying stress. Although this schedule is designed to minimize interference with skills acquisition, at the beginning of training, and to generally bolster the trainee's confidence, it might also teach the trainee to expect a gradually worsening fate, and thus breed despair. Therefore, graduated-intensity training should be employed only if trainees can be given a clear indication of the highest intensity that the stressors to which they are exposed might reach.

An alternative to graduated-intensity training is phased training that dedicates different phases of the training process exclusively to skills training and to stress training. Our investigation of different phased-training procedures revealed that a two-phased process which involves skills training without exposure to any, even mild, stress, in one phase, and a second phase in which trainees practice their newly acquired skills under stress, yielded optimal results.

THE OVERLEARNING DILEMMA

A further dilemma with which designers of training often contend concerns the issue of overlearning. Should trainees be allowed or required to keep exercising their newly acquired skills after their learning curves have reached an asymptote?

There exist some theoretical and empirical grounds to suggest that overlearning is advantageous (cf. Driskell, Willis, & Cooper, 1992):

1. Complex tasks are more likely than simpler tasks to be undermined by high arousal (e.g., Kimble, 1961). Overlearning might simplify the execution of complex tasks and thus make them less vulnerable to stress (see also Weick, 1985).
2. Overlearning creates automatic responses that are less demanding in terms of attentional capacity (Logan, 1985). Because such capacity is limited, reduction of the load carried by the cognitive system might free attentional capacity and improve performance (see also Mandler, 1982).
3. Well-drilled tasks instill a sense of control and predictability. These are likely to reduce the stress experienced in the criterion situation and attenuate its impact on performance (Driskell & Salas, 1991).

Although the grounds justifying the use of overlearning appear sound, there exist equally compelling reasons to doubt its efficiency:

1. Overlearning might constrict the flexibility of response. This effect could become critical if the overlearned response proves suboptimal under criterion conditions, or is altogether impossible. The dominance of the overlearned response might limit people's ability to improvise or to search for alternatives. This could prove fatal in perilous situations or in the face of emergencies.
2. Overlearning might induce boredom, fatigue, and diminished motivation. These can effect suboptimal criterion performance.

3. The overlearning of a number of responses may result in a repertoire of competing dominant responses. This could prove highly detrimental under exposure to actual criterion stressors.

The arguments for and against overlearning create a dilemma that cannot be reconciled easily. Judging by our own experience, overlearned task-relevant responses prove advantageous under stressful criterion conditions. In their absence, persons tend to resort to other, potentially inadequate, overlearned responses that they had acquired in their past. Weick (1985) referred to this phenomenon: "When combat gets serious, we turn off the computers and reach for the grease pencils and the map overlays" (p. 40). Nevertheless, the application of overlearning requires a number of safeguards:

1. One should choose carefully the responses to be overlearned. Responses that are not universally adaptive; that is, responses the usefulness of which is restricted to unique or unusual circumstances and that would have to be changed as circumstances change should not be overlearned. This implies that the responses ought to be chosen on the basis of careful analyses of the criterion situation and on the detailed study of the past performance of individuals in this situation.
2. In choosing behaviors that will undergo overlearning one should avoid competing responses. The overlearning of different responses to similar stimuli should be avoided, as the distinction among the latter might prove difficult under stress.
3. Fatigue and boredom that often accompany overlearning should be minimized. Varied training methods should be used and the maintenance of a high level of motivation should be sought. Thus, for example, elements of training can be introduced in a competitive framework.
4. Some of the disadvantages of overlearning can be alleviated by teaching trainees recipes for improvisation (Crovitz, 1970; Weick, 1979). However, to heighten the likelihood that these recipes would be actually applied, they too ought to be overlearned.

Summary

Overlearning was evaluated as a possible remedy to the adverse effects of stress on performance. Overlearning simplifies complex tasks, automates complex responses, and instills a sense of control and predictability. These heighten the immunity of task performance to the adverse effects of stress. However, overlearning may also have negative side effects. Most notably, it might constrict the flexibility of response. Therefore, overlearning should be limited to instances in which the training environment and training task simulate faithfully and in great detail the criterion situation and actual task. By contrast, overlearning in a training environment that differs in significant ways from conditions which prevail in the criterion situation may actually impede performance.

THE ORIENTATION DILEMMA: SHOULD TRAINING EMPHASIZE AN EMOTION-FOCUSED ORIENTATION OR A PROBLEM-FOCUSED ORIENTATION?

Ways of coping with stress are often classified into problem-focused responses as opposed to emotion-focused responses (Lazarus & Folkman, 1984; Lazarus & Launier, 1978). Problem-focused coping circumvents negative emotions through the emission of behaviors that modify the stressor or minimize its effect, through cognitive activity that fosters the belief that the stressor can be or will be controlled instrumentally, or through both. Emotion-focused coping consists of attempts to alleviate or eliminate dysphoric emotions elicited by a stressor directly, with little attention paid to the characteristics of the situation or to the nature of the threats it poses.

This conceptualization provided grounds for the development of problem-oriented training approaches, as distinct from emotion-focused approaches. The former consist, mostly, of the provision of information about features of the threatening situation, as well as advice concerning solutions to specific problems faced by the individual. These may include, for instance, training better approaches to decision making, suggestions for change of managerial style or for improved communication with the social environment. On the other hand, methods oriented toward emotions seek to train the regulation of the stressful affect without instructing in ways to cope with specific problems. Such methods may include instruction in the control of physiological responses (e.g., biofeedback or muscle relaxation), in the alteration of cognitive strategies (e.g., changes in the inner dialogue or guided imagery), or behavioral strategies (e.g., self-monitoring, modeling, or flooding).

Studies that examined the relative effectiveness of the emotion-focused and problem-focused approaches yielded ambiguous conclusions. Some findings indicated that an emotion-oriented approach is superior (e.g., Strentz & Auerbach, 1988). Other findings showed, by contrast, the emotion-focused training produced the lower adjustment levels (e.g., Martelli, Auerbach, Alexander, & Mercuri, 1987). These inconsistent results can be bridged by the findings of recent studies:

1. Studies conducted in the past few years show that interventions geared at simulating emotion focused modes are more effective than problem oriented interventions in stress-laden situations in which actual control over events is severely curtailed (bereavement, for example). On the other

hand, problem-oriented interventions are likely to be more effective in controllable situations that allow persons to manipulate stressors (Collins, Baum, & Singer, 1983; Kaloupek & Stoupakis, 1985; Kaloupek, White, & Wong, 1984).

2. Emotion-focused training processes appear to be uniquely effective in dealing with brief, stressful incidents. Over time, the effectiveness of such processes tends to diminish (Lazarus, 1983; Roth & Cohen, 1986). As suggested by Strentz & Auerbach (1988), "Few stressors can be effectively dealt with on a sustained basis through largely emotion-focused processes, without increasing attention to environmental cues and execution of appropriate problem solving behaviors" (p. 658). This suggestion was supported by data on parents to children suffering from Leukemia (e.g., Hofer, Wolff, Friedman & Mason, 1972) or on those who were incarcerated in concentration camps (Schmolling, 1984). Specifically, the data show that, over time, persons who do not employ a problem-oriented coping behaviors encountered difficulty in adjusting or surviving.

It follows that the design of training should take into consideration the duration of stressful events that may be expected to occur in the criterion situations. In particular, prolonged stressors require that training include not only emotion-focused methods but also problem-focused ones:

3. A number of studies indicated that individuals often employ a combination of both problem-focused and emotion-focused coping, and that effective coping requires a repertoire which includes both (e.g., Lazarus & Folkman, 1984; Meichenbaum, 1985). Martelli et al. (1987) compared the effectiveness of interventions based on an emotion-focused orientation, those based on a problem-focused orientation, and those based on a mixed-focused orientation. These authors found that the latter approach was the most effective.

Considering this evidence, it appears reasonable to propose that in situations that are moderately controllable and in situations where stress is prolonged, mixed-focus interventions should prove most effective.

Summary

Empirical evidence on the relative merits of the emotion-focused and the problem-focused orientation in stress training is scant. Nevertheless, it does suggest that the emotion-focused orientation is effective in preparing individuals to cope with stressors that they cannot control and which is of brief duration. Thus, the emotion-focused orientation might be effective in preparing a patient to cope with the painful aftermath of surgery. On the other hand, the emotion-focused orientation might prove less effective or relevant, for, say, the training of persons to withstand and cope with combat stressors. These are more prolonged, and yet more controllable. In such conditions, the problem-focused or a mixed-focused orientation should be employed.

THE CUSTOMIZING DILEMMA: SHOULD TRAINING METHODS BE DEVELOPED DIFFERENTIALLY, TO TAKE INTO CONSIDERATION INDIVIDUAL DIFFERENCES?

In recent years, a growing body of research findings suggests that training effective coping under stress can be optimized by establishing a congruency between the training method and the trainee's personality makeup. These findings were obtained when differential methods of training were applied to trainees who varied in their anxiety level (Snow, 1986), their locus of control (Strenz & Auerbach, 1988), confidence expectancy (Keinan, 1988), their tendency to seek information (Martelli et al., 1987; Miller & Mangan, 1983), their tendency to sensitize or repress (Goldstein, 1973; Shipley et al., 1978), and their style of coping with stressors (Fry & Wong, 1991). Thus, for example, Martelli et al. (1987) reported that problem-focused training methods were more effective for persons who habitually seek information, whereas emotion-oriented methods were superior among persons who tend to avoid information. Keinan (1988) found that soldiers with a high level of confidence expectancy; that is, individuals who assign a low probability to the risk of being physically hurt, benefited more from training that exposed them to intense, criterion-like stressors than from training conducted under milder stress. The opposite was found with soldiers whose confidence expectancy was low.

Considering the aforementioned findings, customized training, which attempts to match training methods to personality types, should have become prevalent. In point of fact, such attempts are rare in the context of stress training. Perhaps the implementation of this approach to training is costly and fraught with technical difficulties. Most notably, training programs are usually designed for large groups, the members of which vary along a considerable number of personality attributes. Taking this variability into consideration would entail an onerous screening and selection process and the costly development of numerous training programs.

Given these considerations, the use of customized training appears limited to instances where individuals or very small groups (e.g., intelligence agents, commando units) have to be trained to carry out a critical or highly dangerous mission. In these instances, training should be designed to match personality variables the relevance of which was well established by research (e.g., information seeking), and single out dimensions that can be measured reliably and validly.

The difficulty of implementing customized training ought not rule it out altogether, and alternatives that retain its underlying rationale should be sought. For example, trainees may be offered a number of alternative methods of training and be allowed to choose that which fits them best (see Burton, 1990; Meichenbaum, 1985). Thus, they may choose from among various cognitive, physiological, or behavioral methods of stress management. This approach does away with the need to apply costly screening and selection procedures, the validity and reliability of which is often doubtful. In addition, the opportunity given to trainees to choose a method of training would likely heighten their sense of control and, thus, alleviate their stress (Keinan & Zeidner, 1987).

Concluding Note

Training effective performance under stress presents designers of training with a dual task. First, they have to assure the acquisition of task-relevant skills and bring trainees to a high level of task proficiency. Second, they have to prepare trainees to withstand and cope with high-intensity criterion stressors. Unfortunately, attempts to accomplish the latter objective, during training, might seriously interfere with and impede task acquisition.

The inherent incompatibility of the two training objectives brings to the fore a number of problems and dilemmas. In the present chapter we addressed five: the information dilemma, the intensity dilemma, the overlearning dilemma, the orientation dilemma, and the customizing dilemma. Each of the five dilemmas reflects controversial theoretical positions or contradicting empirical findings that we identified in the literature. Hence, the resolution of these dilemmas is theoretically significant. More important, however, these are dilemmas that often confront the practitioner. Therefore, a considerable emphasis was placed, throughout the chapter, on practical solutions.

The list of dilemmas is by no means exhaustive, and there are additional queries and dilemmas that we did not address. One important issue that is obviously missing concerns generalizability. This issue represents two specific questions that have important theoretical and practical implications.

The first question concerns the generalization of training from one stressor to another. Does the effect of training conducted under a particular stressor generalize or transfer to performance in the presence of a different stressor? This question is particularly relevant to instances in which the stressors characteristic of the criterion situation cannot be reproduced fully in training.

The second question concerns the generalization of training effects from one task to another. Do the effects of training to perform one task, under a particular stressor, generalize to the performance of a different task, in the presence of the same stressor?

Notwithstanding the clear importance of these questions, little research has been devoted to attempts to answer them. An exception is the Terris and Rahhal

(1969) experiment that demonstrated that training under one type of stressor generalized to performance of the same task under a different stressor type.

The questions concerning the generalizability of stress training have to be investigated thoroughly. The answers to these questions carry important implications regarding the practicality of stress training. Clearly, the cost of training might prove prohibitive if research showed that the effectiveness of stress training is limited to specific stressor-task combinations.

We subtitled this chapter "Queries, Dilemmas, and *Possible* Solutions." The choice of terms was meant to indicate that in the area of stress training the number of clear questions exceeds by far that of definitive answers. Many of these questions are centuries-old and many have yet to be answered. It is reasonable to assume that individuals and organizations that are involved in stress training have developed important insights over the years. However, this knowledge remains widely dispersed, and rarely is there an opportunity to evaluate and to integrate it. Much also needs to be done to increase the depth and scope of systematic, controlled research.

The need for answers becomes more pressing with the increasing rate of technological change. The combat soldier will face more lethal weapons while having to exercise more complex skills. New technologies will provide commanders with better intelligence for their decisions, but they might also paralyze them with information overload.

Technological change will also have a marked effect on the training environment. Most notably, new technologies will produce more faithful and more realistic simulations of criterion situations. These create new opportunities for the trainer. However, one should not overlook the fact that the higher the realism of the simulation, the greater its specificity; hence, improved realism could result in lower generalizability and hampered transfer of training.

Technological development and change will not produce answers to old questions. They will just create new questions and will heighten the importance of finding answers to the old ones.

REFERENCES

Auerbach, S. M., Martelli, M. F., & Mercuri, L. G. (1983). Anxiety, information, interpersonal impacts, and adjustment to a stressful health care situation. *Journal of Personality and Social Psychology, 44,* 1284–1296.

Averill, J., & Rosen, M. (1972). Vigilant and nonvigilant coping strategies and psychophysiological stress reactions during anticipation of electric shock. *Journal of Personality and Social Psychology, 23,* 128–141.

Berlyne, D. E. (1960). *Conflict, curiosity and arousal.* New York: McGraw-Hill.

Breznitz, S. (1989). Information induced stress in humans. In S. Breznitz & O. Zinder (Eds.), *Molecular biology of stress* (pp. 253–264). New York: Liss.

Burton, D. (1990). Multimodal stress management in sport: Current status and future directions. In J. G. Jones & L. Hardy (Eds.), *Stress and performance in sport* (pp. 171–201). New York: Wiley.

Cascio, W. F. (1989). *Managing human resources: Productivity, quality of work life, profits*. New York: McGraw-Hill.

Cohen, S. (1978). Environmental load and the allocation of attention. In A. Baum & S. Valins (Eds.), *Advances in environmental psychology* (Vol. 1, pp. 1–29). Hillsdale, NJ: Lawrence Erlbaum Associates.

Cohen, S., Evans, G., Stokols, D., & Krantz, D. (1986). *Behavior, health and environmental stress*. New York: Plenum.

Coleman, J. C. (1976). *Abnormal psychology and modern life*. Glenview, IL: Scott, Foresman.

Collins, D. L., Baum, A., & Singer, J. E. (1983). Coping with chronic stress at Three Mile Island: Psychological and biochemical evidence. *Health Psychology, 2,* 149–166.

Crovitz, H. F. (1970). *Galton's walk*. New York: Harper & Row.

Datel, W. E., & Lifrak, S. T. (1969). Expectations, affect change, and military performance in the army recruit. *Psychological Reports, 24,* 855–879.

Dienstbier, R. A. (1989). Arousal and physiological toughness: Implications for mental and physical health. *Psychological Review, 96,* 84–100.

Driskell, J. E. (1984, August). *Training for a hostile environment*. Paper presented at the annual meeting of the American Psychological Association, Toronto, Canada.

Driskell, J. E., & Salas, E. (1991). Overcoming the effects of stress on military performance: Human factors, training and selection strategies. In R. Gal & A. D. Mangelsdorff (Eds.), *Handbook of military psychology* (pp. 183–193). New York: Wiley.

Driskell, J. E., Willis, R. P., & Cooper, C. (1992). Effect of overlearning on retention. *Journal of Applied Psychology, 77,* 615–622.

Egbert, L. D., Battit, G. E., Welch, C. E., & Bartlett, M. K. (1964). Reduction of prospective pain by encouragement and instruction of patients. *New England Journal of Medicine, 270,* 825–827.

Ellersten, B., Johnsen, T. B., & Ursin, H. (1978). Relationship between the hormonal responses to activation and coping. In H. Ursin, E. Baade, & S. Levine (Eds.), *Psychobiology of stress: A study of coping men* (pp. 105–124). New York: Academic Press.

Epstein, S. (1983). Natural healing processes of the mind: Graded stress inoculation as an inherent mechanism. In D. Meichenbaum & M. E. Jaremko (Eds.), *Stress reduction and prevention* (pp. 39–66). New York: Plenum.

Fenz, W. D., & Epstein, S. (1968). Specific and general inhibitory reactions associated with mastery of stress. *Journal of Experimental Psychology, 77,* 52–56.

Friedland, N., & Keinan, G. (1982). Patterns of fidelity between training and criterion situations as determinants of performance in stressful situations. *Journal of Human Stress, 8,* 41–46.

Friedland, N., & Keinan, G. (1986). Stressors and tasks: How and when should stressors be introduced during training for task performance in stressful situations. *Journal of Human Stress, 12,* 71–76.

Friedland, N., & Keinan, G. (1992). Training effective performance in stressful situations: Three approaches and implications for combat training. *Military Psychology, 4,* 157–174.

Fry, P. S., & Wong, P. (1991). Pain management training in the elderly: Matching interventions with subjects' coping styles. *Stress Medicine, 7,* 93–98.

Glass, D. C., & Singer, J. E. (1972). *Urban stress: Experiments on noise and social stressors*. New York: Academic Press.

Goldstein, M. J. (1973). Individual differences in response to stress. *American Journal of Community Psychology, 1,* 113–137.

Haggard, E. A. (1949). Psychological causes and results of stress. In Committee on Undersea Warfare (Ed.), *Human factors in undersea warfare* (pp. 441–461). Washington, DC: National Research Council.

Hofer, M. A., Wolff, E. T., Friedman, S. B., & Mason, J. W. (1972). A psychoendocrine study of bereavement: Parts I and II. *Psychosomatic Medicine, 34,* 481–504.

Horner, S. O., Meglino, B. M., & Mobley, W. H. (1979). *An experimental evaluation of the effects of a realistic job preview on Marine recruit affect, intentions and behavior* (TR–9). South Carolina: Center for Management and Organizational Research.

Inzana, C. M., Driskell, J. E., Salas, E., & Hall, J. K. (1994). The effects of preparatory information on enhancing performance under stress. Paper presented at the annual meeting of the Society of Industrial and Organizational Psychology, Nashville, TN.

Janis, I. L. (1949). Problems related to the control of fear in combat. In S. A. Stouffer, A. Lumsdaine, M. Lumsdaine, R. Williams, M. Smith, I. L. Janis, S. Star, & L. Cottrell (Eds.), *The American soldier: Combat and its aftermath* (Vol. 2, pp. 192–241). New York: Wiley.

Janis, I. L. (1958). *Psychological stress*. New York: Wiley.

Janis, I. L. (1971). *Stress and frustration*. New York: Harcourt Brace Jovanovich.

Kaloupek, D. G., & Stoupakis, T. (1985). Coping with a stressful medical procedure: Further investigation with volunteer blood donors. *Journal of Behavioral Medicine, 8,* 131–148.

Kaloupek, D. G., White, H., & Wong, M. (1984). Multiple assessment of coping strategies used by volunteer blood donors: Implications for preparatory training. *Journal of Behavioral Medicine, 7,* 35–60.

Keinan, G. (1988). Training for dangerous task performance: The effects of expectations and feedback. *Journal of Applied Social Psychology, 18,* 355–373.

Keinan, G., Carmil, D., & Zomer, E. (in preparation). Stability of stress reactions during the Gulf War.

Keinan, G., & Friedland, N. (1984). Dilemmas concerning the training of individuals for task performance under stress. *Journal of Human Stress, 10,* 185–190.

Keinan, G., Friedland, N., & Sarig-Naor, V. (1990). Training for task performance under stress: The effectiveness of phased training methods. *Journal of Applied Social Psychology, 20,* 1514–1529.

Keinan, G., & Zeidner, M. (1987). The effects of decisional control on state anxiety and achievement. *Personality and Individual Differences, 8,* 973–975.

Kern, R. P. (1966). *A conceptual model of behavior under stress with implications for combat training* (HumRRO Tech. Rep. No. 66–12). Washington, DC: The George Washington University.

Kimble, G. A. (1961). *Hilgard and Marquis' conditioning and learning*. New York: Appleton–Gentury–Crofts.

Langer, E. J., Janis, L., & Wolfer, J. (1975). Reduction of psychological stress in surgical patients. *Journal of Experimental Social Psychology, 11,* 155–165.

Lazarus, R. S. (1966). *Psychological stress and the coping process*. New York: McGraw-Hill.

Lazarus, R. S. (1983). The costs and benefits of denial. In S. Breznitz (Ed.), *The denial of stress* (pp. 1–32). New York: International University Press.

Lazarus, R. S., & Folkman, S. (1984). *Stress, appraisal and coping*. New York: Springer.

Lazarus, R. S., & Launier, R. (1978). Stress-related transactions between person and environment. In L. A. Pervin & M. Lewis (Eds.), *Perspectives in interactional psychology* (pp. 287–327). New York: Plenum.

Leventhal, H., Brown, D., Shachman, S., & Engquist, G. (1979). Effects of preparatory information about sensations, threat of pain, and attention on cold pressor distress. *Journal of Personality and Social Psychology, 37,* 688–714.

Logan, G. D. (1985). Skill and automaticity. *Canadian Journal of Psychology, 9,* 283–286.

Ludwick-Rosenthal, R., & Neufeld, R. W. J. (1988). Stress management during noxious medical procedures: An evaluative review of outcome studies. *Psychological Bulletin, 104*(3), 326–342.

Mandler, G. (1982). Stress and thought processes. In L. Goldberger & S. Breznitz (Eds.), *Handbook of stress: Theoretical and clinical perspectives* (pp. 88–104). New York: The Free Press.

Martelli, M., Auerbach, S. M., Alexander, J., & Mercuri, L. (1987). Stress management in the health care setting: Matching interventions to patient coping style. *Journal of Consulting and Clinical Psychology, 55,* 201–207.

Meichenbaum, D. (1977). *Cognitive behavior modification.* New York: Plenum.
Meichenbaum, D. (1985). *Stress inoculation training.* New York: Pergamon.
Meichenbaum, D., & Novaco, R. W. (1978). Stress inoculation: A preventive approach. In C. Spielberger & I. Sarason (Eds.), *Stress and anxiety* (Vol. 5, pp. 317–330). New York: Halsted.
Miller, S. M. (1981). Predictability and human stress: Towards a clarification of evidence and theory. In L. Berkowitz (Ed.), *Advances in experimental social psychology* (Vol. 14, pp. 203–256). New York: Academic Press.
Miller, S. M., & Mangan, C. E. (1983). Interacting effects of information and coping style in adapting to gynecologic stress: Should the doctor tell all? *Journal of Personality and Social Psychology, 45,* 223–236.
Novaco, R. W., Cook, T. M., & Sarason, I. G. (1983). Military recruit training: An arena for stress-coping skills. In D. Meichenbaum & M. E. Jeremko (Eds.), *Stress reduction and prevention* (pp. 377–418). New York: Plenum.
Orne, M. (1965). Psychological factors maximizing resistance to stress with special reference to hypnosis. In S. Klausner (Ed.), *The quest for self-control* (pp. 286–328). New York: The Free Press.
Perkins, C. C., Jr. (1955). The stimulus conditions which follow learned responses. *Psychological Review, 62,* 341–348.
Perkins, C. C., Jr. (1958). An analysis of the concept of reinforcement. *Psychological Review, 75,* 155–172.
Roth, S., & Cohen, L. J. (1986). Approach, avoidance, and coping with stress. *American Psychologist, 41,* 813–819.
Schmolling, P. (1984). Human reactions to the Nazi concentration camps: A summing up. *Journal of Human Stress, 10,* 108–120.
Seligman, M. E. P., Maier, S. F., & Solomon, R. L. (1971). Unpredictable and uncontrollable aversive events. In F. R. Brush (Ed.), *Aversive conditioning and learning* New York: Academic Press.
Selye, H. (1956). *The stress of life.* New York: McGraw-Hill.
Shipley, R. H., Butt, J. H., Horwitz, B., & Farbry, J. E. (1978). Preparation for a stressful medical procedure: Effect of amount of stimulus preexposure and coping style. *Journal of Consulting and Clinical Psychology, 46,* 499–507.
Snow, R. E. (1986). Individual differences and the design of educational programs. *American Psychologist, 41,* 1029–1039.
Strentz, T., & Auerbach, S. M. (1988). Adjustment to the stress of simulated captivity: Effects of emotion-focused versus problem-focused preparation on hostages differing in locus of control. *Journal of Personality and Social Psychology, 55,* 652–660.
Taylor, S. E., & Clark, L. F. (1986). Does information improve adjustment to noxious medical procedures? In M. J. Saks & L. Saxe (Eds.), *Advances in applied social psychology* (Vol. 3, pp. 1–28). Hillsdale, NJ: Lawrence Erlbaum Associates.
Terris, W., & Rahhal, D. K. (1969). Generalized resistance to the effects of psychological stressors. *Journal of Personality and Social Psychology, 13,* 93–97.
Vernon, D. T. A. (1971). Information seeking in a natural stress situation. *Journal of Applied Psychology, 55,* 359–363.
Vernon, D. T. A., & Bigelow, D. A. (1974). Effect of information about a potentially stressful situation on response to stress impact. *Journal of Personality and Social Psychology, 29,* 50–59.
Vossell, G., & Laux, L. (1978). The impact of stress experience on heart rate and task performance in the presence of a novel stressor. *Biological Psychology, 6,* 193–201.
Vroom, V. H. (1964). *Work and motivation.* New York: Wiley.
Weick, K. E. (1979). *The social psychology of organizing.* Reading, MA: Addison-Wesley.
Weick, K. E. (1985). A stress analysis of future battlefields. In J. G. Hunt (Ed.), *Leadership and future battlefields* (pp. 32–46). Washington: Pergamon-Brassey's.

West, L. J. (1958). Psychiatric aspects of training for honorable survival as prisoner of war. *American Journal of Psychiatry, 15*, 329–336.

Wexley, K. N., & Latham, G. P. (1981). *Development and training human resources in organizations*. Glenview, IL: Scott, Foresman.

Willis, M. P. (1967). Stress effects on skill. *Journal of Experimental Psychology, 74*, 460–465.

Worchel, S., & Yohai, S. (1979). The role of attribution in the experience of crowding. *Journal of Experimental Social Psychology, 15*, 91–104.

Zakay, D., & Wooler, S. (1984). Time pressure, training, and decision effectiveness. *Ergonomics, 27*, 273–284.

9 Designing for Stress

Christopher D. Wickens
University of Illinois at Urbana–Champaign

The proposal must be postmarked no later than 5 p.m., but as the copying is being done frantically an hour before the deadline, the machine ceases to function, displaying a series of confusing error messages on its computer-driven display. With panic gripping the unfortunate victim, he finds himself unable to decipher the complex and confusing instructions. Meanwhile, in another building on campus, the job candidate giving a talk has fielded a few difficult (some might say nasty) questions, and now turns to the video demo that should answer the questions. Nervous and already upset, she now finds that the VCR will not function, and as she fiddles with the various buttons, no one lifts a hand to assist her. Finally, in the sky above the city, the pilot of the advanced commercial airliner is descending toward a landing, having programmed the flight management system to proceed according to ATC clearances. Suddenly, the message arrives from ATC that the plane has dangerously descended in conflict with another airline. The system must be reprogrammed to an alternative heading and level.

Difficulties in a human–system interface, like those described here, characterizing stressful performance with a complex system, are typically addressed by three generic means: training, selection, and design (Huey & Wickens, 1993). Design issues themselves may be categorized into those that focus on task design, (e.g., workload reduction through task sharing) and on interface design. Many of the chapters in this book have discussed training issues, whereas Hogan and Lesser (chap. 6, this volume) have considered the implications for selection. In the present chapter we consider the implications for design.

It is obvious that design changes, intended to lessen the unfortunate consequences of stress on human–system performance, must be based on an accurate (and therefore valid) model of stress effects on performance. Unfortunately, however, we find that there are several factors that stand in the way of creation of such a model. We discuss these factors first, before describing the most important (validated) components of the model, and then discussing some of its implications for design.

A VALID STRESS MODEL

There are a variety of sources of information that can be used to derive a model of human performance under stress: accident analysis, incident analysis, operator protocols, and laboratory experiments. Unfortunately, all of these suffer certain limitations, as I have discussed elsewhere (Wickens, 1995). Accident analysis suffers from the problems of low sample size and small *N*. For example, the case study of the USS Vincennes (Klein, chap. 2, this volume) has received tremendous scrutiny in its analysis, but, without the benefit of hindsight, it is very difficult to draw conclusions on the critical causal factors in the incident, and to attribute the manner in which stress may or may not have been a responsible agent. A single accident case study is simply incapable of revealing which of several causal factors might have been most responsible.

Incidents and case studies have the advantage of occurring more frequently (than accidents) and therefore providing a larger, and more stable data base on which to draw conclusions. Nevertheless, many inferences must be made in order to establish the degree of stress that may have been present at the time of the incident, and most incidents may have multiple causes as well. However, the advantages of large numbers is that, through statistical amalgamation, true causal factors tend to remain, whereas random, noncausal, or "noise" factors diminish in their influence. In this manner, for example, McKinney (1993) inferred that inflight emergencies of solo combat pilots would be a source of stress, and, through analysis of a large number of such incidents, he was able to generalize some characteristics of problem solving degradation in these stressful conditions.

Protocol analysis of users performing an operational system, a useful tool in many instances, is less than fully satisfactory for several aspects of stress research, because it is unlikely that analyses will be taken while stressful conditions are present. (However, this is not invariably the case. For example, it is known that the student pilot's initial check ride is quite stressful, and these could be favorable conditions for passive, nonverbal protocols to be collected.)

The fourth, best controlled technique for the analysis of stress effects is through laboratory experiments. These may be done in relatively realistic surroundings, such as the Army combat training arena (Berkun, 1964) or the real-mission flight simulator (Ruffle-Smith, 1979), or the environment may be somewhat more simplified (and, therefore, often more controlled; Keinan & Friedland, chap. 8, this volume; Wickens, Stokes, Barnett, & Hyman, 1993).

However, two statistical factors often limit the generalizability of conclusions from such studies. First, the actual effect size (in terms of variance accounted for) may be quite small, despite its statistical reliability. For example, a change in decision strategy of .08 probability units, or a change in response time of 150 msec may be observed as a statistically significant effect, given a sufficiently large *N* experiment. However, great caution must be exercised before assuming that this effect would be manifest in the single operator that operates a complex system without benefit of experimental control over all of the extraneous variables. If such effects are not likely to be shown, it hardly seems wise to base extensive and expensive design changes on them.

Reflecting the other side of the statistical coin, the second concern is with insufficient sample size that is sometimes a deficiency of experimental design (particularly experiments carried out in complex environments). Here, the low *N* can so reduce the power of an experiment as to make finding an effect very unlikely, even if the effect is substantial, so long as the "traditional" .05 criterion is accepted for effect reporting. As Harris (1991) has eloquently pointed out, in doing experiments with design implications (i.e., human factors research), it is incumbent on the researcher to consider the potential costs of type-2 statistical errors (reporting no effect when there really is one), as well as the standard guarding against type-1 errors (the traditional .05 significance test). Sometimes the cost of rejecting an effect when there is one there may be large, particularly if a low power design (i.e., with small subject sample size), makes the probability of such an error fairly high.

These two statistical concerns lead logically to the importance of meta-analysis, (Rosenthal, 1991) for pooling what data are available from multiple experiments that have provided common manipulations. Some of the chapters in this book have provided valuable contributions in this direction (see also Cannon-Bowers, Salas, & Converse, 1990; Hockey, 1986). The data base of research on which to draw, however, remains woefully inadequate—the result, in part, of the difficult roadblocks that confront the investigator of stress research. These roadblocks of course have to do with the legitimate concerns for the health and safety of human subjects. Whether or not the need for the human factors community to better understand human performance under stress is sufficiently balanced against the limited availability of data is a judgment that protocol review committees must decide.

A SIMPLIFIED STRESS MODEL

As a consequence of the limitations of all methodologies, and the small experimental data base available, it is prudent to provide a stress model with design implications that only considers well-documented and relatively robust effects. It is also prudent to restrict the operational definition of stress in the current writing to what I term *emergency stress*. These are the circumstances in which the operator believes that external conditions create a potential for serious damage either in bodily harm, or psychological state, and are normally coupled by some degree of time pressure (Svenson & Maule, 1993). Given the assumption of a correlation with time pressure, which will indeed be experienced in most emergency situations, it is possible to state with some confidence that the effects discussed in the next sections will often be observed.

Attentional Tunneling

The idea that the scope or "breadth" of an attentional spotlight will narrow under stress has been documented by a number of studies (Hockey, 1986). For example, Weltman, Smith, and Egstrom (1971) compared the performance of two groups of divers on a central and peripheral signal-detection task. One group was led to believe that it was under conditions of a 60-foot dive in a pressure chamber; the other was not. In fact, there was no change in pressure for either group. Both groups showed similar performance on the central task, but performance on the peripheral task was significantly degraded for the pressure group. This group also showed greater anxiety-related increases in heart rate, substantiating the increased level of stress. Stress may also cause a focusing of attention on one particular task or cognitive activity, as well as on a particular perceptual channel. In the analysis of cockpit voice recordings at times of high stress prior to accidents, Helmreich (1984) found evidence of severe breakdowns in the pilots' ability to handle multiple tasks or multiple concerns.

Although perceptual or cognitive tunneling produced by stress usually degrades performance, it is also possible to envision circumstances in which this tunneling may actually facilitate performance, in which focused attention on critical task aspects is desired. Indeed, this positive effect was observed in a study by Houston (1969), who found that noise stress improved the focus of attention on the relevant aspect of a stimulus and reduced the distracting effect of irrelevant aspects.

Data provided by Houston's experiment and others suggest that the attentional or cognitive tunneling resulting from stress is defined in terms of subjective importance or priority. That is, performance of those tasks or processing of that information thought to be most important remains unaffected or perhaps is enhanced (through arousal), whereas processing information with lower perceived priority is filtered (Bacon, 1974; Broadbent, 1971). In one sense, this kind of tunneling is optimal, but it will produce undesirable effects if the subjective importance that defines the attended channel proves to be unwarranted. Such was the case, for example, in the incident of the Three Mile Island nuclear power plant (Rubinstein & Mason, 1979). Immediately after the crisis in the plant developed and under the high stress caused by the initial failure, the operators appeared to be fixated on a single faulty indicator, supporting an incorrect belief that the water level in the reactor was too high, thereby preventing their attention from focusing on more reliable indicators that supported the opposite (and correct) hypothesis. As suggested by this example and elaborated later, this narrowing effect can be directly related to biases in decision making. For the operator who has a well-structured and accurate model of task demands and a well-developed skill in discriminating sources of useful (versus trivial) information, however, it can be expected that stress should lead to little degradation of performance. This buffering of stress effects was demonstrated by the flight crew of United Airlines flight 232, which suffered a life threatening total failure of the hydraulics system on approach to Sioux City Airport, yet was able to focus attention effectively on implementing a solution to the problem at hand, (Predmore, 1991).

Working Memory Loss

Mandler (1979) has discussed the degrading effects of anxiety on working memory. Correspondingly, Berkun (1964) performed a study in which U.S. Army personnel were placed under very realistically simulated combat stress conditions in which they believed that their life was at risk. Much of the degraded performance in procedures that Berkun observed can be attributed to reduced working memory capacity. Logie and Baddeley (1983) and Lewis and Baddeley (1981) have noted similar working memory decrements of divers performing at depth. Idzikowski and Baddeley (1983a, 1983b) observed an anxiety-related working memory loss in speakers waiting in the wings to give their first public speech. Noise, as well as danger and anxiety, has also produced consistent effects on working memory (Hockey, 1986). Although it is intuitively evident that the presence of noise would disrupt the ability to rehearse verbal information in working memory (Poulton, 1976), it also appears that the combined stress effects of noise and anxiety may disrupt spatial working memory systems as well (Stokes, Belger, & Zhang, 1990; Stokes & Raby, 1989). Indeed, in a simulation study of pilot decision making, Wickens, Stokes, Barnett, and Hyman (1988) observed that the effects of noise stress were greatest on decision problems that relied on spatial visualization for their successful resolution.

Long-Term Memory

Although stress appears to disrupt working memory, it appears to have less of an effect on the retrieval of information from long-term memory to the extent that information is well rehearsed and memorized. For example, in their study of pilot judgment, Wickens et al. (1988) found that those judgments requiring direct retrieval of facts from long-term memory were relatively unimpaired by stress.

In Berkun's study of Army soldiers under stress, he found a difference between more- and less-skilled soldiers in the extent to which performance degraded. The more experienced soldiers were less disrupted, a finding which is also consistent with this view assuming that the more skilled soldiers had better learned procedures stored in long-term memory. Stokes et al. (1990) found that the decision-making performance of novice pilots deteriorated under stress, whereas performance of a group of highly trained pilots in the same circumstances was not affected. Presumably, the latter group was more able to rely on direct retrieval of information from long-term memory in making their decisions.

Similar to the narrowing effect that stress exerts on perception and selective attention, however, stress appears to restrict the information retrieved from long-term memory more specifically to those habits that are well learned or over-learned (Eysenck, 1976). Although there do not appear to be much experimental data supporting this claim, at least one study shows that increased stress actually eliminates some of the benefits of expertise and training in decision making (Ben Zur & Breznitz, 1981). Studies by Fitts and Seeger (1953) and Fuchs (1962) and analyses of aircraft accidents carried out by Allnutt (1987) have all suggested that stress will lead to a regression to earlier learned and more compatible response patterns when these patterns may conflict with incompatible (but appropriate) ones. Collectively, these findings suggest the importance of extensive training in procedures and actions that may need to be taken in emergency situations. They further emphasize that such procedures should require only actions of high compatibility. The findings, however, emphasize the extent to which creative innovative problem solving may be degraded under stress because, by definition, such problem solving will not have been accomplished in the same way in the past and hence repetitive practice of the same steps cannot be achieved.

Strategic Shifts

There is some evidence from behavioral studies that stress leads to consistent shifts in processing strategy. In a study of the anxiety brought on by the first parachute jump, for example, Simonov, Frolov, Evtushenko, and Suiridov (1977) observed a shift in performance of a signal detection task that can be characterized by a riskier criterion setting. The paratroopers were simply more likely to respond "yes" and hence made more hits and more false alarms. Hockey (1986) concluded that there is a general effect of noise, and/or anxiety stress on the speed–accuracy tradeoff, shifting performance to a less accurate but not slower level. In their study of pilot judgment, Wickens et al. (1988) found that judgments were less accurate but not necessarily slower under the combined stress effects of noise, time pressure, and threat of loss of income.

Dorner and Pfeifer (1993) observed systematic shifts in strategic thinking in operators of a video firefighting game that result from unpredictable bursts of noise stress. On the one hand, stressed subjects respond more rapidly and give more tactical commands to the simulated firefighting units under their control; on the other hand, however, these commands are simpler, somewhat less effective, and lead to more tactical errors than the commands issued by nonstressed subjects, hence illustrating a form of speed–accuracy tradeoff that may be described as "adaptive" in the sense of minimizing the disruptive effects of the noise bursts.

The tendency of those under the stress of an emergency to shift performance from accurate to fast (but error-prone) responding has been cited as a concern in operator response to complex failures in nuclear power control rooms. The operator often has a desire to do something rapidly, when in fact this impatience is often counterproductive until the nature of the failure is well understood. In the Three Mile Island incident the hasty action of the control room operators was to shut down an automated device that had in fact been properly doing its job. Fischer, Orasanu, and Montalvo (1993) noted that good performing crews under the time stress of a simulated inflight emergency were actually slower, but more careful in the selection of emergency procedures, than poorer performing crews.

Decision Making

Understanding the effects of stress on decision making has always been of great interest to the human factors profession. The importance of this knowledge has been enhanced by the analysis of the faulty decisions made in the Three Mile Island incident and, more recently, by concerns over the decisions made in the U.S.S. Vincennes incident (U.S. Navy, 1988; Klein, chapter 2, this volume). The concern that decisions degrade under stress is reinforced by anecdotes, and case studies of poor pilot judgments that have occurred under stressful conditions of bad weather, spatial disorientation, or aircraft failure (Jensen, 1982; Nagel, 1988; Simmel, Cerkovnik, & McCarthy, 1987). However, it is often difficult to tell whether stress was itself a causal factor in the poor decision or whether the conditions that produced the stress were also those that, for example, degraded the information available in such a way that the poorer decision became more likely.

To predict the effects of stress on decision making, one approach is to adopt a componential approach (Wickens & Flach, 1988; see also Orasanu, chap. 3, this volume). Because different decisions may involve varying dependence on such components as working memory, attention, and long-term memory retrieval, each decision may be affected differently by stress as a function of the components on which it depends and of the differential effects of stress on those components as described earlier (Wickens et al., 1988).

An alternative approach is to examine the results of experiments that impose stress on decision making, diagnosis, and problem-solving tasks. Although few studies exist, their results are consistent with the picture of a stress-sensitive decision-making process. Such studies have not only shown that decisions of various sorts degrade under stress but have also concluded that this degradation takes specific forms. Thus, for example, Cowen (1952) found that subjects perseverated longer with inappropriate or rigid problem solutions under the stress produced by threat of shock, a sort of action tunneling that is consistent with the idea of attentional tunneling. Keinan, Friedlan, and Benporat (1987) found that the allocation of attention to a word problem became increasingly nonoptimal and unsystematic as stress was imposed by the threat of an electrical shock. The investigators also observed that this stressor produced a premature closure: Subjects terminated their decision before all alternatives had been considered. Ben Zur and Breznitz (1981) found that stress, although leading to some filtering of information, also led subjects to give more weight to negative task dimensions and, as a consequence, make less risky decisions.

Driskell and Salas (1991) observed that stress actually increased the receptivity of members of a dyad to judgments provided by the other member in reaching a problem solution. They imposed stress on U.S. Navy personnel who believed they were one component of a cooperative decision-making dyad. In fact, unknown to the subjects, the other component of the dyad was always the same source of computer-generated information for all subjects. The investigators were interested in the extent to which perceived stress influenced the receptivity to information received from a team member when that member was perceived to be either above or below them in the military chain of command. Not surprisingly, greater receptivity to information was shown when the member was perceived of higher rank. Whether the rank was higher or lower, however, both groups were more willing to accept the information under the conditions of perceived stress.

As we have discussed earlier, Wickens et al. (1988) observed that the combined stress of noise, time pressure, risk, and task loading produced a general degradation of pilot judgments on a computer-based flight simulation. The stress effect, however, was selectively observed only on decision problems that were difficult in terms of their spatial memory demand. As noted previously, their data indicated that decisions did not degrade under stress when long-term memory retrieval was the primary mechanism. These conclusions are related to those drawn by Klein (1989; see also chap. 2, this volume), who argued that expert firefighters use direct long-term memory retrieval to make their decisions. Given this characteristic, experts making decisions about familiar courses of action should be less likely to suffer degrading effects of stress.

The conclusion drawn by Klein (1989), however, is partially contradicted by a study carried out by Koehler and McKinney (1991), who evaluated the decision-making performance of Air Force pilots in 195 aircraft malfunction mishaps. Their analysis of decision quality revealed that experts (long-time pilots) did not necessarily perform better than relatively new pilots under the high-stress conditions of an inflight malfunction. Although this conclusion must be tempered by the fact that experts were more often flying a lead aircraft in formation, thereby adding to their attention demands, a second finding is consistent with the picture presented here; that is, expert performance was particularly disrupted when the malfunctions were novel and unique, more so than the performance of the novices. This is consistent with the view that experts had well-learned solutions available in long-term memory for the routine problems that were not available for the unique ones. The novices had fewer such solutions available for either problem type, and the loss in their performance from routine to unique would therefore be less.

Of course a finding that is, by now, very well documented is that different people adapt very differently to stressors. Fischer, Orasanu, and Montalvo (1993) and Orasanu and Strauch (1994) reported a finding for a full-mission flight simulation of how aircrews responded to the stressful time pressured circumstances of inflight emergencies. Although some crews behaved "optimally," others clearly made some poor choices on their handling of the emergency. In contrast to the more effective crews, those who were less effective and harmed more by stress (a) tended to make certain key decisions more rapidly, without careful considerations of the complexity of the options (a form of speed–accuracy tradeoff), and (b) tended to be reactive, and procrastinating rather than proactive and foresightful in scheduling subtasks that needed to be done in advance of the simulated landing.

A FEW DESIGN PRINCIPLES

The previous section has documented several relatively well validated "stress effects." Many of these, of course, have direct implications for training. This includes not only training of the skill itself, but training on strategies to deal with stress (see chaps. 7 and 8, this volume), such as avoiding the tendency to perform premature actions. The direct implications for design are perhaps somewhat less clear cut. To some extent it is evident from the simplified stress model presented earlier that design decisions to simplify operations under normal procedures will also better support performance under stress, and here we discuss three such principles:

1. *Reduction of working memory load.* Working memory is a vulnerable bottleneck for routine as well as stressed performance. Hence, design decisions made to allow continuous visual display of information that may need to be entered or computed will surely benefit performance of both kinds of tasks. An example here would be the visual display of computer-coded digital data link messages from controller to pilot (Kerns, 1991). Such visual digital coding should be able to augment (rather than replace) auditory channels, and would seemingly offer a substantial advantage for a stressed pilot disoriented in bad weather.
2. *Principle of pictorial realism and ecological interface design.* Much can be said about the potential "intuitive" advantages of presenting information about physical systems, as a "picture" that mimics many aspects of the

system in question. Roscoe (1968) has described this as the principle of pictorial realism (PPR). Indeed, to the extent that the operator is familiar with the underlying structure of the information presented, presenting this information as a picture would appear to support more "natural" and "automatic" processing, and would thereby capitalize on and exploit the well-learned habits that are stress resistant.

In an aircraft, this might involve representing guidance for flight path control as a forward-looking "highway in the sky" (Haskell & Wickens, 1993). A derivative of the PPR is the principle of ecological interface design (Bennett, Toms, & Woods, 1993; Rasmussen & Vicente, 1989), in which pictorial graphic displays attempt to mimic or display prominently the physical constraints and variables relations necessary for control that are inherent in the operator's mental model of complex processes, even as these variables may not be embodied in a true visible "picture." Applications of this principle have primarily been in the area of nuclear and energy conversion processes (Bennett, et al., 1993; Moray, et al., 1994). At an even simpler level, the PPR advises the depiction of continuously changing quantities as pictures (analog), rather than as symbols (digital), and the advantages of analog over digital displays in high-stress environments is well documented (Sanders & McCormick, 1993; Wickens, 1992). Indeed, one of the clear design deficiencies that could have been responsible for the misclassification of the Iranian Airlines jet as a hostile fighter in the USS Vincennes incident was the depiction of the aircraft's altitude by a digital readout (see Klein, this volume).

Nevertheless, the PPR can be oversold as a design improvement. There are clearly times, for example, when symbolic depictions or distortions of the true "picture" might actually improve the user's performance. For example, in a number of investigations, we have found that depiction of 3-D space is more effective by a set of planar displays, than by "pictorially real" 3-D displays (Wickens, Liang, Prevett, and Olmos, in press), because of inherent perceptual ambiguities caused by the latter.

3. *Proximity compatibility principle (PCP).* A fundamental part of our stress model is the assertion that information access, through visual search or other retrieval mechanisms, is disrupted and sometimes curtailed in times of stress (particularly when that stress has a time component). The major bias of perceptual or cognitive tunneling certainly represents one important aspect of this disruption. Therefore, it may be expected that performance of tasks which require accessing two sources of information to be integrated or compared might be more disrupted by the degrading effects of stress (Pamperin & Wickens, 1987). Good design will thereby present in close display proximity those sources of information that must be processing in close "mental proximity" (Wickens & Carswell, 1995). For example, it is useful to have the error codes explaining a malfunction available

to be viewed right next to the computer-driven display that annunciates a particular error symbol. Here again, the Vincennes incident was potentially illustrative. The altitude reading of the target aircraft was displayed quite separately from the radar depiction of its ground track, making it more difficult for operators to integrate the lateral and vertical trend information of the single aircraft (and therefore understand it as climbing and not descending).

The PCP naturally has a number of derivative implications for design (Wickens & Carswell, 1995). All have in common the design principle to make it easier to compare and relate sources of information that must be used together in the service of a single task. By so doing, the stress-related damage to cue sampling can be lessened (because the travel of attention between samples is reduced). Furthermore, the implications of stress damage to working memory will be attenuated, because working memory is often involved in the comparison of displayed information sources.

The principles listed here are all generally helpful for design and, for the most part, have more "upsides" than downsides. Applying them for design in normal use should also facilitate performance under stressful conditions in emergencies.

However, design for stress is also characterized by a certain number of trade-offs: What is good in one situation (i.e., stress) is not necessarily good for normal operation, and it is to these trade-offs we shall now turn.

DESIGN TRADE-OFFS

Guidance Versus Awareness. A characteristic of routine operation is that it is just that—it is routine and predictable. Therefore, the need to make the operator aware of a broad set of alternative actions, or a wide range of the information space in which performance takes place, is less important than the need to provide continuous feedback that the actions performed are correct and correctly carried out by the system. These generic information requirements transcend quite different system types: In aviation navigation during routine flight, for example, the pilot primarily needs to know if the aircraft flight parameters are maintained near the target values, and that other systems are normal; in process control operation, the operator needs to know that the system is responding to control inputs, and, again, that other system components are normal; in routine computer usage, the operator simply needs feedback that commands have been registered from keystrokes. As Landeweerd (1979) has noted in the context of process control in normal operation, primary awareness must be given to the forward causal flow of events: Do the operator's operations CAUSE the appropriate system response to occur? Using a term borrowed from the domain of assessment, this is *convergent thinking*.

In marked contrast, the emergencies and abnormalities that often trigger stress have a very different set of information requirements—now reasoning must often work backward, from effect to possible cause (diagnostic reasoning), and the operator may need to be aware of a much broader range of task domain information in order to support divergent thinking and problem solving. Two examples illustrate the difference. First, unlike the pilot on routine flight, the pilot who suddenly finds himself lost or spatially disoriented can no longer afford to consider only what is directly in front of the aircraft (the information for routine local guidance), but now must be aware of the full terrain surrounding the aircraft in a 360° sphere because the desired flight path could now be anywhere, and any number of hazards could intervene between current location and heading and the desired levels of these parameters (Wickens, Liang, Prevett, & Olmos, in press; Wickens & Prevett, 1995). Second, the process control operator confronting a malfunction must now try to make sense of a myriad of confusing lights and warning signals, tracing these back to the cause through a combination of symptomatic interpretations and trouble-shooting tests. It is clear that neither narrowly focused information displays nor routine procedural actions are appropriate for either of these situations. Roth and Woods (1988), for example, have explicitly pointed out the problems with "brittle" procedural checklists to be rigidly followed in times of failure and fault. Each fault may be so unique that subtle departures from a proceduralized action sequence may need to be taken based on the symptoms and consequences emerging from previous actions.

In sum, the challenge to the designer is how to support the very different information needs in the stressful times of failure recovery, and in the nonstressful times of routine operation, without overwhelming the operator at the moment of emergency. A key design concept here seems to be the development of ecological displays that can efficiently provide a single context within which both routine and nonroutine information can be depicted in integrated functional terms that match the operator's mental model of the process under supervision. These design efforts for single display frameworks have been examined in both process control (Bennett et al., 1993) and Aviation (Wickens et al., in press; Wickens & Prevett, 1995) contexts.

Command Versus Status Displays. A related trade-off is inherent in the choice between displaying what is (a status display) an displaying what to do (a command display). There is a certain irony in this choice. When life is routine, the operator is not under stress and there is ample time to absorb and interpret a great deal of information; then presentation of status information makes a great deal of sense. The operator can take the time to fully understand the status of the system under his or her purview, and chose actions appropriately. However, under the crisis of an emergency, the operator may have neither the time nor the resources available to figure out what to do, and hence may need a command display that instructs the most appropriate action. This is certainly the philosophy of the

TCAS display for pilots that under the crisis of a potential midair conflict, commands the pilot to perform certain maneuvers (turn, climb, etc.; Chappell, 1990).

The potential problem with this situation is that effective command displays depend critically on a certain amount of computer intelligence to forecast the future system states, and to make assumptions about how different elements in the system will respond. TCAS logic, for example, must make assumptions about the future flight path of the aircraft and about the likelihood that both aircraft involved in a potential conflict will or will not perform evasive maneuvers in the timely and accurate fashion recommended. Where such intelligence may be imperfect, there may be occasions in which an operator should see the current status on which the computer's command recommendations are based. This double need for both status and command information requires design considerations of *foreground–background redundancy.* That is, redundant presentation of both command and status information is desirable; however it is CRITICAL that the nature of the two sources be clearly distinguished so that there can be no confusion regarding what is status information ("you are too far *left*") and what is command information ("turn *right*"). Display design should also make the command information more salient, because that is what may need to be processed in the times of stress when cognitive tunneling serves to filter out less important channels of information. Therefore, a viable approach would seem to be one that presents both command and status information within the same framework, but highlights the command information (presents in the foreground) while deemphasizing the background status information, allowing it to be plainly visible, but not the primary source of attention capture as the display is quickly viewed.

Flexibility Versus Consistency. With increasing sophistication of automation and artificial intelligence, it is progressively more feasible to structure "intelligent interfaces (Hancock & Chignell, 1989) that assume configurations which are most appropriate for a given operation, phase, or set of user requirements. For example, "tailored logic" in a multifunction display would yield those displays that are appropriate for a current phase of flight most easily through a menu system (e.g., takeoff, cruise, landing; Reising & Curry, 1987). Within the context described in the previous section, status displays could appear in routine circumstances and command displays could be made to appear during emergencies. A corresponding adaptive presentation could contingently present local guidance versus global awareness information. Such flexibility has the advantage of saving the operator the time of engaging in information search, allowing the maximum amount of display space devoted to the most valuable information for a given function, or both.

However, flexible display (or control) design is purchased at the cost of inconsistency, and there are severe dangers of inconsistently configured interfaces, particularly in times of stress (Braune, 1989). Consider, for example, the dangers inherent when a given digital symbol has a very different meaning, depending on the setting of another mode switch. Such confusion was thought to underlie a recent crash of an Airbus A320, when accident investigators inferred that the pilots might have confused a setting of 3.3° flight path angle (what they thought it meant) with 3.3 × 1000 vertical feet/minute descent rate (what the automation had actually commanded and was therefore depicted on the display). The digital indicator read "3.3" in both cases, and only a relatively subtle indicator suggested the critical differences in modes. The inconsistent meaning of similarly appearing displays or controls appears to be a particularly lethal invitation for stress-induced confusion, and is a force that should temper a designer's enthusiasm for overly flexible, changeable, and modifiable interfaces. Design for stress, in short, should strive for the creation of consistency.

CONCLUSION

In conclusion, we have tried to offer some advice, although it was not often very concrete, on how to address these trade-offs between what is best suited for normal nonemergency situations, and what is best suited for the stressful operation during emergency. In many cases these recommendations are based on extrapolations from models of stress in simpler situations, rather than concrete scientific evaluations of emergency response in real-world tasks. However, given the importance of both efficient stress management in crisis and appropriate display-control interfaces in preventing human error, it would appear to be of some urgency to proceed with research evaluations of the suggestions made. As other chapters in this book have noted, training can be an important and necessary remediation for stress, but it is far from sufficient (Huey & Wickens, 1993).

REFERENCES

Allnutt, M. F. (1987). Human factors in accidents. *British Journal of Anaesthesia, 59,* 856–864.

Bacon, S. J. (1974). Arousal and the range of cue utilization. *Journal of Experimental Psychology, 102,* 81–87.

Baddeley, A. D. (1972). Selective attention and performance in dangerous environments. *British Journal of Psychology, 63,* 537–546.

Bennett, K. B., Toms, M. L., & Woods, D. D. (1993). Emergent features and graphical elements: Designing more effective configural displays. *Human Factors, 35,* 71–98.

Ben Zur, H., & Breznitz, S. J. (1981). The effect of time pressure on risky choice behavior. *Acta Psychologica, 47,* 89–104.

Berkun, M. M. (1964). Performance decrement under psychological stress. *Human Factors, 6,* 21–30.

Braune, R. J. (1989). The common/same type rating: Human factors and other issues. *SAE Technical Paper Series 892229,* 1–5. Warrendale, PA: Society of Automotive Engineers.

Broadbent, D. (1971). *Decision and stress,* NY: Academic Press.

Cannon-Bowers, J. A., Salas, E., & Converse, S. A. (1990). Cognitive psychology and team training: Training shared mental models of complex systems. *Human Factors Bulletin, 33,* 1–4.

Chappell, S. L. (1990). Pilot performance research for TCAS. *Third Human Error Avoidance Techniques Conference Proceedings* 6 (pp. 51–68). Warrendale, PA: Society of Automotive Engineers.

Cowen, E. L. (1952). The influence of varying degrees of psychosocial stress on problem-solving rigidity. *Journal of Abnormal and Social Psychology, 47,* 512–519.

Dorner, D., & Pfeifer, E. (1993). Strategic thinking and stress. *Ergonomics, 36*(11), 1345–1360.

Driskell, J. E., & Salas, E. (1991). Group decision making under stress. *Journal of Applied Psychology, 76*(3), 473–478.

Eysenck, M. W. (1976). Arousal, learning, and memory. *Psychological Bulletin, 83,* 389–404.

Fischer, V., Orasanu, J. & Montalvo, M. (1993). Efficient strategies on the flight deck. In R. Jensen (Ed.), *Proceedings for Symposium on Aviation Psychology* (pp. 238–243). Columbus, Ohio: Ohio State University.

Fitts, P. M., & Seeger, C. M. (1953). S–R compatibility: Spatial characteristics of stimulus and response codes. *Journal of Experimental Psychology, 46,* 199–210.

Fuchs, A. (1962). The progressive–regressive hypothesis in perceptual-motor skill learning. *Journal of Experimental Psychology, 63,* 177–181.

Hancock, P. A., & Chignell, M. H. (Eds.). (1989). *Intelligent interfaces: Theory, research and design.* New York: Elsevier.

Harris, D. (1991). The importance of the type 2 error in aviation safety research. In E. Farmer (Ed.), *Stress and error in aviation* (pp. 151–157). Brook Feld, VT: Auebury.

Haskell, I. D., & Wickens, C. D. (1993). Two- and three-dimensional displays for aviation: A theoretical and empirical comparison. *International Journal of Aviation Psychology, 3*(2), 87–109.

Helmreich, R. (1984). Cockpit management attitudes. *Human Factors, 26,* 583–589.

Hockey, G. R. J. (1986). Changes in operator efficiency as a function of environmental stress, fatigue, and circadian rhythms. In K. R. Boff, L. Kaufman, & J. P. Thomas (Eds.), *Handbook of perception and human performance* (Vol. II, pp. 44–1/44–49). New York: Wiley.

Houston, B. K. (1969). Noise, task difficulty, and Stroop color–word performance. *Journal of Experimental Psychology, 82,* 403–404.

Huey, M. B., & Wickens, C. D. (Eds.) (1993). *Workload transition: Implications for individual and team performance.* Washington, DC: National Academy Press.

Idzikowski, C., & Baddeley, A. D. (1983a). Fear and dangerous environments. In G. R. J. Hockey (Ed.), *Stress and fatigue in human performance* (pp. 123–144). London: Wiley.

Idzikowski, C., & Baddeley, A. D. (1983b). Waiting in the wings: Apprehension, public speaking, and performance. *Ergonomics, 26,* 575–583.

Jensen, R. S. (1982). Pilot judgment: Training and evaluation. *Human Factors, 24,* 61–74.

Keinan, G., Friedland, N., & Benporat, Y. (1987). Decision making under stress: Scanning of alternatives under physical threat. *Acta Psychologica, 64,* 219–228.

Kerns, K. (1991). Data-link communication between controllers and pilots: A review and synthesis of the simulation literature. *The International Journal of Aviation Psychology, 1*(3), 181–204.

Klein, G. A. (1989). Recognition-primed decision. In W. Rouse (Ed.), *Advances in man–machine systems research, Vol. 5* (pp. 47–92). Greenwich, CT: JAI.

Koehler, J. J., & McKinney, E. H. (1991). *Uniqueness of task, experience, and decision making performance: A study of 176 U.S. Air Force mishaps.* Unpublished manuscript.

Landeweerd, J. A. (1979). Internal representation of a process fault diagnosis and fault correction. *Ergonomics, 22,* 1343, 1351.

Lewis, V. J., & Baddeley, A. D. (1981). Cognitive performance, sleep quality, and mood during deep oxyhelium diving. *Ergonomics, 24,* 773–793.

Logie, R. H., & Baddeley, A. D. (1983). A trimex saturation dive to 660 m: Studies of cognitive performance, mood, and sleep quality. *Ergonomics, 26,* 359–374.

Mandler, G. (1979). Thought processes, consciousness, and stress. In V. Hamilton & D. M. Warburton (Eds.), *Human stress and cognition: An information processing approach.* Chichester, England: Wiley.

McKinney, E., Jr. (1993). Flight leads and crisis decision-making. *Aviation, Space, and Environmental Medicine, 64*(5), 359–362.

Moray, N. (1986). Monitoring behavior and supervisory control. In K. R. Boff, L. Kaufman, & J. P. Thomas (Eds.), *Handbook of perception and performance* (Vol. II, pp. 40–1/40–51). New York: Wiley.

Moray, N., Lee, J., Vicente, K., Jones, B., & Rasmussen, J. (1994). A direct perception interface for nuclear power plants. In *Proceedings 38th meeting of the Human Factors Society.* Santa Monica: Human Factors.

Nagel, D. C. (1988). Human error in aviation operations. In E. Wiener & D. Nagel (Eds.), *Human factors in aviation* (pp. 263–303). New York: Academic Press.

Orasanu, J., & Strauch, B. (1994). Temporal factors in aviation decision making. In *Proceedings of the 38th Annual Meeting of the Human Factors and Ergonomics Society* (935–939). Santa Monica, CA: Human Factors and Ergonomics Society.

Pamperin, K. L., & Wickens, C. D. (1987). The effects of modality stress across task type on human performance. *Proceedings of the 31st Annual Meeting of the Human Factors Society.* Santa Monica, CA: Human Factors Society.

Poulton, E. C. (1976). *Tracking skill and manual control.* New York: Academic Press.

Predmore, S. C. (1991). Micro-coding of cockpit communications in accident analyses: Crew coordination in the United Airlines flight 232 accident. In R. S. Jensen (Ed.), *Proceedings of the Sixth International symposium on Aviation Psychology* (pp. 353–358). Columbus: Dept. of Aviation, Ohio State University.

Rasmussen, J., & Vicente, K. J. (1989). Coping with human errors through system design: Implications for ecological interface design. *International Journal of Man–Machine Studies, 31,* 517–532.

Reising, J. M., & Curry, D. G. (1987). A comparison of voice and multifunction controls: Logic design is the key. *Ergonomics, 7,* 1063–1077.

Roscoe, S. N. (1968). Airborne displays for flight and navigation. *Human Factors, 10,* 321–332.

Rosenthal, R. (1991). *Meta analysis for the Social Sciences* (2nd ed.). Beverly Hills: Sage.

Roth, E. M., & Woods, D. D. (1988). Aiding human performance: I. Cognitive analysis. *Le Travail Humain, 51,* 39–64.

Rubinstein, T., & Mason, A. F. (1979). The accident that shouldn't have happened: An analysis of Three Mile Island. *IEEE Spectrum, 16,* 33–57.

Ruffle-Smith, H. P. (1979). *A simulator study of the interaction of pilot workload with errors, vigilance, and decision* (NASA Tech. Memorandum 78482). Washington, DC: NASA Technical Information Office.

Sanders, M. S., & McCormick, E. J. (1993). *Human factors in engineering and design* (7th ed.). New York: McGraw-Hill.

Simmel, E. C., Cerkovnik, M., & McCarthy, J. E. (1987). Sources of stress affecting pilot judgment. In R. Jensen (Ed.), *Proceedings of the 4th International Symposium on Aviation Psychology* (pp. 190–194). Columbus: Department of Aviation, Ohio State University.

Simonov, P. V., Frolov, M. V., Evtushenko, V. F., & Suiridov, E. P. (1977). *Aviation, Space, and Environmental Medicine, 48,* 856–858.

Stokes, A. F., Belger, A., & Zhang, K. (1990). *Investigation of factors comprising a model of pilot decision making: Part II. Anxiety and cognitive strategies in expert and novice aviators.* University of Illinois Institute of Aviation Tech. Rep. (ARL–90–8/SCEEE–90–2). Savoy, IL: Aviation Research Laboratory.

Stokes, A. F., & Raby, M. (1989). Stress and cognitive performance in trainee pilots. In *Proceedings of the 33rd Annual Meeting of the Human Factors Society.* Santa Monica, CA: Human Factors Society.

Svenson, O., & Maule, J. (Eds.). (1993). *Time pressure and stress in human judgment and decision making.* New York: Plenum.

U.S. Navy (1988). *Investigation report: Formal investigation into the circumstances surrounding the downing of Iran Air Flight 655 on 3 July 1988.* Washington, DC: Department of Defense.

Weltman, G., Smith, J. E., & Egstrom, G. H. (1971). Perceptual narrowing during simulated pressure-chamber exposure. *Human Factors, 13*(2), 99–107.

Wickens, C. D. (1992). *Engineering psychology and human performance* (2nd ed.). New York: HarperCollins.

Wickens, C. D. (1995). Aerospace techniques. In J. Weimer (Ed.), *Research techniques in human engineering.* Englewood Cliffs, NJ: Prentice Hall.

Wickens, C. D., & Carswell, C. M. (1995). The proximity compatibility principle: Its psychological foundation and its relevance to display design. *Human Factors, 37.*

Wickens, C. D., & Flach, J. (1988). Human information processing. In E. Wiener & D. Nagel (Eds.), *Human factors in aviation* (pp. 111–155). New York: Academic Press.

Wickens, C. D., & Prevett, T. (1995). Exploring predimensions of egocentricity in aircraft navigation displays. *Journal of Experimental Psychology: Applied, 1,* 110–135.

Wickens, C. D., Liang, C., Prevett, T., & Olmos, O. (in press). *Egocentric and exocentric displays for terminal area navigation. International Journal of Aviation Psychology.*

Wickens, C. D., Stokes, A. F., Barnett, B., & Hyman, F. (1988). Stress and pilot judgment: An empirical study using MIDIS, a microcomputer-based simulation. *Proceedings of the 32nd Annual Meeting of the Human Factors Society.* Santa Monica, CA: Human Factors Society.

Wickens, C. D., Stokes, A. F., Barnett, B., & Hyman, F. (1993). The effects of stress on pilot judgment in a MIDIS simulator. In O. Svenson & J. Maule (Eds.), *Time pressure and stress in human judgment and decision making* (pp. 271–292). New York: Plenum.

Author Index

A

B

C

D

Subject Index

A

C

D

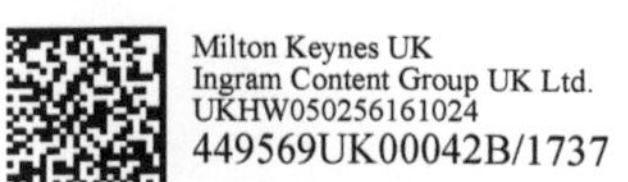
Milton Keynes UK
Ingram Content Group UK Ltd.
UKHW050256161024
449569UK00042B/1737

9 781138 983045